Dynamic Sun

Dynamic Sun presents a modern, comprehensive, and authoritative overview of the Sun, from its deep core to the outer corona, and the solar wind. Each chapter is written by eminent scientists in the field of solar physics. Chapters deal with solar models and neutrinos, seismic Sun, rotation of the solar interior, helioseismic tomography, solar dynamo, spectro-polarimetry, solar photosphere and convection, dynamics and heating of the solar chromosphere, solar transition region, solar MHD, solar activity, particle acceleration, radio observations of explosive energy releases on the Sun, coronal seismology, coronal heating, VUV solar plasma diagnostics, and the solar wind. Solar observing facilities are presented in the last chapter. With a foreword by eminent astrophysicist Eugene Parker, the twenty chapters of this book are all fully illustrated and have comprehensive reference lists. The book covers all major topics in solar physics, and is suitable for graduate students and researchers in solar physics, astrophysics, and astronomy.

BHOLA N. DWIVEDI is a Reader in Applied Physics at Banaras Hindu University, India, and a visiting scientist at the Max-Planck-Institut für Aeronomie, Germany. He has over twenty-two years teaching experience, and broad experience in Solar Physics, with involvement in almost all the major solar space experiments, including Skylab, Yohkoh, SOHO, and TRACE. His current research interests include physics and diagnostics of solar X-ray and EUV emission processes, and waves and oscillations in the solar atmosphere.

Dynamic Sun

Edited by
B. N. Dwivedi
Banaras Hindu University, India

Foreword by
E. N. Parker

CAMBRIDGE
UNIVERSITY PRESS

CAMBRIDGE UNIVERSITY PRESS
Cambridge, New York, Melbourne, Madrid, Cape Town, Singapore, São Paulo

Cambridge University Press
The Edinburgh Building, Cambridge CB2 8RU, UK

Published in the United States of America by Cambridge University Press, New York

www.cambridge.org
Information on this title: www.cambridge.org/9780521810579

First published 2003
This digitally printed version (with corrections) 2007

A catalogue record for this publication is available from the British Library

Library of Congress Cataloguing in Publication data

Dynamic sun / edited by B. N. Dwivedi.
   p.   cm.
  Includes bibliographical references and index.
  ISBN 0 521 81057 4
  1. Sun. I. Dwivedi, B. N., 1950–

QB521 .D96 2003
523.7–dc21   2002073928

ISBN 978-0-521-81057-9 hardback
ISBN 978-0-521-03808-9 paperback

Contents

12   Solar Magnetohydrodynamics 217
E.R. Priest

13   Solar activity 238
Z. Švestka

16   **Coronal oscillations** 314
      V.M. Nakariakov

17   **Probing the Sun's hot corona** 335
      K.J.H. Phillips and B.N. Dwivedi

Foreword

The Sun has posed a challenge to science since the telescope was first turned on it by Galileo, Schreiner, and others around 1610, and the Sun is no less mysterious today for all of our extensive knowledge of its structure. For the improving sensitivity and resolution of the observations, aimed at understanding the old mysteries, have led to the discovery of new mysteries. We have come a long way since 1610, with the overall static structure of the Sun evidently now firmly established. On the other hand, we have a long way to go to understand solar variability and magnetic activity in terms of the basic principles of physics. Observations and theory have progressed to a detailed description of the surface activity down to scales of the order of a hundred km. Unfortunately much of the action lies at still smaller scales, and the magnetic activity deep under the surface cannot be observed directly, so that inference replaces direct observation. Much of what we see at the surface defies theoretical explanation, e.g. the intense fibril structure of the magnetic field, the formation of sunspots, the remarkable penumbral structure of the sunspot, etc. We can describe these phenomena, but we cannot show yet why the Sun is compelled by the basic laws of physics (Newton, Maxwell, Boltzmann, Lorentz, *et al*) to produce them.

The review articles that collectively make up this book are intended as a survey of existing knowledge, which is substantial. It is the starting point for addressing the formidable scientific tasks that lie ahead. We should take heart, then, from the formidable scientific challenges that have already been overcome. For instance, a hundred years after Newton propounded the theory of mechanics and gravitation, the laboratory measurement of the gravitational constant G by Cavendish in 1797 provided the mass of the Sun. Avogadro's number was determined only in 1811. Then, followed by the development of thermodynamics and the kinetic theory of gases, the first self - gravitating polytropic models of the Sun were constructed in the late 19th century, indicating multi-million degree temperatures in the central region.

It must be appreciated that the elemental composition of the Sun was baffling for several decades, with the solar spectrum dominated by the lines of C,O,Ca,Na,Si,Fe, etc. on the one hand, while the theoretical models of the Sun required a molecular weight closer to H on the other. Theory of atomic physics and radiative transfer eventually made it clear that the Sun is mostly hydrogen and helium, with a photosphere too cool to excite their emission. The theoretical recognition of the negative hydrogen ion in the period 1940 - 1950 then explained the high photospheric opacity.

The advent of nuclear physics in the 1930's led to the realization that the thermal energy of the Sun is supplied by thermonuclear reactions in the core, dominated by the proton-proton chain and the carbon cycle. This laid to rest the traditional speculations that the energy was supplied by continuing gravitational contraction or by the infall of comets and asteroids.

Thus by 1950 the essential physics was in hand and theoretical models of the internal structure of the Sun could move forward, with continuing improvement in the calculation of the opacity of the solar gas as a function of density and temperature. Fortunately the recognition of helioseismology by 1980 provided a comprehensive precision test of the theoretical models of the solar interior. With the inclusion of such subtleties as the gravitational settling of the heavier ions, and the accumulation of He in the thermonuclear core, the theoretical model of the solar interior now provides the speed of sound as a function of radial distance that agrees everywhere with the speed of sound inferred from helioseismology to within the observational uncertainties of about one part in 500. So in the last decade we have achieved a firm standard model for the solar interior, based on the simple assumption that the original Sun was chemically homogeneous.

This state of affairs has proved essential in pursuing the observed solar neutrino emission from the thermonuclear core, the observed flux being only 0.3 - 0.5 of the theoretical value. Given that the internal structure of the Sun is now known accurately, it would appear that the discrepancy lies with the physics of the neutrino, initially assumed to be a stable particle. Neutrino oscillations between the e, μ, τ neutrino states are presently under intense experimental and observational study, with the rest mass of the neutrino already established experimentally as nonvanishing.

This brief but heroic history of solar physics is the platform from which we attack the contemporary array of mysteries of the Sun, originating in the vigorous and erratic generation of magnetic field in the convective zone. The convective zone is an unavoidable feature of a star like the Sun. It constitutes the outer 2/7 of the solar radius, across which the temperature falls from 2×10^6 K to 5.6×10^3 K (at the visible surface). The convection arises from the fact that below 2×10^6 K the radiation cannot handle the outward heat flux without the temperature declining outward so fast that the hot gas below continually changes places with the cool gas above.

The hydrodynamic antics of the convection have yet to be fully understood. Presumably the convection is responsible for the nonuniform rotation of the Sun, with a rotation period of 25 days at the equator and something in excess of 30 days at the poles. However, helioseismology shows the surprising fact that the surface rotation extends vertically downward to the bottom of the convective zone, with the radiative interior rotating approximately rigidly with an intermediate period of about 27 days. The best numerical hydrodynamic models of the convection have yet to duplicate this peculiar internal rotation profile, showing instead an angular velocity that is primarily a function of distance from the spin axis of the Sun.

The convective hydrodynamics becomes vastly more complicated and baffling when the magnetic fields are included. The essential point is that the gas is ionized and, therefore, on the large scale of the Sun, the gas cannot support any significant electric field in its own moving frame of reference. Consequently on all but the smallest scales the magnetic field is obliged to move bodily with the convecting gas, becoming enormously stretched and tangled. The magnetic fields appear to be sustained by stretching and winding in the nonuniform rotation and the convective cells.

One of the first puzzles to confront the theoretician is the very small resistive diffusion provided by the resistivity of the gas ($\sim 10^4$ cm^2/sec) and the rapid diffusion ($\sim 10^{11}$ cm^2/sec) required to understand the generation of magnetic fields over dimensions greater than 10^{10} cm in only a few years. One turns to the concept of turbulent diffusion, of the order of $0.1\lambda v$ for eddies with scale λ and velocity v. This automatically supplies diffusivities of the desired order of magnitude. However, with present estimates of the mean azimuthal magnetic field of 3×10^3 Gauss or more in the lower convective zone, it appears that the magnetic field is far too strong to be carried about at random by the turbulent convection. Further it appears that the general magnetic field of the Sun is in an intensely fibril state throughout the convective zone. Estimates of the fibril intensity range from 1.5×10^3 Gauss at the visible surface to thirty or more times as much at the base of the convective zone. The fibril form of the field implies an enhanced magnetic energy for a given total magnetic flux, and the question is why the field exists in this elevated energy state.

The vigorous interaction of neighboring fibrils in the tenuous atmosphere above the visible surface involves rapid reconnection and dissipation of magnetic energy in that tenuous atmosphere. Million degree temperatures are the rule at coronal levels and fast particle populations, occasionally up to 10^{10}eV per particle, are created in the larger flares. The background population of microflares and nanoflares generates an ambient and rapidly varying suprathermal particle population. The solar X-ray corona is one manifestation of this magnetic dissipation, while coronal holes and the fast solar wind streams are another. Coronal mass ejections and large flares arise from the large-scale convective distortion of the magnetic fields arching above the visible surface.

To reiterate the present state of solar physics, we understand enough of the basic principles of magnetohydrodynamics and plasma physics to describe the gross features of the observed magnetic activity. However, we do not understand enough to show how the activity follows from first principles. So we do the best we can, using magnetohydrodynamics in the large and more complicated plasma physical processes in the small, which can be very complex indeed in the intense thin current sheets of a flare, large or small. This volume provides a brief review of the intellectual properties presently in hand.

E.N. Parker
30 October 2001

Dynamic Sun: an introduction

B.N. Dwivedi

Banaras Hindu University, India

1.1   Introduction

"A gaze blank and pitiless as the Sun", was how W.B. Yeats evoked faceless doom in his poem *The Second Coming*. The glare of the Sun may still seem pitiless, but it is blank no longer - at least not to solar physicists. In spite of the stability of the Sun's deep interior, the solar atmosphere is extremely active and dynamic. Observation reveals a wide-ranging repertoire of phenomena occurring at all times. Such phenomena are the consequence of magnetic flux emerging through the Sun's surface from its interior, our understanding of which has been built up by theoretical modelling and observations, since they are beyond simulation in a terrestrial laboratory.

Although many exotic astronomical objects are available for study, the seemingly pedestrian Sun is the object of special study by large-scale ground-based telescopes and other facilities as well as by major spacecraft which were launched over the past twenty years by several space agencies around the world. One reason for this is of course the Sun's proximity, which makes it a fundamental testing ground for virtually all astrophysical techniques. The signal-to-noise associated with the collection in one second of solar photons is comparable to that from a similar source at one parsec in 1000 years. Hence we are able to analyse solar data (in, e.g., the polarimetric, spectral, temporal, or spatial domain) to a very considerable extent. While we do not see surface details on any other star, we can resolve regions on the Sun as small as 150 km across, the size of a large city, using the latest ground-based instruments, and 700 km with spacecraft instrumentation, though recognizing it is likely to be even better soon. The success of solar observations has spawned studies in a wider context, e.g. atomic and nuclear spectroscopy in astrophysics, cosmic magnetometry, neutrino astrophysics, and asteroseismology. Another example of the Sun's uniqueness for observational astronomers is the way in which it is possible, using particle detectors and magnetometers on spacecraft in orbits that take them far from the Earth's magnetosphere, to sample the solar wind which is the dynamic extension of the solar corona consisting of outward-streaming fully ionised plasma.

Despite the success of recent solar physicists in elucidating the workings of the Sun, there remain many challenges for our complete understanding. Some of the items include the filamentary structure of the photospheric magnetic field, the origin and behaviour of small magnetic bipoles continually emerging in supergranules, magnetic diffusion (which is so essential for understanding the solar dynamo), the Sun's peculiar internal rotation inferred from helioseismology, the energy source of the hot solar corona and the generation and acceleration of the solar wind, as well as the small but very important solar brightness variations with level of activity.

1.2   Main contents

In this book we present a modern, comprehensive and authoritative overview of the Dynamic Sun from its deep core to the outer corona, and the solar wind (see Figure 1.1), including a chapter on solar observing facilities. All the chapters have been refereed. They present an up-to-date account of the subject and list extensive references for further study. The main contents of each chapter is as follows:

Chapter 2: Solar models: structure, neutrinos, and helioseismological properties (Bahcall, Basu and Pinsonneault): Solar models remain at the frontiers of two different scientific disciplines, solar neutrino studies and helioseismology. After presenting the details of some state-of-the-art solar models, this chapter gives an overview of solar neutrino physics in some detail (helioseismology is covered in Chapters 3-5 of this book). The neutrino predictions from the set of solar models discussed have been contrasted with the results of the solar neutrino experiments. Finally, the structure of the solar models are compared with helioseismic results obtained using different data sets.

Chapter 3: Seismic Sun (Chitre and Antia): Helioseismology probes the internal structure and dynamics of the Sun with high precision. Frequencies of nearly half a million resonant modes of oscillations have been measured by the ground-based Global Oscillation Network Group (GONG) project and space-based Michelson Doppler Imager (MDI) on the SOHO spacecraft. Each of these modes is trapped in a different region of the solar interior and hence its frequency is sensitive to structure and dynamics in the corresponding region. Conversely, by combining the information from these large number of independedent modes of solar oscillations, the inference is made of the structure and dynamics of the solar interior to unprecedented precision. These seismic data provide a test for solar models and theories of stellar structure and evolution.

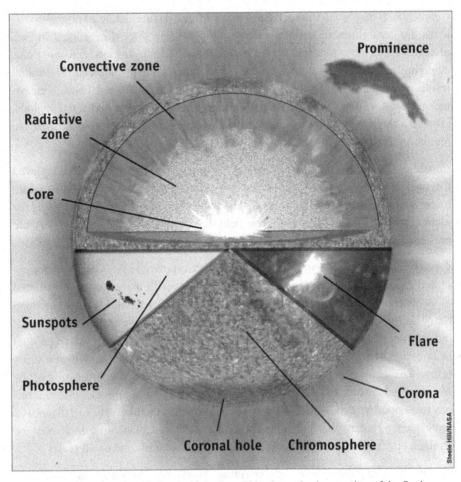

Figure 1.1. The chief parts of the Sun: This gives a basic overview of the Sun's structure. The three major interior zones are the core (the innermost part of the Sun where energy is generated by nuclear reactions), the radiative zone (where energy travels outward by radiation through about 70% of the Sun), and the convection zone (in which convection circulates the Sun's energy to the surface). The flare, sunspots and photosphere, chromosphere, corona, coronal holes, and the prominence are all clipped from actual SOHO images of the Sun. Courtesy of Steele Hill and SOHO/ESA-NASA.

Chapter 4: Rotation of the solar interior (Christensen-Dalsgaard and Thompson): Helioseismology allows us to infer the rotation in the greater part of the solar interior with high precision and resolution. The results show interesting conflicts with earlier theoretical expectations, indicating that the Sun is host to complex dynamical phenomena, so far hardly understood. This has important consequences for our ideas about the evolution of stellar rotation, as well as for models for the generation of the solar magnetic field. An overview of our current knowledge about solar rotation is given, much of it obtained from the SOHO spacecraft, and the broader implications are discussed.

Chapter 5: Helioseismic tomography (Kosovichev): Helioseismic tomography extends the capabilities of helioseismology by providing three-dimensional images of sound-speed variations and mass flows associated with sunspots, active regions, emerging magnetic flux, convective cells and other solar phenomena. The initial results reveal the structure of supergranulation and meridional flows beneath the solar surface as well as large-scale mass motions around sunspots and active regions, provide a clue for the mechanism of sunspots, and even show the presence of active regions on the far side of the Sun.

Chapter 6: The solar dynamo as a model of the solar cycle (Choudhuri): It is believed that the Sun's magnetic field is produced by the dynamo process, which involves non-linear interactions between the solar plasma and the magnetic field. Summarising the main characteristics of solar magnetic field, the basic ideas of dynamo theory are presented and its current status is discussed.

Chapter 7: Spectro-polarimetry (Stenflo): Spectro-polarimetry is our tool for remotely diagnosing the Sun's magnetic field. It deals with the wavelength variation of an observable vector quantity, the Stokes vector. The observational task is to map the Stokes vector both in the spectral and spatial domain with highest possible resolutions (spatial, spectral, temporal) and polarimetric accuracy. The interpretation or inversion of Stokes vector data to derive the magnetic and thermodynamic structure of the solar atmosphere must take into account the extreme structuring of the magnetic field, which extends to scales far smaller than we can resolve with present-day telescopes. With novel imaging Stokes polarimeters qualitatively new diagnostic tools like the Hanle effect and optical pumping are now available to complement the Zeeman effect in the exploration of the magnetized solar plasma on all scales.

Chapter 8: Solar photosphere and convection (Nordlund): An abrupt transition from convective to radiative energy transport at the solar surface results in a spatially and temporally very complex photosphere. The properties of the solar photosphere as well as its importance for both the sub-surface layers and for the chromosphere and corona above are now beginning to be understood in some detail. Progress has been made largely through the use and interpretation of numerical simulations of this region. Comparisons are made in a forward sense; synthetic observational data are generated from the numerical models, and are compared directly with corresponding observational data.

Chapter 9: The dynamics of the quiet solar chromosphere (Kalkofen, Hasan and Ulmschneider): Wave propagation in the nonmagnetic chromosphere is described for plane and spherical waves, and excitation by means of impulses in small source regions in the photosphere; excitation for flux tube waves in the magnetic network is described for large, single impulses and for a fluctuating velocity field. Observational signatures of the various wave types and their effect on chromospheric heating are considered. It is concluded that calcium bright points in the nonmagnetic chromosphere are due to spherical acoustic waves, and that for the oscillations in the magnetic network, transverse waves are more important than longitudinal waves; they may penetrate into the corona, giving rise to some coronal heating.

Chapter 10: Heating of the solar chromosphere (Ulmschneider and Kalkofen): Overlying the photosphere is the chromosphere, a layer that is dominated by mechanical and magnetic heating. By simulating the chromospheric line and continuum emission, empirical models can be constructed that allow the energy balance to be evaluated. Several possible heating processes are discussed as well as the search is made for the actual heating mechanisms. It is found that dissipation by acoustic waves is the basic heating mechanism for nonmagnetic regions of the chromosphere, and MHD tube waves for magnetic regions.

Chapter 11: The solar transition region (Kjeldseth-Moe): What is the solar transition region like? The view of a static, thin transition region has long been left behind. Modern concepts are emerging, but a new model is not generally agreed upon. The observational facts and theoretical considerations, however, consistently point towards a strongly dynamic solar plasma. A comprehensive account of all this is presented here.

Chapter 12: Solar magnetohydrodynamics (Priest): The magnetic field exerts a force, stores energy, acts as a thermal blanket, channels plasma, drives instabilities, and supports waves. For many purposes the behaviour of the magnetic field and its interaction with plasma is governed by the equations of magnetohydrodynamics (MHD). This chapter gives a brief account of some of the basics of MHD, and summarises the simple properties of the different kinds of waves that are present in ideal MHD.

Chapter 13: Solar activity (Švestka): What is the active Sun which is a very important factor in our life ? Observations from SOHO and TRACE reveal the highly turbulent nature of Sun's surface and its atmospheric layers: all the time and everywhere we see brightness variations, loop formations and decays, plasma flows and ejections of gas. However, this is not what we call solar activity. The real processes called solar activity appear only in limited parts of the solar surface, and their occurrence varies quasi-periodically with time, creating 11-year cycles of solar activity whose main characteristics are described in this chapter. Particular attention is paid to coronal mass ejections, as the most important phenomenon affecting the Earth.

Chapter 14: Particle acceleration (Emslie and Miller): The acceleration of particles to high energies is a ubiquitous phenomenon at sites throughout the universe. Despite decades of observations in X-rays and gamma-rays, the mechanism for particle acceleration in solar flares remains an enigma. A comprehensive account of the Sun as a very efficient particle accelerator is presented in this chapter.

Chapter 15: Radio observations of explosive energy releases on the Sun (Kundu and White): This chapter is devoted to a discussion of the radio observations of explosive energy releases (normal flares and small-scale energy releases) on the Sun. Radio imaging observations of solar flares and coronal transients and the relationship of radio phenomena with those observed in hard and soft X-rays, and underlying physics are discussed.

Chapter 16: Coronal oscillations (Nakariakov): The detection of coronal waves provides us with a new tool for the determination of the unknown parameters of the corona - MHD seismology of the corona. The method is similar to helioseismology. But MHD coronal seismology is much richer as it is based upon three different wave modes – Alfvén, slow and fast magnetoacoustic modes. These MHD modes have quite different dispersive, polarization and propagation properties, which make this approach even more powerful. The delicate interplay of MHD wave theory and the observations of coronal waves and oscillations are presented, illustrating it with several examples.

Chapter 17: Probing the Sun's hot corona (Phillips and Dwivedi): The mega-Kelvin temperature of the solar corona has been recognized since the 1940s. While it is generally realized that the magnetic field is the underlying reason, the detailed heating mechanism still eludes solar physicists. This chapter reviews the main historical developments and discoveries right up to those from currently operating satellites such as SOHO and TRACE as well as the chief theoretical problems. An account of the two main competing ideas for coronal heating, nanoflares and MHD wave dissipation, is then given.

Chapter 18: Vacuum-ultraviolet emission line diagnostics for solar plasmas (Dwivedi, Mohan and Wilhelm): Observations of the solar vacuum-ultraviolet emission lines obtained by SUMER/*SOHO* and their interpretation in terms of atomic physics concepts are given. Electron temperature and density diagnostics of the low corona are described. Doppler line-of-sight measurements demonstrate an outflow at the base of the corona in the dark areas of coronal holes, which are seen as the source of the solar wind. Some aspects of the dynamics of the upper solar atmosphere, such as explosive events and sunspot oscillations, are mentioned as examples of the quiet-Sun activity, but spectral observations during solar flare are also shown with indications of plasmas with temperatures of several million Kelvins.

Chapter 19: Solar wind (Marsch, Axford and McKenzie): There are three major types of solar wind – the steady fast wind, the unsteady slow wind, and the variable transient wind. The fast streams are the normal modes of the solar wind. Their basic properties can be reproduced by multi-fluid models involving waves. After briefly reviewing the history of the subject and describing some of the modern theories of the fast wind, the boundary conditions and *in-situ* constraints are discussed which are imposed on the models, in particular by Ulysses at high latitudes.

Some of the results are then presented from SOHO observations that have brought a wealth of new information on the state of the wind in the inner corona as well as the plasma source conditions prevailing in the transition region and solar chromosphere. Finally, problem areas are identified and future research perspectives are outlined.

Chapter 20: Solar observing facilities (Fleck and Keller): An overview is given of current and planned ground-based solar telescopes and instruments, balloon-borne and suborbital solar telescopes, and solar and heliospheric space missions. These observing facilities operate in all areas of solar physics, from the solar interior to interplanetary space and from regimes of high energy to observations requiring high resolution. The next generation of solar telescopes and instruments promise us the ability to investigate solar processes on their fundamental scales, whether sub-arc second or global in nature.

1.3   Concluding remarks

SOHO and TRACE have produced a host of high-resolution observations that have already substantially improved our insights into the physics of the Sun itself as well as how the solar wind and coronal mass ejections influence the near-Earth environment. A new generation of satellites is expected to unravel further solar mysteries and to monitor space weather in a similar way to its terrestrial counterpart. The scientific future of solar physics thus offers exciting prospects for the simple reason that the Sun presents more and more mysteries giving opportunities to learn new physics. And as Yeats says elsewhere, "I'll . . . pluck till time and times are done . . . the golden apples of the Sun", I hope the intended readership (graduate students, and researchers in solar physics, astrophysics, and astronomy) will find each chapter of this book, a 'golden apple' of the Dynamic Sun and as a whole an indispensable guide. This has been possible with the kind support of all my co-authors, and I can hardly thank them enough.

2

Solar models: Structure, neutrinos, and helioseismological properties

J.N. Bahcall

Institute for Advanced Study, Einstein Drive, Princeton, NJ 08540, USA

S. Basu

Astronomy Department, Yale University
P.O. Box 208101, New Haven, CT 06520-8101, USA

M.H. Pinsonneault

Department of Astronomy, Ohio State University, Columbus, Ohio 43210, USA

2.1 Introduction

Why are new calculations of standard solar models of interest? After all, solar models have been used to calculate neutrino fluxes since 1962 (Bahcall *et al.* 1963) and solar atmospheres have been used to calculate p-mode oscillation frequencies since 1970 (Ulrich 1970; Leibacher and Stein 1971). Over the past four decades, the accuracy with which solar models are calculated has been steadily refined as the result of increased observational and experimental information about the input parameters (such as nuclear reaction rates and the surface of abundances of different elements), more accurate calculations of constituent quantities (such as radiative opacity and equation of state), the inclusion of new physical effects (such as element diffusion), and the development of faster computers and more precise stellar evolution codes.

Solar models nevertheless remain at the frontiers of two different scientific disciplines, solar neutrino studies and helioseismology. In an era in which many major laboratory studies are underway to study neutrino oscillations with the aid of very long baselines, $\sim 10^3$ km, between accelerator and detector, solar neutrinos have a natural advantage, with a baseline of 10^8 km (Pontecorvo 1968). In addition, solar neutrinos provide unique opportunities for studying the effects of matter upon neutrino propagation, the so-called MSW effect (Wolfenstein 1978; Mikheyev and Smirnov 1985), since on their way to terrestrial detectors they pass through large amounts of matter in the Sun and, at night, also in the earth.

The connection with ongoing solar neutrino research imposes special requirements on authors carrying out the most detailed solar modeling. Precision comparisons between neutrino measurements and solar predictions are used by many physicists to refine the determination of neutrino parameters and to test different models of neutrino propagation. Since the neutrino experiments and the associated analysis of solar neutrino data are refined at frequent intervals, it is appropriate to reevaluate and refine the solar model predictions as improvements are made in the model input parameters, calculational techniques, and descriptions of the microscopic and macroscopic physics.

In this paper, we provide new information about the total solar neutrino fluxes and the predicted neutrino event rates for a set of standard and non-standard solar models. We also present the number density of scatterers of sterile neutrinos. These quantities are important for precision studies of neutrino oscillations using solar neutrinos.

At the present writing, the Sun remains the only main-sequence star for which p-mode oscillations have been robustly detected. Thus only for the Sun can one measure precisely tens of thousands of the eigenfrequencies for stellar pressure oscillations. The comparison between the sound speeds and pressures derived from the observed p-mode frequencies and those calculated with standard solar models has provided a host of accurate measurements of the interior of the nearest star. The solar quantities determined by helioseismology include the sound velocity and density as a function of solar radius. The excellent agreement between the helioseismological observations and the solar model calculations has shown that the large discrepancies between solar neutrino measurements and solar model calculations cannot be due to errors in the solar models (cf. Figure 2.3). In this paper, we present a refined comparison between our best standard solar model and measurements of the solar sound speeds obtained using oscillation data from a number of different sources.

The interested reader may wish to consult the following works that summarize the solar neutrino aspects of solar models (Bahcall 1989; Bahcall and Pinsonneault 1992, 1995; Berezinsky, Fiorentini, and Lissia 1996; Castellani et al. 1997; Richards et al. 1996; Turck-Chièze et al. 1993; Bahcall, Basu, and Pinsonneault 1998) and the helioseismologic aspects of solar models (Bahcall and Ulrich 1988; Bahcall and Pinsonneault 1995; Christensen-Dalsgaard et al. 1996; Guenther and Demarque 1997; Guzik 1998; Turck-Chièze et al. 1998; Brun, Turck-Chièze, and Zahn 1999; Ricci and Fiorentini 2000).

2.2   Standard solar model

By 'the Standard solar model' (henceforth SSM), we mean the solar model which is constructed with the best-available physics and input data. All of the solar models we consider, standard or 'deviant' models, (see below) are required to fit the observed luminosity and radius of the Sun at the present epoch, as well as the observed heavy element to hydrogen ratio at the surface of the Sun. No helioseismological constraints are used in defining the SSM.

Naturally, Standard models improve with time, as the input data are made more accurate, the calculational techniques become faster and more precise, and the physical description is more detailed.

Our SSM[†] is constructed with the OPAL equation of state (Rogers, Swenson, and Iglesias 1996) and OPAL opacities (Iglesias and Rogers 1996), which are supplemented by the low temperature opacities of Alexander and Ferguson (1994). The model was calculated using the usual mixing length formalism to determine the convective flux.

The principal change in the input data is the use of the Grevesse and Sauval (1998) improved standard solar composition in the OPAL opacities (see http://www. phys.llnl.gov/Research/OPAL/index.htm) and in the calculation of the nuclear reaction rates. The refinements in this composition redetermination come from many different sources, including the use of more accurate atomic transition probabilities in interpreting solar spectra. The OPAL equation of state and the Alexander and Ferguson opacities are not yet available with the composition recommended by Grevesse and Sauval 1998.

The nuclear reaction rates were evaluated with the subroutine exportenergy.f (cf. Bahcall and Pinsonneault 1992), using the reaction data in Adelberger et al. (1998) and with electron and ion weak screening as indicated by recent calculations of Gruzinov and Bahcall (1998); see also Salpeter (1954)[‡]. The model incorporates helium and heavy element diffusion using the exportable diffusion subroutine of Thoul (cf. Thoul, Bahcall and Loeb, 1994; Bahcall and Pinsonneault 1995)[§]. An independent and detailed treatment of diffusion by Turcotte et al. (1998) yields results for the impact of diffusion on the computed solar quantities that are very similar to those obtained here. We have used the most recent and detailed calculation (Marcucci et al. 2000) for the S_0-factor for the ^{3}He$(p,e^+ + \nu_e)^4$He reaction: S_0(hep) $= 10.1 \times 10^{-20}$ keV b, which is a factor of 4.4 times larger than the previous best-estimate [see §2.4.2, Bahcall and Krastev (1998), and Marcucci et al. (2001) for a discussion of the large uncertainties in calculating S_0(hep)]. For values of S_0(hep)

[†] To simplify the language of the discussion, we will often describe characteristics of the Standard model as if we knew they were characteristics of the Sun. We will sometimes abbreviate the reference to this Standard model as BP2000 or Bahcall-Pinsonneault 2000.

[‡] Other approximations to screening are sometimes used. The numerical procedures of Dzitko et al. (1995) and Mitler (1977) predict reaction rates that are too slow for heavy ions because they assumed that the electron charge density near a screened nucleus is the unperturbed value, $en_e(\infty)$. This assumption seriously underestimates the charge density near heavy ions. For example, it is known that a screened beryllium nucleus under solar interior conditions has charge density near the nucleus $\approx -3.85 en_e(\infty)$ (Gruzinov and Bahcall 1997; Brown and Sawyer 1997; all quantum mechanical calculations give similar results, see Bahcall 1962, and Iben, Kalata, and Schwartz 1967).

[§] Both the nuclear energy generation subroutine, exportenergy.f, and the diffusion subroutine, diffusion.f, are available at the Web site www.sns.ias.edu/~jnb, menu item: neutrino software and data.

in the range of current estimates, the assumed rate of the *hep* reaction only affects in a noticeable way the calculated flux of *hep* neutrinos and does not affect the calculated fluxes of other neutrinos, the helioseismological characteristics, or other physical properties of the Sun.

For the standard model, the evolutionary calculations were started at the main-sequence stage. The model has a radius of 695.98 Mm. We do not study the pre-main sequence evolution in our standard model. We do consider one pre-main sequence model, which differs very little from the corresponding model started at the zero-age main sequence.

The ratio of heavy elements to hydrogen (Z/X) at the surface of the model is 0.0230, which was chosen to be consistent with the value obtained by Grevesse and Sauval (1998). A Krishna-Swamy T-τ relationship for the atmosphere was used. We adopt a solar luminosity $L_\odot = 3.844 \times 10^{33}$ erg s^{-1} and a solar age of 4.57×10^9 yr (see Bahcall and Pinsonneault 1995). For the calculations of uncertainties in neutrino flux predictions, we assume a 1σ uncertainty of 0.4%. The uncertainty in the predicted solar neutrino fluxes due to the luminosity is an order of magnitude smaller than the total uncertainty in predicting the neutrino fluxes. For all the other quantities we calculate, the uncertainty in the solar luminosity has an even smaller effect.

We have created from the output of the present calculations two exportable computer files that contain in easily readable form the details of our best SSM (BP2000). Physical variables are given at 875 separate radial shells, which is the same number of shells used to calculate the solar interior model. In addition to the variables cited above, this file contains the pressure, electron number density, the mass fractions of ^{3}He, ^{7}Be, ^{12}C, ^{14}N, and ^{16}O, as well as the source densities of all eight of the most important solar neutrino fluxes. These files are accessible at http://www.sns.ias.edu/~jnb. Previous standard solar models in this series (published in 1982, 1988, 1992, 1995, as well as 1998) are available at the same URL and can be used to test the robustness of conclusions that depend upon models of the solar interior.

2.3   Variant and deviant solar models

In this section, we describe eleven solar models, seven of which are slight variants on the theme of the SSM, and four of which are deficient in one or more significant aspects of the physics used in their construction.

Since the first report (Davis, Harmer, and Hoffman 1968) that the solar neutrino event rate in the chlorine experiment was less than the predicted rate (Bahcall, Bahcall, and Shaviv 1968) obtained by using the then SSM and the then standard electroweak theory, there have been many studies of deviant solar models that were designed to 'explain' the solar neutrino problem. With the advent of precise measurements of solar p-modes that extend deep into the solar interior, many non-standard models have been explicitly shown to be inconsistent with the inferred solar sound speeds (Bahcall,

Pinsonneault, Basu, and Christensen-Dalsgaard 1997) or the p-mode frequencies (Guenther and Demarque 1997).

We explore here the range of solar parameters predicted by various non-standard models, even those that are strongly inconsistent with helioseismological data. Our purpose is to set extreme limits on predicted solar parameters, such as the luminosity evolution or neutrino emission, rather than the traditional goal of avoiding new neutrino physics.

Table 2.1 and Table 2.2 summarize some of the important physical characteristics of the complete set of solar models studied.

Models NACRE, AS00, GN93, "Pre-M.S.", "Rotation", "Radius$_{78}$" and "Radius$_{508}$" are what we called "variant" solar models. Models "No Diffusion", "Old physics", $S_{34} = 0$, and "Mixed" are the deviant models.

The NACRE model was constructed using the same input physics as our Standard model except that we use for the NACRE model the charged particle fusion cross sections recommended in the NACRE compilation (Angulo *et al.* 1999). The model GN93 was considered our Standard model in BPB2000 and differs only in the adopted value of $Z/X = 0.0245$ (Grevesse and Noels 1993) from the current Standard model (see §2.2) which has $Z/X = 0.0230$ (Grevesse and Sauval 1998). The model AS00 is the same as the two models described above except that it has lower heavy element abundance $Z/X = 0.0226$ (Asplund 2000). As a consequence of a more detailed calculation of the solar atmosphere, Asplund (2000) suggests that all meteoritic

Table 2.1. *Some interior characteristics of the solar models. The quantities T_c (in units of 10^6 K), ρ_c (gm cm^{-3}), and P_c (10^{17} erg cm^{-3}) are the current epoch central temperature, density, and pressure; Y and Z are the helium and heavy element mass fractions, where the subscript 'init' denotes the zero-age main sequence model, and the subscript 'c' denotes the center of the solar model*

Model	T_c	ρ_c	P_c	Y_{init}	Z_{init}	Y_c	Z_c
Standard	15.696	152.7	2.342	0.2735	0.0188	0.6405	0.0198
NACRE	15.665	151.9	2.325	0.2739	0.0188	0.6341	0.0197
AS00	15.619	152.2	2.340	0.2679	0.0187	0.6341	0.0197
GN93	15.729	152.9	2.342	0.2748	0.02004	0.6425	0.02110
Pre-M.S.	15.725	152.7	2.339	0.2752	0.02003	0.6420	0.02109
Rotation	15.652	148.1	2.313	0.2723	0.01934	0.6199	0.02032
Radius$_{78}$	15.729	152.9	2.342	0.2748	0.02004	0.6425	0.02110
Radius$_{508}$	15.728	152.9	2.341	0.2748	0.02004	0.6425	0.02110
No Diffusion	15.448	148.6	2.304	0.2656	0.01757	0.6172	0.01757
Old physics	15.787	154.8	2.378	0.2779	0.01996	0.6439	0.02102
$S_{34} = 0$	15.621	153.5	2.417	0.2722	0.02012	0.6097	0.02116
Mixed	15.189	90.68	1.728	0.2898	0.02012	0.3687	0.02047

Table 2.2. *Some characteristics of the convective zones of solar models at the current epoch. Here Y_s and Z_s are the surface helium and heavy element abundances, α is the mixing length parameter, $R(CZ)$ and $T(CZ)$ are the radius and temperature at the base of the convective zone, and $M(CZ)$ is the mass included within the convective zone*

Model	Y_s	Z_s	α	$R(CZ)$ $(R_\odot)$	$M(CZ)$ $(M_\odot)$	$T(CZ)$ $(10^6$ K)
Standard	0.2437	0.01694	2.04	0.7140	0.02415	2.18
NACRE	0.2443	0.01696	2.04	0.7133	0.02451	2.19
AS00	0.2386	0.01684	2.05	0.7141	0.02394	2.18
GN93	0.2450	0.01805	2.06	0.7124	0.02457	2.20
Pre-M.S.	0.2455	0.01805	2.05	0.7127	0.02443	2.20
Rotation	0.2483	0.01797	2.03	0.7144	0.02388	2.15
Radius$_{78}$	0.2450	0.01806	2.06	0.7123	0.02461	2.20
Radius$_{508}$	0.2450	0.01806	2.06	0.7122	0.02467	2.20
No Diffusion	0.2655	0.01757	1.90	0.7261	0.02037	2.09
Old physics	0.2476	0.01796	2.04	0.7115	0.02455	2.21
$S_{34}=0$	0.2422	0.01811	2.03	0.7151	0.02309	2.17
Mixed	0.2535	0.01782	1.85	0.7315	0.01757	2.02

abundances should be adjusted downward by 0.04 dex. All of the models described below use the Grevesse and Noels (1993) composition mix with $Z/X = 0.0245$.

Model Pre-M.S. is evolved from the pre-main sequence stage, but otherwise is the same as our Standard model. The model Rotation incorporates mixing induced by rotation and is a reasonable upper bound to the degree of rotational mixing which is consistent with the observed depletion of lithium in the Sun[†] (Pinsonneault *et al.* 1999). The prescriptions for calculating this model are described in §5 of Pinsonneault (1997) and in BPB2000. There has been considerable discussion recently regarding the precise value of the solar radius (cf. Antia 1998; Schou *et al.* 1997; Brown and Christensen-Dalsgaard 1998) and some discussion of the effects of the uncertainty in radius on the quantities inferred from the helioseismological inversions (cf. Basu 1998). We have therefore considered two models which were constructed with the same input physics as BP2000, but which have model radii which differ from the radius assumed in constructing the Standard model. Radius$_{78}$ has a radius of 695.78 Mm, which is the radius that has been determined from the frequencies of f-modes (cf. Antia 1998) and Radius$_{508}$ has a radius of 695.508 Mm, which is the

[†] The Rotation model discussed here differs somewhat from the rotation model analyzed in BPB2000 in that the metals heavier than CNO were inadvertently not mixed in the previous version of this model. The rotation profile computed from this model does not match precisely the best current estimates of the rotation profile in the inner regions of the Sun. The case considered here corresponds to the maximum amount of mixing.

solar radius as determined by Brown and Christensen-Dalsgaard (1998), who used the measured duration of solar meridian transits during the 6 years 1981–1987 and combined these measurements with models of the solar limb-darkening function to estimate the value of the solar radius.

All of these variant models are approximately as consistent with the helioseismological evidence as the Standard model (see BPB2000).

The model 'Old physics' was constructed using the old Yale equation of state (cf. Guenther *et al.* 1992), supplemented with the Debye-Hückel correction (cf. Bahcall, Bahcall, and Shaviv 1968) and older OPAL radiative opacities (Iglesias, Rogers, and Wilson 1992; Kurucz 1991). The model includes helium and heavy element diffusion and nuclear reaction cross section data in the same way as our Standard model. The $S_{34} = 0$ model was calculated assuming that the rate of the $^3He(\alpha, \gamma)^7Be$ reaction is zero, which implies that no ^7Be or ^8B neutrinos are produced. In the Standard solar model, about 12% of the terminations of the $p - p$ chain involve the $^3He(\alpha, \gamma)^7Be$ reaction, whose rate is proportional to S_{34}. The No Diffusion model does not include helium or heavy-element diffusion and therefore represents the state-of-the art in solar modeling prior to 1992. The model Mixed has an artificially mixed core, with the inner 50% by mass (25% by radius) required to be chemically homogeneous at all times. This model was constructed to be similar to the prescription of Cumming and Haxton (1996), who changed by hand the ^3He abundance as a function of radius in the final BP95 (Bahcall and Pinsonneault 1995) solar model in order to minimize the discrepancy between measurements of the total event rates in neutrino experiments and the calculated event rates. Cumming and Haxton did not calculate the time evolution of their model.

We showed in BPB2000 that the Mixed, No Diffusion, and $S_{34} = 0$ models are strongly disfavored by helioseismological data. We use these deviant (or deficient) models here to test the robustness of the discrepancies between solar model predictions and solar neutrino measurements.

2.4   Neutrino physics

We present at http://www.sns.ias.edu/~jnb a detailed numerical table which gives the fraction of each of the eight important neutrino fluxes that is produced in each spherical shell. These neutrino production fractions are important for calculating the effect of MSW (matter) oscillations in the Sun, but will not be discussed further here.

The Solar Neutrino Problem

The Sun is powered by nuclear reactions, the chief of these being the so called p-p chain, through which four protons combine to form a helium nucleus (an α particle) releasing energy. By-products of these reactions include neutrinos. Since neutrinos are very weakly interacting particles, they penetrate through the Sun and reach the Earth. If these neutrinos can be detected on Earth, we can get direct evidence of nuclear reactions that take place inside the Sun. The main steps in the pp-chain are given below, with the neutrino emitting reactions given in italics. The neutrino energy in MeV is given in parenthesis. As indicated, all neutrinos are electron-type neutrinos.

$$p + p \rightarrow {}^2H + e^+ + \nu_e \text{ (0 to 0.4); } \textbf{or } p + e^- + p \rightarrow {}^2H + \nu_e \text{ (1.4)}$$

$${}^2H + p \rightarrow {}^3He + \gamma$$

$${}^3He + {}^3He \rightarrow \alpha + 2p; \quad \textbf{or} \quad {}^3He + \alpha \rightarrow {}^7Be + \gamma$$

$${}^7Be + e^- \rightarrow {}^7Li + \nu_e \text{ (0.38, 0.86)}$$

$${}^7Li + p \rightarrow 2\alpha$$

or

$${}^7Be + p \rightarrow {}^8B + \gamma$$

$${}^8B \rightarrow {}^8Be + e^+ + \nu_e \text{ (0 to 15)}$$

$${}^8Be \rightarrow 2\alpha$$

A small fraction of solar energy is produced by the CNO cycle reactions which also give rise to neutrinos.

The first solar neutrino detector was proposed in 1964 by Raymond Davis Jr. and John Bahcall and consisted basically of a tank containing 615 tons of a chlorine-based fluid, which can interact with neutrinos via the reaction

$${}^{37}Cl + \nu_e \rightarrow {}^{37}Ar + e^-$$

The argon atoms produced in this reaction can be detected. The first results showed that the experiment detected far fewer neutrinos than were predicted by solar-structure and nuclear reaction theories. This discrepancy is what is usually referred to as the solar neutrino problem.

The chlorine experiment has an energy threshold just below that of the 0.86 MeV ^{7}Be neutrinos and hence could only detect the high and intermediate energy ^{7}Be, pep, ^{8}B and CNO neutrinos. Subsequently, other neutrino detectors have been built, water based ones (Kamiokande and Superkamiokande) as well as heavy-water based ones (the Sudbury Neutrino Observatory, SNO). These have very high energy threshold (≈ 5 MeV) and hence can only detect the high energy ^{8}B neutrinos. There are also low threshold (≈ 0.233 MeV) Gallium based detectors (GALLEX, SAGE and GNO) that can detect neutrinos from all reactions. The results from these experiments have shown that the solar neutrino problem is not just that we detect fewer neutrinos, but that the relative number of neutrinos detected by each type of experiment is inconsistent with theoretical expectations. These results thus give rise to additional solar neutrino problems.

There are two possible solutions of the solar-neutrino problems: either that the standard model of the Sun is wrong, or that the standard model of particle physics is wrong. The first option implies that fewer neutrinos are detected because fewer are actually produced in the Sun and that our calculated values are wrong. The second option implies that the Sun produces the predicted number of neutrinos, but the electron type neutrinos change into some other type and hence cannot be detected by the solar neutrino experiments which can, with the exception of Super-kamiokande and SNO, only detect electron type neutrinos. Super-kamiokande detects other flavours with very low efficiency.

Helioseismology has shown that our current solar models are very close to the real Sun, with the sound-speed profile being within fractions of a percent of solar values. Changing the core-temperature (and hence sound speed) in the models will worsen the agreement between the Sun and the model. Besides, while changing the temperature profile can change the overall number of neutrinos predicted, it has been demonstrated that it is not possible to change the numbers so that they agree with all neutrino experiments simultaneously. Hence this is not a solution to the solar neutrino problem.

The second solution, i.e., that the neutrinos change flavour is the more attractive because with that assumption, it is possible to get neutrino fluxes in all experiments close to the observed values. In fact, recent results from the SNO detector (Ahmad et al. 2002, PRL, **89**, 011301), which can detect other flavours of neutrinos, implies that this is indeed the preferred solution to the solar neutrino problem.

2.4.1   Standard model

Table 2.3 gives the neutrino fluxes and their uncertainties for our SSM. In order to obtain the most precise values that we can for the predicted fluxes, we have recomputed the Standard model discussed elsewhere in this paper. We use in this subsection the most recently determined absolute value for the solar luminosity, 3.842×10^{33} ergs^{-1} (Fröhlich and Lean 1998, Crommelynck, Fichot, Domingo, and Lee 1996), which is 0.2% smaller than the value used in the model calculations discussed elsewhere in this paper. The largest changes due to the 0.2% decrease in the solar luminosity are a 2% decrease in the ^{8}B neutrino flux and a 1% decrease in the ^{7}Be neutrino flux. All other quantities calculated in this paper are changed by negligible amounts.

The adopted uncertainties in different input parameters are given in Table 2 of Bahcall, Basu, and Pinsonneault (1998) (hereafter BP98). We also present in Table 2.3 the calculated event rates in the chlorine, gallium, and lithium experiments. The rate in the Super-Kamiokande experiment is usually quoted as a fraction of the best-estimate theoretical flux of ^{8}B neutrinos, assuming an undistorted (standard) energy spectrum.

Table 2.4 compares the predictions of the combined standard model, i.e., the SSM (BP2000) and the standard electroweak theory (no neutrino oscillations), with the results of the chlorine, GALLEX + GNO, SAGE, Kamiokande, and

Super-Kamiokande solar neutrino experiments. The observed rate in the chlorine experiment is 2.56 ± 0.23 SNU (Lande 2000; Davis 1994; Cleveland *et al.* 1998), which is to be compared to the calculated value of $7.6^{+1.3}_{-1.1}$ SNU. This discrepancy between calculated and observed neutrino capture rates has been approximately the same for

Table 2.3. *Standard Model Predictions (BP2000): solar neutrino fluxes and neutrino capture rates, with 1σ uncertainties from all sources (combined quadratically). The tabulated fluxes correspond to a present-day solar luminosity of 3.842×10^{33} ergs^{-1}. The cross sections for neutrino absorption on chlorine are from Bahcall et al. (1996); the cross sections for gallium are from Bahcall (1997); the cross sections for ^{7}Li are from Bahcall (1989) and Bahcall (1994)*

Source	Flux $\left(10^{10}\ \mathrm{cm}^{-2}\mathrm{s}^{-1}\right)$	Cl (SNU)	Ga (SNU)	Li (SNU)
pp	$5.95\left(1.00^{+0.01}_{-0.01}\right)$	0.0	69.7	0.0
pep	$1.40 \times 10^{-2}\left(1.00^{+0.015}_{-0.015}\right)$	0.22	2.8	9.2
hep	9.3×10^{-7}	0.04	0.1	0.1
^{7}Be	$4.77 \times 10^{-1}\left(1.00^{+0.10}_{-0.10}\right)$	1.15	34.2	9.1
^{8}B	$5.05 \times 10^{-4}\left(1.00^{+0.20}_{-0.16}\right)$	5.76	12.1	19.7
^{13}N	$5.48 \times 10^{-2}\left(1.00^{+0.21}_{-0.17}\right)$	0.09	3.4	2.3
^{15}O	$4.80 \times 10^{-2}\left(1.00^{+0.25}_{-0.19}\right)$	0.33	5.5	11.8
^{17}F	$5.63 \times 10^{-4}\left(1.00^{+0.25}_{-0.25}\right)$	0.0	0.1	0.1
Total		$7.6^{+1.3}_{-1.1}$	128^{+9}_{-7}	$52.3^{+6.5}_{-6.0}$

Table 2.4. *Solar Neutrino Rates: Theory versus Experiment. The units are SNU (10^{-36} interactions per atom per sec) for the radiochemical experiments: Chlorine, GALLEX + GNO, and SAGE. The unit for the ^{8}B and hep fluxes are, respectively, 10^{6} cm^{-2} s^{-1} and 10^{3} cm^{-2} s^{-1}. The errors quoted for Measured/BP2000 are the quadratically combined uncertainties for both BP2000 and the Measured rates. For simplicity in presentation, asymmetric errors were averaged*

Experiment	BP2000	Measured	Measured/ BP2000
Chlorine	$7.6^{+1.3}_{-1.1}$	2.56 ± 0.23	0.34 ± 0.06
GALLEX + GNO	128^{+9}_{-7}	$74.1^{+6.7}_{-7.8}$	0.58 ± 0.07
SAGE	128^{+9}_{-7}	$75.4^{+7.8}_{-7.4}$	0.59 ± 0.07
^{8}B-Kamiokande	$5.05\left[1.00+^{+0.20}_{-0.16}\right]$	$2.80\left[1.00 \pm 0.14\right]$	0.55 ± 0.13
^{8}B-Super-Kamiokande	$5.05\left[1.00+^{+0.20}_{-0.16}\right]$	$2.40\left[1.00+^{+0.04}_{-0.03}\right]$	0.48 ± 0.09
hep-Super-Kamiokande	9.3	$11.3(1 \pm 0.8)$	~ 1

more than three decades (cf. Bahcall, Bahcall, and Shaviv 1968; Davis, Harmer, and Hoffman 1968; Bahcall 1989). The average of the SAGE (Abdurashitov et al. 1999; Gavrin 2000) and the GALLEX (Hampel et al. 1999) plus GNO (Bellotti 2000) results is 74.7 ± 5.0 SNU, which is more than 6σ away from the calculated standard rate of 128^{+9}_{-7} SNU. After 1117 days of data acquisition, the flux of ^{8}B neutrinos measured by Super-Kamiokande is $[2.40 \pm 0.03(\text{stat}) ^{+0.08}_{-0.07}(\text{syst.})]10^{-6}$ cm^{-2}s^{-1} (Suzuki 2001; Fukuda et al. 1998), which corresponds to 0.475 of the BP2000 predicted flux.

Comparing the second and third columns of Table 2.4, we see that the predictions of the combined standard model differ by many standard deviations from the results of the solar neutrino experiments.

The flux of *hep* neutrinos was calculated for BP2000 using the most recent theoretical evaluation by Marcucci et al. (2000, 2001) of the cross section factor $S_0(\text{hep})$, which is 4.4 times larger than the previous best estimate. The most recent report (after 1117 days of data) of the Super-Kamiokande collaboration (Suzuki 2001) is that the *hep* flux observed in their $\nu_e - e$ scattering experiment is 5.4 ± 4.5 times the best-estimate from BP98. Since the BP2000 estimate of the *hep* flux is a factor of 4.4 times larger than the flux quoted in BP98, the best-estimate theoretical *hep* flux now agrees with the best-estimate experimental *hep* flux measurement, although we do not attach much significance to this agreement since we cannot quote an uncertainty on the theoretical estimate (see discussion of the *hep* reaction in §2.4.2).

The event rates predicted by BP2000 for the chlorine, gallium, and Super-Kamiokande solar neutrino experiments are within two percent of the rates predicted for the BP98 SSMs. As far as these experiments are concerned, the effects of using the improved heavy element composition essentially cancels the effect of correcting the error in the opacity interpolation. The difference in the ^{7}Be flux predicted by BP98 and BP2000 is only 0.6%. The ^{7}Be flux will be measured by the BOREXINO solar neutrino experiment.

2.4.2   Calculated uncertainties

We have calculated the uncertainties in the neutrino fluxes and in the experimental event rates by including the published errors in all experimental quantities and by taking account of the correlated influence of different input parameters using the results of detailed solar model calculations. The procedure for calculating the uncertainties has been developed over the past three decades and is described in detail in Chapter 7 of Bahcall (1989) (see also Bahcall and Pinsonneault 1992, 1995, and Bahcall, Basu, and Pinsonneault 1998).

In order that the reader can see the specific implementation of the uncertainty calculations, we are making available the exportable fortran code that evaluates the rates in different neutrino experiments and also calculates the uncertainties in the individual neutrino fluxes and experimental rates. The code, exportrates.f can be downloaded from the web page: http://www.sns.ias.edu/~jnb/SNdata/sndata.html.

The uncertainties in the nuclear fusion cross sections (except for *hep*, see below) were taken from Adelberger *et al.* (1998), the neutrino cross sections and their uncertainties are from Bahcall (1994, 1997) and Bahcall *et al.* (1996), the luminosity and age uncertainties were adopted from Bahcall and Pinsonneault (1995), the 1σ fractional uncertainty in the diffusion rate was taken to be 15% (Thoul, Bahcall, and Loeb 1994) and the opacity uncertainty was determined by comparing the results of fluxes computed using the older Los Alamos opacities with fluxes computed using the modern Livermore opacities (Bahcall and Pinsonneault 1992).

We follow the discussion in Bahcall and Pinsonneault (1995) and adopt a 1σ uncertainty in the heavy element abundance of

$$\sigma(Z/X) = \pm 0.061 \times (Z/X). \tag{2.1}$$

This uncertainty spans the range of values recommended by Grevesse (1984), Grevesse and Noels (1993), and Grevesse and Sauval (1998) over the fourteen year period covered by the cited Grevesse *et al.* review articles. The uncertainty adopted here is about twice as large as the uncertainty recommended by Basu and Antia (1997) based on their helioseismological analysis. In support of the larger uncertainty used in this paper, we note that the difference between the Grevesse and Noels (1993) values of $Z/X = 0.0230$ is 1σ according to Eq. (2.1).

We include the uncertainty in the small ^{17}F neutrino flux due to the uncertainty in the measured S_0-factor for the reaction ^{16}O$(p, \gamma)^{17}$F. We use the 1σ uncertainty 18.1% estimated by Adelberger *et al.* (1998). It was an oversight not to include this uncertainty in our previous calculations.

The only flux for which we do not quote an estimated uncertainty is the *hep* flux (see Table 2.3). The difficulty of calculating from first principles the nuclear cross section factor S_0(hep) is what has caused us not to quote in this series of papers an uncertainty in the *hep* flux (see discussion in Bahcall 1989 and in Bahcall and Krastev 1998). The *hep* reaction is uniquely difficult to calculate among the light element fusion reactions since the one-body and two-body contributions to the reaction rate are comparable in magnitude but opposite in sign, so that the net result is sensitive to a delicate cancellation. Also, two-body axial currents from excitations of Δ isobars are model dependent. In addition, the calculated rate is sensitive to small components in the wave function, particularly D-state admixtures generated by tensor interactions. These complications have been discussed most recently and most thoroughly by Marcucci *et al.* (2001).

The calculated errors are asymmetric in many cases. These asymmetries in the uncertainties in the neutrino fluxes and experimental event rates result from asymmetries in the uncertainties of some of the input parameters, for example, the important pp, ^{7}Be + p, and ^{14}N + p fusion reactions and the effect of excited states on neutrino absorption cross sections. To include the effects of asymmetric errors, the code exportrates.f was run with different input representative uncertainties and the different higher (lower) rates were averaged to obtain the quoted upper (lower) limit uncertainties.

2.4.3 NACRE charged particle fusion rates

In order to estimate the systematic uncertainties associated with different treatments of the nuclear fusion reactions, we have constructed a solar model that is the same as the Standard Model discussed in §2.4.1, except that we have used the charged particle fusion cross sections recommended in the NACRE compilation (Angulo *et al.* 1999) rather than the fusion cross sections determined by Adelberger *et al.* (1998). We will refer to this solar model as the NACRE model.

The low energy cross section factors, S_0, that are recommended by the NACRE collaboration and by Adelberger *et al.* agree within their stated 1σ uncertainties for all of the fusion reactions that are important for constructing a solar model. The only important solar nuclear reactions for which the NACRE collaboration did not recommend interaction rates are the electron capture reactions that produce the ^{7}Be and the *pep* neutrinos; the NACRE collaboration also did not provide energy derivatives for the cross section factors of the CNO reactions. Wherever the data necessary for computing solar fusion rates was not available in the NACRE compilation, we continued to use the Adelberger *et al.* (1998) recommended values in computing the NACRE model.

Table 2.5 gives the calculated neutrino fluxes and capture rates predicted by the NACRE solar model. In all cases, the fluxes for the NACRE solar model agree

Table 2.5. *Neutrinos with NACRE reaction rates. The cross sections for neutrino absorption on chlorine are from Bahcall et al. (1996); the cross sections for gallium are from Bahcall (1997); the cross-sections for ^{7}Li are from Bahcall (1989) and Bahcall (1994)*

Source	Flux $(10^{10}\ \mathrm{cm}^{-2}\mathrm{s}^{-1})$	Cl (SNU)	Ga (SNU)	Li (SNU)
pp	$5.96\left(1.00^{+0.01}_{-0.01}\right)$	0.0	69.8	0.0
pep	$1.39 \times 10^{-2}\left(1.00^{+0.015}_{-0.015}\right)$	0.22	2.8	9.1
hep	9.4×10^{-7}	0.04	0.1	0.1
^{7}Be	$4.81 \times 10^{-1}\left(1.00^{+0.10}_{-0.10}\right)$	1.15	34.5	9.1
^{8}B	$5.44 \times 10^{-4}\left(1.00^{+0.20}_{-0.16}\right)$	6.20	13.1	21.2
^{13}N	$4.87 \times 10^{-2}\left(1.00^{+0.21}_{-0.17}\right)$	0.08	2.9	2.1
^{15}O	$4.18 \times 10^{-2}\left(1.00^{+0.25}_{-0.19}\right)$	0.28	4.7	10.3
^{17}F	$5.30 \times 10^{-4}\left(1.00^{+0.25}_{-0.25}\right)$	0.0	0.1	0.1
Total		$8.0^{+1.4}_{-1.1}$	128^{+9}_{-7}	$52.0^{+6.5}_{-5.9}$

with the fluxes calculated with the SSM to well within the 1σ uncertainties in the Standard Model fluxes. The ^{7}Be flux from the NACRE model is 1% larger than for the Standard model and the ^{8}B flux is 8% higher. The chlorine capture rate predicted by the NACRE model is 5% higher than for the Standard model; the predicted rates for the NACRE model and the Standard model differ by less than 1% for the gallium and lithium experiments.

We conclude that this estimate of the systematic uncertainties due to the relative weights given to different determinations of nuclear fusion cross sections suggests likely errors from this source that are significantly smaller than our quoted 1σ errors for the Standard model neutrino fluxes and event rates (cf. Table 2.3). Similar conclusions have been reached by Morel, Pichon, Provost, and Berthomieu (1999) in an independent investigation (see also Castellani *et al.* 1997).

2.4.4 Variant and deviant models

Table 2.6 compares the calculated neutrino fluxes for the seven variant solar models four deficient solar models described in §2.3 with the fluxes obtained for the SSM.

The deviant models listed in Table 2.6 are all deficient in some important aspect of the physics used in their calculation.

2.4.5 The electron number density

The probability of converting an electron type neutrino to a muon or tau neutrino in the Sun depends upon the profile of the electron number density as a function of solar radius. For particular values of the electron density, neutrino energy, and neutrino

Table 2.6. *Neutrino fluxes from twelve solar models discussed in §2.3*

Model	pp (E10)	pep (E8)	hep (E3)	^{7}Be (E9)	^{8}B (E6)	^{13}N (E8)	^{15}O (E8)	^{17}F (E6)	Cl (SNU)	Ga (SNU)
Standard	5.96	1.40	9.3	4.82	5.15	5.56	4.88	5.73	7.7	129
NACRE	5.97	1.39	9.4	4.85	5.54	4.93	4.24	5.39	8.1	129
AS00	5.99	1.41	9.4	4.62	4.70	5.25	4.56	5.33	7.1	126
GN 93	5.94	1.39	9.2	4.88	5.31	6.18	5.45	6.50	8.0	130
Pre-M.S.	5.95	1.39	9.2	4.87	5.29	6.16	5.43	6.47	7.9	130
Rotation	5.98	1.40	9.2	4.68	4.91	5.57	4.87	5.79	7.4	127
Radius$_{78}$	5.94	1.39	9.2	4.88	5.31	6.18	5.45	6.50	8.0	130
Radius$_{508}$	5.94	1.39	9.2	4.88	5.31	6.18	5.45	6.50	8.0	130
No Diffusion	6.05	1.43	9.6	4.21	3.87	4.09	3.46	4.05	6.0	120
Old physics	5.95	1.41	9.2	4.91	5.15	5.77	5.03	5.92	7.8	130
$S_{34} = 0$	6.40	1.55	10.1	0.00	0.00	6.47	5.64	6.70	0.8	89
Mixed	6.13	1.27	6.2	3.57	4.13	3.04	3.05	3.61	6.1	115

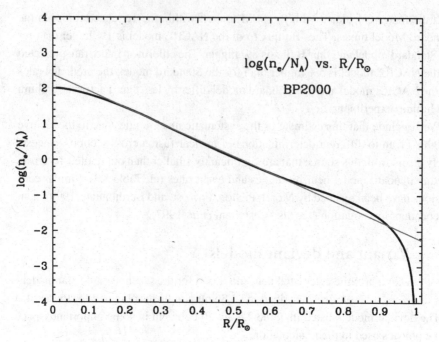

Figure 2.1. The electron number density, n_e, versus solar radius for the SSM (BP2000). The straight-line fit shown in Figure 2.1 is an approximation given by Bahcall (1989). This has been used previously in many analyses of matter effects on solar neutrino propagation. Precise numerical values for n_e are available at www.sns.ias.edu/~jnb.

mass, neutrinos can be resonantly converted from one type of neutrino to another. The Mikheyev-Smirnov resonance occurs if the electron density at a radius r satisfies

$$\frac{n_{e,res}(r)}{N_A} \approx 66 \cos 2\theta_V \left(\frac{|\Delta m^2|}{10^{-4} \, \text{eV}}\right) \left(\frac{10 \, \text{MeV}}{E}\right), \qquad (2.2)$$

where n_e is the electron number density measured in cm^{-3}, N_A is Avogadro's number, θ_V is the neutrino mixing angle in vacuum, $|\Delta m^2|$ is the absolute value of the difference in neutrino masses between two species that are mixing by neutrino oscillations, and E is the neutrino energy.

Figure 2.1 and Table 2.7 give the electron number density as a function of solar radius for the SSM (BP2000). A much more extensive numerical file of the electron number density versus radius is available at www.sns.ias.edu/~jnb; this file contains the computed values of the electron number density at 2493 radial shells.

We see from Figure 2.1 that for typical values of the neutrino parameters that allow the so-called LMA and SMA MSW solutions which fit all of the currently available solar neutrino data (e.g., Bahcall, Krastev, and Smirnov 1998), the experimentally most important ^{8}B neutrinos ($E \geq 5 \, \text{MeV}$) satisfy the resonance condition, Eq. (2.2), at radii that are smaller than the radius of the convective zone. For the so-called LOW

MSW solutions and for all MSW solutions with $\theta_V \sim \pi/4$, the resonance radius falls in the outer part of the Sun.

We have not previously published accurate values for the electron density in the outer parts of the Sun, $r \geq 0.8 R_\odot$. The straight line in Figure 2.1 is an approximation to the electron number density in the SSM of Bahcall and Ulrich (1988) (see Bahcall 1989). This approximation, which has been used for over a decade by different groups analyzing solar neutrino data, is

$$n_e/N_A = 245 \exp(-10.54 R/R_\odot) \text{ cm}^{-3}. \tag{2.3}$$

Figure 2.1 shows that the approximation for n_e given in Eq. (2.3) fails badly in the outer regions of the Sun. Recently, several different analyses have been published using the electron number density shown in Figure 2.1 (or, more precisely, the computer file for $n_e(r)$ which is available at www.sns.ias.edu/~jnb).

The effective density of particles for interacting with sterile (right handed) neutrinos is not the electron number density discussed in the previous subsection, but rather n_{sterile} (see Mikheyev and Smirnov 1986; Lim and Marciano 1988; Barger et al. 1991), where

$$n_{\text{sterile}} = n_e - 0.5 \times n_{\text{neutrons}}, \tag{2.4}$$

and n_{neutrons} is the number density of neutrons. Since nearly all of the neutrons in the Sun are either in ^{4}He or in heavier elements with $Z \simeq A/2$, it is easy to derive an analytic expression that relates n_{sterile} and n_e. One obtains

$$n_{\text{sterile}} = n_e \left(\frac{1+3X}{2(1+X)} \right). \tag{2.5}$$

Table 2.7. *The electron number density versus radius for the standard solar model. The quantity n_e is measured in number per cm³ and N_A is Avogadro's number*

$R/R_\odot$	$\log(n_e/N_A)$	$R/R_\odot$	$\log(n_e/N_A)$
0.01	2.008E+00	0.55	−1.527E−01
0.05	1.956E+00	0.60	−3.605E−01
0.10	1.827E+00	0.65	−5.585E−01
0.15	1.662E+00	0.70	−7.428E−01
0.20	1.468E+00	0.75	−9.098E−01
0.25	1.249E+00	0.80	−1.099E+00
0.30	1.012E+00	0.85	−1.330E+00
0.35	7.687E−01	0.90	−1.642E+00
0.40	5.269E−01	0.95	−2.164E+00
0.45	2.914E−01	1.00	−6.806E+00
0.50	6.466E−02		

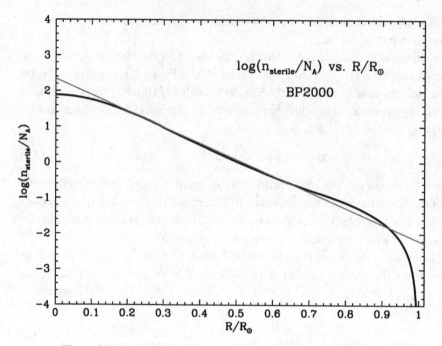

Figure 2.2. The number density, n_{sterile}, of scatterers of sterile neutrinos versus solar radius for the Standard solar model (BP2000). The straight line in Figure 2.2 is given by an equation of the same form as Eq. (2.3) except that the coefficient for n_{sterile}/N_A is 223 (instead of 245 for n_e/N_A).

Figure 2.2 and Table 2.8 gives the radial distribution of n_{sterile} in the SSM. The functional form of $n_{\text{sterile}}(r)$ is similar to the function form of $n_e(r)$.

The number density n_{sterile} is about 25% smaller than the electron number density, n_e, in the center of the Sun, where helium is most abundant. In the central and outer regions of the Sun, n_{sterile} is about 9% less than n_e. The neutrino survival probabilities are the same for sterile and for active neutrinos as long as the adiabatic approximation is valid (see, e.g., §9.2 of Bahcall 1989).

2.5   Sound speeds

In this section we compare the sound-speed profile of our solar models with that of the Sun. Some authors (see, e.g., Guenther and Demarque 1997) compare their solar models directly with the p-mode oscillation frequencies rather than with inverted quantities such as the sound speed. The reader is referred to that paper for a discussion of the direct comparison method, its application, and additional references.

We have chosen to use the sound speed profile because the inversion process that produces the inferred sound speeds allows one to remove the uncertainties, common to all p-mode oscillation frequencies, that arise from the near-surface regions of the Sun. These common uncertainties are due to the treatment of convection, turbulence, and non-adiabatic effects. Inversion techniques are designed to minimize the effects of these outer layers (see, e.g., Basu *et al.* 1996) Moreover, the sound speed profile summarizes in a robust way the results obtained for many thousands of oscillation frequencies. Finally, the inversion procedure allows one to isolate different regions of the Sun, which is important in the context of discussions regarding solar neutrinos. The neutrinos are produced deep in the solar interior. In BPB2000, we have discussed in detail the systematic uncertainties and assumptions related to the inversion for the sound speeds and for the less-accurately determined density.

Figure 2.3 shows the fractional differences between the calculated sound speeds for the Standard model and what may be the most accurate available sound speeds measured by helioseismology, the LOWL1 + BiSON Measurements presented in Basu *et al.* (1997). These sound speeds are derived from a combination of the data obtained by the Birmingham Solar Oscillation Network (BiSON; cf. Chaplin *et al.* 1996) and the Low-ℓ instrument (LOWL; cf. Tomczyk *et al.* 1995a,b).

The rms fractional difference between the calculated and the measured sound speeds is 10.4×10^{-4} over the entire region in which the sound speeds are well measured, $0.05R_\odot \leq r \leq 0.95R_\odot$. In the solar core, $0.05R_\odot \leq r \leq 0.25R_\odot$ (in which about 95% of the solar energy and neutrino flux are produced in an SSM), the rms fractional difference between measured and calculated sound speeds is 6.3×10^{-4}. The Standard model sound speeds agree with the measured sound speeds to 0.1% whether or not one limits the comparison to the solar interior or averages over the

Table 2.8. *The sterile number density,* n_{sterile}, *versus radius for the standard solar model. The quantity* n_{sterile} *is measured in number per* cm^3 *and* N_A *is Avogadro's number. A more extensive numerical file of* n_{sterile} *containing values of* n_{sterile} *at 2499 radial shells, is available at www.sns.ias.edu/~jnb*

$R/R_\odot$	$\log(n_e/N_A)$	$R/R_\odot$	$\log(n_e/N_A)$
0.01	1.885E+00	0.55	-1.901E-01
0.05	1.853E+00	0.60	-3.978E-01
0.10	1.757E+00	0.65	-5.956E-01
0.15	1.611E+00	0.70	-7.777E-01
0.20	1.425E+00	0.75	-9.436E-01
0.25	1.209E+00	0.80	-1.133E+00
0.30	9.731E-01	0.85	-1.364E+00
0.35	7.303E-01	0.90	-1.676E+00
0.40	4.887E-01	0.95	-2.198E+00
0.45	2.536E-01	1.00	-6.839E+00
0.50	2.701E-02		

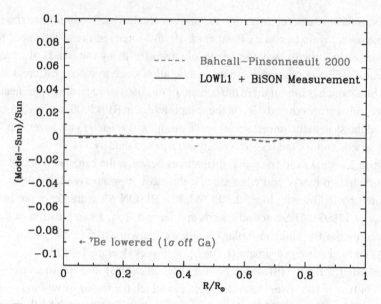

Figure 2.3. Predicted versus measured sound speeds. The figure shows the excellent agreement between the calculated sound speeds for the SSM (BP2000) and the helioseismologically measured (Sun) sound speeds. The horizontal line at 0.0 represents the hypothetical case in which the calculated sound speeds and the measured sound speeds agree exactly everywhere in the Sun. The rms fractional difference between the calculated and the measured sound speeds is 0.10% for all solar radii between between $0.05R_\odot$ and $0.95R_\odot$ and is 0.08% for the deep interior region, $r \leq 0.25R_\odot$, in which neutrinos are produced.

entire Sun. Systematic uncertainties $\sim 3 \times 10^{-4}$ are contributed to the sound speed profile by each of three sources: the assumed reference model, the width of the inversion kernel, and the measurement errors (see BPB2000).

The vertical scale of Figure 2.3 was chosen so as to include the arrow marked "^{7}Be lowered (1σ off Ga)." This arrow indicates the typical difference between solar model speeds and helioseismological measurements that would be expected if the discrepancy between the gallium solar neutrino measurements and the predictions in Table 2.3 were due to errors in the solar physics of the SSM (see discussion in BP98). An extensive table of the Standard model sound speeds is available at www.sns.ias.edu/~jnb .

Figure 2.4 compares the results of six precise observational determinations of the solar sound speed with the results of our SSM. The vertical scale has been expanded by a factor of 21 relative to Figure 2.3 in order to show the small but robust discrepancies between the calculations and the observations and to indicate the size of the differences between the various measurements. In the deep solar solar interior where neutrinos are produced, $R/R_\odot \leq 0.25$, the differences between the various observational determinations of the sound speed are comparable to the differences between BP2000 and any one of the measured sets of sound speeds.

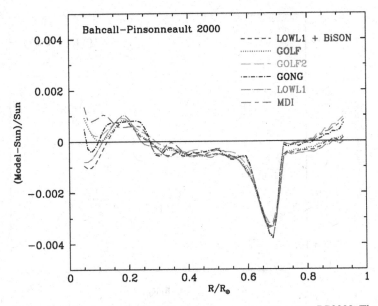

Figure 2.4. Six precise helioseismological measurements versus BP2000. The figure compares the fractional difference between the sound speeds calculated for the Standard solar model (BP2000) and the sound speeds in six helioseismological experiments. The references to the helioseismological data are given in the text. Systematic uncertainties due to the assumed reference model and the width of the inversion kernel are each ~ 0.0003 (see BPB2000).

The p−mode frequencies used in deriving the observed sound speeds shown in Figure 2.4 were obtained from a number of different sources. In addition to the LOWL1 + BiSON data described above, we have used data from a number of other sources. (1) Data from the first year of LOWL observations. The sound speed inversions are described in Basu *et al.* (1997) and are referred to as LOWL1 in this paper. (2) Data from the Michelson Doppler Imager (MDI) instrument on board the Solar and Heliospheric Observatory (SOHO) during the first 144 days of its operation (cf. Rhodes *et al.* 1997). The results of the sound speed inversions using these data are from Basu (1998). (3) The frequencies obtained from the data obtained by the Global Oscillation Network Group (GONG) between months 4–14 of its observations. The solar sound-speeds are from Basu (1998). (4) Initial observation taken by the Global Oscillations at Low Frequencies (GOLF) instrument on board SOHO, combined with intermediate data from MDI. We have labeled the sound speeds obtained as GOLF1 and the solar sound speed results can be found in Turck Chièze *et al.* (1997). (5) More recent data from GOLF (Thiery *et al.* 2000), combined with intermediate degree data obtained from the first 360 days observations by the MDI instrument (Schou *et al.* 1997). The sound-speed results are described in Basu *et al.* (2000). These results have been labeled as GOLF2. There are other helioseismological data sets that could have been used. All give very similar results in the outer layers and fairly similar results in the core. See Basu *et al.* (2000) for a discussion.

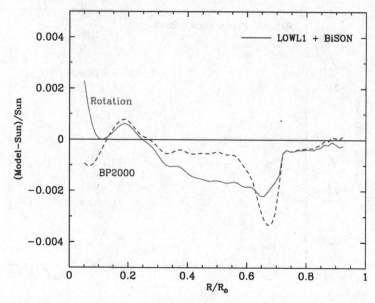

Figure 2.5. BP2000 versus the Rotation model. The figure compares the fractional difference between the sound speeds determined from the LOWL1 + BiSON data with the sound speeds calculated for the BP2000 solar model and the Rotational solar model. The BP2000 model agrees slightly better with the measured sound speeds in the intermediate region between $0.3R_\odot$ and $0.6R_\odot$. The rotation mpodel was developed by Pinsonneault and his associates to explain the depletion of lithium in the Sun.

The MDI and GONG sets have good coverage of intermediate degree modes. The MDI set has p-modes from $\ell = 0$ up to a degree of $\ell = 194$ while the GONG set has modes from $\ell = 0$ up to $\ell = 150$. However, both these sets are somewhat deficient in low degree modes. The LOWL1 + BiSON combination, on the other hand, has a better coverage of low degree modes, but has modes from $\ell = 0$ only up to $\ell = 99$. The GOLF data sets only contain low-degree modes ($l = 0, 1, 2$) and hence have to be combined with other data before they can be used to determine the solar sound speed profile.

Figure 2.5 compares both the Rotation and the Standard model with just the LOWL1 + BiSON sound speeds, This figure shows that the Rotation model gives marginally better agreement with the measured sound speeds right at the base of the convective zone, comparable agreement in the deep interior, $r \leq 0.25R_\odot$, and slightly less good agreement in the intermediate region between $0.3R_\odot$ and $0.6R_\odot$.

The neutrino fluxes calculated with the Rotation model lie well within the estimated errors in the Standard model fluxes, as can be seen easily by comparing the fluxes and the errors given in Table 2.3 and Table 2.6. The ^{7}Be flux for the Rotation model is 3% less than the Standard model ^{7}Be and the ^{8}B flux is 5% lower than the corresponding Standard model value. The Rotation model predicts a capture rate by ^{37}Cl that is 0.3 SNU less than the Standard model rate and a ^{71}Ga capture rate that is 2 SNU less than the Standard model rate.

2.6   Discussion and summary

This chapter provides new information about four topics: 1) the characteristics of the Standard solar model (SSM) at the current epoch (see §2.6.1 below); 2) neutrino fluxes and related quantities for standard and variant solar models (§2.6.2); and 3) measured versus calculated solar sound speeds (§2.6.3). Extensive numerical data that are useful for applications are available at www.sns.ias.edu/~jnb .

2.6.1   Standard solar model: current epoch

We present detailed numerical tabulations of the computed characteristics of our SSM, which is defined and discussed in §2.2. These tables include, as a function of the solar radius, the enclosed mass fraction, the temperature, mass density, electron number density, the pressure, and the luminosity fraction created in a given spherical shell, as well as the mass fractions of ^{1}H, ^{3}He, ^{4}He, ^{7}Be, ^{12}C, ^{14}N, and ^{16}O. Over the years, previous numerical versions of our Standard model have been used for a variety of purposes that range from comparisons with other stellar evolution codes, estimating the importance in the Sun of newly considered physical effects, searching for possible instabilities in the Sun, comparison with helioseismological measurements, and the calculation of processes (especially the MSW effect) that influence the propagation of solar neutrinos.

In the past, we have published in hard copy form increasingly more detailed and precise numerical tables of the characteristics of the solar interior. The capabilities of current calculations and the requirements of some of the most interesting applications have made complete hard copy publication no longer appropriate. We have therefore limited ourselves in §2.2 to describing briefly the ingredients we use in calculating the current Standard model. We present the numerical results in exportable data files that are available at www.sns.ias.edu/~jnb.

2.6.2   Neutrino fluxes and related quantities

Figure 2.1 and Table 2.7 give the electron number density as a function of position in the Sun for the Standard solar model. The distribution of the electron density is required to compute the probability for matter-induced oscillations between active neutrinos. Similarly, Table 2.8 gives the radial distribution of the number density of scatterers of sterile neutrinos, n_{sterile}, in the SSM. We have not previously published precise values for the electron number density, or of n_{sterile}, in the outer regions of the Sun. The outer regions are relevant for large mixing angle neutrino oscillations with relatively low neutrino mass differences ($\Delta m^2 < 10^{-8} \text{eV}^2$).

Table 2.3 presents the neutrino fluxes and the event rates in the chlorine, gallium, lithium, and electron-scattering neutrino experiments that are predicted by the SSM. These predictions assume that nothing happens to solar neutrinos after they are produced. The table also gives estimates of the uncertainties in the fluxes and the event rates.

How do the predictions of solar neutrino event rates compare with experiment? Table 2.4 compares the predictions of BP2000 with the results of the chlorine, GALLEX + GNO, SAGE, Kamiokande, and Super-Kamiokande solar neutrino experiments. This table assumes nothing happens to the neutrinos after they are created in the Sun. The standard predictions differ from the observed rates by many standard deviations. Because of an accidental cancellation, the predicted solar neutrino event rates for BP2000 and BP98 are almost identical (see §2.4.1).

Table 2.6 compares the neutrino fluxes and the experimental event rates for all nine of the solar models whose helioseismological properties were investigated in BPB2000, plus two additional standard-like models considered here which have somewhat different heavy-element to hydrogen ratios. The seven standard-like models (the first seven models in Table 2.6) all produce essentially the same neutrino predictions; the spreads in the predicted pp, ^{7}Be, and ^{8}B fluxes are $\pm 0.7\%$, $\pm 3\%$, and $\pm 6.5\%$ respectively. The calculated rates for the seven standard-like solar models have a range of ± 0.45 SNU for the chlorine experiment and ± 2 SNU for the gallium experiments.

The estimated total errors from external sources (see Table 2.3), such as nuclear cross section measurements and heavy element abundances, are about a factor of three larger than the uncertainties resulting from the solar model calculations.

We have investigated one possible source of systematic errors, the relative weights assigned to different determinations of nuclear fusion cross sections. We calculated the neutrino fluxes and predicted event rates using the NACRE (Angulo *et al.* 1999) fusion cross sections rather than the Adelberger *et al.* (1998) cross sections. The NACRE parameters lead to slightly higher predicted event rates in solar neutrino experiments. However, all changes in the neutrino fluxes and event rates between the NACRE-based predictions and the Standard predictions (based upon Adelberger *et al.* nuclear parameters) are much less than the 1σ uncertainties quoted for the Standard model (cf. Table 2.3 and Table 2.5).

The neutrino event rates predicted by all seven of the standard-like solar models considered here are inconsistent at the 5σ level (combined theoretical and experimental errors) with the results of the two gallium experiments, GALLEX and SAGE, assuming no new physics is occurring. The inconsistency with the chlorine experiment is similar but more complex to specify, since the largest part of the theoretical uncertainty in the calculated standard capture rate is due to the electron-type neutrinos from ^{8}B beta-decay. The fractional uncertainty in the ^{8}B flux depends upon the magnitude of the flux created in the solar interior. Moreover, the amount by which this ^{8}B flux is reduced depends upon the adopted particle physics scenario.

A similar level of inconsistency persists even for the *ad hoc* deficient models, such as the $S_{34} = 0$ and Mixed models, that were specially concocted to minimize the discrepancy with the neutrino measurements. For example, the calculated rates for the Mixed model are $6.15^{+1.0}_{-0.85}$ SNU for the chlorine experiment and $115^{+6.8}_{-5.1}$ SNU for gallium experiments. The Mixed model is 4.2σ below the measured chlorine rate and 6.3σ below the measured gallium rate. In addition, the deficient models are strongly disfavored by the helioseismology measurements.

2.6.3 Sound speeds

Figure 2.3 shows the excellent agreement between the helioseismologically determined sound speeds and the speeds that are calculated for the SSM. The scale on this figure was chosen so as to highlight the contrast between the excellent agreement found with the Standard model and the two orders of magnitude larger rms difference for a solar model that could reduce significantly the solar neutrino problems. One would expect characteristically a 9% rms difference between the observations and the predictions of solar models that significantly reduce the conflicts between solar model measurements and solar model predictions. Averaged over the entire Sun, the rms fractional difference is only 0.10% between the SSM sound speeds and the helioseismologically-determined sound speeds. The agreement is even better, 0.06%, in the interior region in which the luminosity and the neutrinos are produced.

There are small but robust discrepancies between the measured and the calculated solar sound speeds. Figure 2.4 shows the fractional differences between the Standard model sound speeds and the speeds measured in each of six different determinations (using *p*-mode data from LOWL1, BiSON, GONG, MDI, GOLF, and GOLF2). The vertical scale for Figure 2.4 is expanded 21 times compared to the vertical scale of Figure 2.3. As can be seen from Figure 2.4, the differences between observed and measured sound speeds are comparable over much of the Sun to the differences between different measurements of the sound speed, but there is a clear discrepancy near the base of the convective zone that is independent of which observational data set is used. There may also be a less prominent discrepancy near $0.2R_\odot$. The agreement between the BP2000 sound speeds and the measured values is improved over what was found earlier with the BP98 model.

The Rotation model includes a prescription for element mixing that was designed to explain the depletion of lithium. The calculated differences between the Rotation and the Standard models represent a reasonable upper limit to the effects that rotation, sufficient to explain lithium depletion, might produce. The overall agreement between the sound speeds of the Rotation model and the helioseismologically determined sound speeds is slightly worse, 0.12% rather than 0.10%, than the agreement obtained with the SSM. The differences between the solar neutrino fluxes predicted between the Rotation model and the Standard model is typically about $0.3\sigma_\nu$, where σ_ν represents the uncertainties given in Table 2.3 for the Standard model neutrino fluxes. We conclude that further improvements of the theoretical calculations motivated by refinements of $p-$ mode oscillation measurements are unlikely to significantly affect the calculated solar neutrino fluxes.

Acknowledgements

We are grateful to many colleagues in solar physics, nuclear physics, and particle physics for valuable discussions, advice, criticism, and stimulation. JNB is supported in part by an NSF grant #PHY-0070928.

This chapter is edited in a shortened form from *ApJ*, 555, 990–1012 (2001). The total ^{8}B neutrino flux calculated in this paper is in good agreement with the subsequently-measured ^{8}B neutrino flux, see Q. R. Ahmad *et al.*, *Phys. Rev. Lett.*, 87, 071301 (2001).

References

Abdurashitov, J. N. *et al.* (SAGE Collaboration) (1999). *Phys. Rev.*, **C60**, 055801.

Adelberger, E. C. *et al.* (1998). *Rev. Mod. Phys.*, **70**, 1265.

Alexander, D. R. and Ferguson, J. W. (1994). *Astrophys. J.*, **437**, 879.

Angulo, C. *et al.* (1999). *Nucl. Phys.*, **A656**, 3.

Antia, H. M. (1998). *Astron. Astrophys.*, **330**, 336.

Asplund, M. (2000). *Astron. Astrophys.*, **359**, 755.

Bahcall, J. N. (1962). *Phys. Rev.*, **128**, 1297.

Bahcall, J. N. (1989). *Neutrino Astrophysics* (Cambridge University Press, Cambridge).

Bahcall, J. N. (1994). *Phys. Rev.*, **D49**, 3923.

Bahcall, J. N. (1997). *Phys. Rev.*, **C56**, 3391.

Bahcall, J. N., Bahcall, N. A., and Shaviv, G. (1968). *Phys. Rev. Lett.*, **20**, 1209.

Bahcall, J. N., Basu, S., and Pinsonneault, M. H. (1998). *Phys. Lett.*, **B433**, 1.

Bahcall, J. N., Fowler, W. A., Iben, I., and Sears, R. L. (1963). *Astrophys. J.*, **137**, 344.

Bahcall, J. N. and Krastev, P. I. (1998). *Phys. Lett.*, **B436**, 243.

Bahcall, J. N., Krastev, P. I., and Smirnov, A. Yu. (1998). *Phys. Rev.*, **D58**, 096016-1.

Bahcall, J. N., Lisi, L., Alburger, D. E., De Braeckeleer, L., Freedman, S. J., and Napolitano, J. (1996). *Phys. Rev.*, **C54**, 411.

Bahcall, J. N. and Pinsonneault, M. H. (1992). *Rev. Mod. Phys.*, **64**, 885.

Bahcall, J. N. and Pinsonneault, M. H. (1995). *Rev. Mod. Phys.*, **67**, 781.

Bahcall, J. N., Pinsonneault, M. H., Basu, S., and Christensen-Dalsgaard, J. (1997). *Phys. Rev. Lett.*, **78**, 171.

Bahcall, J. N. and Ulrich, R. K. (1988). *Rev. Mod. Phys.*, **60**, 297.

Barger, V., Deshpande, N., Pal, P. B., Phillips, R. J. N., and Whisnant, K. (1991). *Phys. Rev.*, **D43**, 1759.

Basu, S. (1998). *Mon. Not. Roy. Astron. Soc.*, **298**, 719.

Basu, S. and Antia, H. M. (1997). *Mon. Not. Roy. Astron. Soc.*, **287**, 189.

Basu, S. *et al.* (1997). *Mon. Not. Roy. Astron. Soc.*, **292**, 243.

Basu, S., Christensen-Dalsgaard, J., Hernandez, F. P., and Thompson, M.J. (1996). *Mon. Not. Roy. Astron. Soc.*, **280**, 651.

Basu, S., Pinsonneault, M. H., and Bahcall, J. N. (2000). *Astrophys. J.*, **529**, 1084.

Basu, S. *et al.* (2000). *Astrophys. J.*, **535**, 1078.

Bellotti, E., *et al.* (GNO Collaboration) (2001). in *Neutrino 2000*, Proc. of the XIXth International Conference on Neutrino Physics and Astrophysics, 16–21 June 2000, ed. J. Law, R.W. Ollerhead, and J.J. Simpson, *Nucl. Phys. Proc. Suppl.*, **B91**, 44.

Berezinsky, V., Fiorentini, G., and Lissia, M. (1996). *Phys. Lett.*, **B365**, 185.

Brown, T. M. and Christensen-Dalsgaard, J. (1998). *Astrophys. J.*, **500**, L195.

Brown, L. S. and Sawyer, R. F. (1997). *Astrophys. J.*, **489**, 968.

Brun, A. S., Turck-Chièze, S., and Zahn, J. P. (1999). *Astrophys. J.*, **525**, 1032.

Castellani, V., Degl'Innocenti, S., Fiorentini, G., Lissia, M., and Ricci, B. (1997). *Phys. Rep.*, **281**, 309.

Chaplin, W. J. *et al.* (1996). *Solar Phys.*, **168**, 1.

Christensen-Dalsgaard, J. *et al.* (1996). *Science*, **272**, 1286.

Cleveland, B. T., Daily, T., Davis, R., Jr., Distel, J. R., Lande, K., Lee, C. K., Wildenhain, P. S., and Ullman, J. (1998). *Astrophys. J.*, **496**, 505.

Crommelynck, D., Fichot, A., Domingo, V., and Lee, R. III (1996). *Geophys. Res. Lett.*, **23**, No. 17, 2293.

Cumming, A. and Haxton, W. C. (1996). *Phys. Rev. Lett.*, **77**, 4286.

Davis, R. (1994). *Prog. Part. Nucl. Phys.*, **32**, 13.

Davis, R., Jr., Harmer, D. S., and Hoffman, K. C. (1968). *Phys. Rev. Lett.*, **20**, 1205.

Dzitko, H., Turck-Chièze, S., Delbourgo-Salvador, P., and Lagrange, C. (1995). *Astrophys. J.*, **447**, 428.

Fröhlich, C. and Lean, J. (1998). *Geophys. Res. Lett.*, **25**, No. 23, 4377.

Fukuda, Y., *et al.* (Kamiokande Collaboration) (1996). *Phys. Rev. Lett.*, **77**, 1683.

Fukuda, Y. *et al.* (Super-Kamiokande Collaboration) (1998). *Phys. Rev. Lett.*, **81**, 1158; erratum, **81**, 4279.

Gavrin, V. (SAGE Collaboration) (2001). *Nuclear Phys. Proc. Suppl.*, **B91**, 36.

Grevesse, N. (1984). *Phys. Scripta*, **T8**, 49.

Grevesse N., and Noels A. (1993). in *Origin and Evolution of the Elements*, ed. N. Prantzos, E. Vangioni-Flam, and M. Cassé (Cambridge University Press, Cambridge), 15.

Grevesse, N. and Sauval, A. J. (1998). *Space Sci. Rev.*, **85**, 161.

Gruzinov, A. V. and Bahcall, J. N. (1997). *Astrophys. J.*, **490**, 437.

Gruzinov, A. V., and Bahcall, J. N. (1998). *Astrophys. J.*, **504**, 996.

Guenther, D. B. and Demarque, P. (1997). *Astrophys. J.*, **484**, 937.

Guenther, D. B., Demarque, P., Kim, Y.-C., and Pinsonneault, M. H. (1992). *Astrophys. J.*, **387**, 372.

Guzik, J. A. (1998). in Proc. SOHO 6/GONG 98 workshop: *Structure and*

Dynamics of the Sun and Sun-like Stars, ed. S. G. Korzennik, and A. Wilson (ESA, Noordwijk), 417.

Hampel, W. *et al.* (GALLEX Collaboration) (1999). *Phys. Lett.*, **B447**, 127.

Iben, I., Jr., Kalata, K., and Schwartz, J. (1967). *Astrophys. J.*, **150**, 1001.

Iglesias, C. A. and Rogers, F. J. (1996). *Astrophys. J.*, **464**, 943.

Iglesias, C. A., Rogers, F. J., and Wilson, B. G. (1992). *Astrophys. J.*, **397**, 717.

Kurucz, R. L. (1991). *Stellar Atmospheres: Beyond Classical Models*, ed. L. Crivellari, I. Hubeny, and D. G. Hummer (Kluwer, Dordrecht), 441.

Lande, K. (Homestake) (2001). in *Neutrino 2000*, Proc. of the XIXth International Conference on Neutrino Physics and Astrophysics, 16–21 June 2000, ed. J. Law, R.W. Ollerhead, and J.J. Simpson, *Nucl. Phys. Proc. Suppl.*, **B91**, 50.

Leibacher, J. W. and Stein, R. F. (1971). *Astrophys. J.*, **7**, L191.

Lim, C. S. and Marciano, W. (1988). *Phys. Rev.*, **D37**, 1368.

Marcucci, L. E., Schiavilla, R., Viviani, M., Kievsky, A., and Rosati, S. (2000). *Phys. Rev. Lett.*, **84**, 5959.

Marcucci, L. E., Schiavilla, R., Viviani, M., Kievsky, A., Rosati, S., and Beacom, J. F. (2001). *Phys. Rev.*, **C63**, 015801.

Mikheyev, S. P. and Smirnov, A. Yu. (1985). *Sov. J. Nucl. Phys.*, **42**, 913.

Mikheyev, S. P. and Smirnov, A. Yu. (1986). in *'86 Massive Neutrinos in Astrophysics and in Particle Physics*, Proc. of the Sixth Moriond Workshop, ed. O. Fackler, and Trân Thanh Vân (Editions Frontières, Gif-sur-Yvette), 355.

Mitler, H. E. (1977). *Astrophys. J.*, **212**, 513.

Morel, P., Pichon, B., Provost, J., and Berthomieu, G. (1999). *Astron. Astrophys.*, **350**, 275.

Pinsonneault, M. H. (1997). *Annu. Rev. Astron. Astrophys.*, **35**, 557.

Pinsonneault, M. H., Steigman, G., Walker, T. P., and Narayanan, V. K. (1999). *Astrophys. J.*, **527**, 180.

Pontecorvo, B. (1968). *Sov. Phys. JETP*, **26**, 984.

Rhodes, E. J., Jr., Kosovichev, A. G., Schou, J., Scherrer, P. H., and Reiter, J. (1997). *Solar Phys.*, **175**, 287.

Ricci, B. and Fiorentini, G. (2000). *Nucl. Phys. Proc. Suppl.*, **B81**, 95.

Richard, O., Vauclair S., Charbonnel, C., and Dziembowski, W. A. (1996). *Astron. Astrophys.*, **312**, 1000.

Rogers, F. J., Swenson, F. J., and Iglesias, C. A. (1996). *Astrophys. J.*, **456**, 902.

Salpeter, E. E. (1954). *Aust. J. Phys.*, **7**, 373.

Scherrer, P. H., Wilcox, J. M., Christensen-Dalsgaard, J., and Gough, D. O. (1983). *Solar Phys.*, **82**, 75.

Schou, J., Kosovichev, A. G., Goode, P. R., and Dziembowski, W. A. (1997). *Astrophys. J.*, **489**, L197.

Suzuki, Y. (2001). in *Neutrino 2000*, Proc. of the XIXth International Conference on Neutrino Physics and Astrophysics, 16–21 June 2000, ed. J. Law, R.W. Ollerhead, and J.J. Simpson, *Nucl. Phys. Proc. Suppl.*, **B91**, 29.

Thiery, S., *et al.* (2000). *Astron. Astrophys.*, **355**, 743.

Thoul, A. A., Bahcall, J. N., and Loeb, A. (1994). *Astrophys. J.*, **421**, 828.

Tomczyk, S., Schou, J., and Thompson, M. J. (1995a). *Astrophys. J.*, **448**, L57.

Tomczyk, S., Streander, K., Card, G., Elmore, D., Hull, H., and Caccani, A. (1995b). *Solar Phys.*, **159**, 1.

Turck Chièze, S., Vignaud, D., Däppen, W., Fossat, E., Provost, J., and Schatzman, E. (1993). *Phys. Rep.*, **230**, 57.

Turck Chièze, S. *et al.* (1997). *Solar Phys.*, **175**, 247.

Turck Chièze, S. *et al.* (1998). in Proc. SOHO 6/GONG 98 workshop: *Structure and Dynamics of the Sun and Sun-like Stars*, ed. S. G. Korzennik, and A. Wilson (ESA, Noordwijk), 555.

Turcotte, S., Richer, J., Michaud, G., Iglesias, C. A., and Rogers, F. J. (1998). *Astrophys. J.*, **504**, 539.

Ulrich, R. K. (1970). *Astrophys. J.*, **162**, 993.

Wolfenstein, L. (1978). *Phys. Rev.*, **D17**, 2369.

3

Seismic Sun

S.M. Chitre

Department of Physics, University of Mumbai, Mumbai 400098, India

H.M. Antia

Tata Institute of Fundamental Research, Homi Bhabha Road, Mumbai 400005, India

3.1   Introduction

The Sun has been widely described as the Rosetta Stone of astronomy and has clearly contributed in a major way to the development of physics during the past two centuries. This is a very apt description since our star has provided a ready-made cosmic laboratory for studying a variety of processes and phenomena operating both within and outside this object. In fact, the study of the Sun has served as a very valuable guide for theory of structure and evolution of stars in general and pulsating stars in particular.

The internal layers of the Sun are clearly not directly accessible to observations. Nevertheless, it is possible to construct a plausible picture of the interior with the help of mathematical equations governing its mechanical and thermal equilibrium, along with the boundary conditions provided by observations. The basic question confronting solar physicists was how to ascertain the correctness of theoretically evolved solar models and in particular, to examine if there was any way of probing the inner regions of the Sun. It turns out the Sun is, indeed, transparent to neutrinos released in the nuclear reaction network operating in the energy-generating core and also to the waves generated through bulk of the solar body. These valuable probes complement each other and enable us "to see" inside the Sun for deducing the sound speed, thermal and composition profiles and rotation and magnetic fields prevailing in the solar interior. This information can be gainfully employed to calculate the neutrino fluxes in various experiments operating here on the Earth and furthermore, to constrain the physical properties of neutrinos. The internal layers of our Sun thus furnish an ideal laboratory for testing atomic and nuclear physics, high-temperature plasma physics and magneto-hydrodynamics, neutrino physics and even general relativity.

3.2   Structure equations and the Standard Solar Model

The central problem of solar structure is to obtain the 'march' of thermodynamic quantities (e.g., pressure, density, temperature, composition) with depth. The early investigations of the Sun were largely concerned with the collection and study of spectroscopic data for determining the surface temperature and chemical composition. Later analytical efforts were mainly concentrated on the study of polytropic models for inferring the physical conditions in the solar interior. With the advent of high-speed computers, extensive numerical computations were performed for integrating the structure equations with the auxiliary input of physics, supplemented by appropriate boundary conditions. The conventional approach is to assume at zero-age, the Sun to have a homogeneous chemical composition and the mass, $M_\odot = (1.9889 \pm 0.0002) \times 10^{33}$ gm, and then to evolve it over the solar lifetime to yield the luminosity, $L_\odot = (3.846 \pm 0.006) \times 10^{33}$ erg s^{-1} and the radius, $R_\odot = (6.9599 \pm 0.0007) \times 10^{10}$ cm at the present age, $t_\odot = (4.6 \pm 0.1) \times 10^9$ yrs, assuming a couple of adjustable parameters. For this purpose, the Standard Solar Model (SSM) is constructed (cf., Bahcall and Ulrich 1988; Christensen-Dalsgaard *et al.* 1996) using a variety of simplifying assumptions. The Sun is assumed to be spherically symmetric, maintaining mechanical and thermal equilibrium, with negligible effects due to rotation, magnetic field, mass loss or accretion of material and tidal forces on its overall structure. The mechanical equilibrium ensures that the pressure gradient balances the gravitational forces (Cox and Giuli, 1968),

$$\frac{dP(r)}{dr} = -\frac{Gm(r)}{r^2}\rho(r), \tag{3.1}$$

$$\frac{dm(r)}{dr} = 4\pi r^2 \rho(r). \tag{3.2}$$

Here $P(r)$, the pressure, $\rho(r)$, the density and $m(r)$, the mass within a radius of r, are all functions of the radial coordinate, r, for a spherically symmetric star.

The energy generation takes place in the central regions by thermonuclear reactions converting hydrogen into helium mainly by the proton-proton chain. The energy generated by the thermonuclear reactions is transported from the centre to the surface of the Sun from which it is radiated into the space outside. In about two-thirds of the solar interior the energy flux is carried by radiative processes and the flux of radiation, F_{rad} is related to the temperature gradient by,

$$F_{\text{rad}} = -\frac{4acT^3}{3\kappa\rho}\frac{dT}{dr}. \tag{3.3}$$

Here a is the Stefan-Boltzmann constant, c the speed of light and κ the opacity of the solar material caused by a host of atomic processes involving many elements and ionisation stages (cf., Iglesias and Rogers 1996). In the zone extending approximately a third of the solar radius below the photosphere, the radiative temperature gradient becomes unstable to convection and the energy flux is carried largely by convection modelled in the framework of a local mixing-length formalism (Böhm-Vitense 1958) and is expressed as,

$$F_{\text{conv}} = -\kappa_t \rho T \frac{dS}{dr},$$ (3.4)

where κ_t, the turbulent diffusivity is given by $\kappa_t = \alpha w l$, w being the mean vertical velocity, l the mixing length, S the entropy and α is a parameter of order unity.

For thermal equilibrium, the energy lost by the Sun must be balanced by the nuclear energy generation, throughout the solar interior,

$$\rho T \frac{\partial S}{\partial t} = -\frac{1}{4\pi r^2} \frac{\partial L(r, t)}{\partial r} + \rho \epsilon,$$ (3.5)

where $L(r) = 4\pi r^2 (F_{\text{rad}} + F_{\text{conv}})$ is the luminosity and ϵ is the energy-generation rate per unit mass.

In addition, a knowledge of the thermodynamic state of matter inside the Sun is required. For most part except for the outermost layers, the solar material is essentially completely ionised and the perfect gas law is an adequate description of the equation of state which expresses the gas pressure as

$$P_g = \frac{k_B}{m_H \mu} \rho T,$$ (3.6)

where k_B is the Boltzmann constant, m_H the mass of hydrogen atom and μ the mean molecular weight (cf., Schwarzschild 1958). There are corrections (amounting to a few per cent) to this ideal gas law arising from effects due to electron degeneracy, plasma screening, Coulomb free energy between charged particles and pressure ionisation (cf., Eggleton et al. 1973; Däppen et al. 1988; Christensen-Dalsgaard and Däppen 1992; Rogers et al. 1996).

In the standard solar model there is supposed to be no mixing of material outside the convection zone, except for a slow gravitational diffusion of helium and heavy elements relative to hydrogen into the radiative interior because of the momentum exchange between heavier and lighter elements (cf., Guzik and Cox 1993). SSM also assumes there is no wave transport of energy or material in the solar interior. Furthermore, the standard nuclear and neutrino physics is adopted for constructing theoretical models satisfying the observed constraints, namely, mass, radius, luminosity and ratio of chemical abundance by mass, Z/X. Here, X and Z refer respectively, to the fractional abundance by mass of hydrogen and elements heavier than helium.

The early attempts to probe the Sun were mainly confined to the spectroscopic studies of surface layers. Since the 1960s, there have been valiant efforts to measure the flux of neutrinos generated by the nuclear reaction network operating in the solar core (see Chapter 2 of this book). The neutrino flux is sensitive to the temperature and composition profiles in the interior. It was, therefore, expected that the steep temperature dependence of high-energy neutrino fluxes will help in determining the Sun's central temperature to better than a few percent. However, the persistent discrepancy between the measured solar neutrino counting rates and the predictions of standard solar model raised serious doubts about the reliability of structure calculations, based on the veracity of input physics, in particular, the assumption of standard physical properties of neutrinos. This had prompted solar physicists to look for some independent diagnostic tool to explore conditions prevailing inside the Sun. With the observations of solar oscillations, the extensive techniques developed in geo-seismology come very handy with a view to infer the structure of the Sun to a great accuracy by using the precisely measured eigenfrequencies of global solar oscillations.

3.3   Seismology of the Sun

The solar surface is observed to undergo a series of mechanical vibrations which manifest themselves as Doppler shifts oscillating with a period centred around 5 minutes (Leighton *et al.* 1962). These have now been identified as acoustic (p-) modes of pulsation of the entire Sun (Ulrich 1970; Leibacher and Stein 1971; Deubner 1975) oscillating in a number of characteristic modes whose frequencies are determined by the internal solar structure and dynamics. These oscillation modes represent a superposition of millions of standing waves with amplitude of an individual mode of the order of a few centimetres per second. The frequencies of many of these modes have been determined to an accuracy of better than 0.01%. In much the same manner as geophysicists study the internal layers of Earth from seismic events, the helioseismic data of the rich spectrum of velocity fields observed at the solar surface furnish a valuable tool to probe the Sun's internal layers to a remarkable degree of precision. The extraordinarily accurate measurements of oscillations frequencies provide very stringent constraints on the admissible solar models.

A determination of the mode frequencies to a high accuracy, of course, requires continuous observations extending over very long periods of time. From most terrestrial observatories it is not possible to track the Sun continuously for more than 15 hours owing to the day-night cycle. A number of strategies have, therefore, been adopted for getting a longer coverage of the Sun, which include observations from the geographic south pole, from a network of sites located around the globe and from a suitably located satellite. There are several ground-based networks observing the Sun almost continuously with a variety of instruments. The most prominent amongst these is the Global Oscillation Network Group (GONG) which comprises six stations located in contiguous longitudes around the world (Harvey *et al.* 1996). GONG has been observing the Sun since October 1995 and frequencies of close to half a million modes have been calculated for different periods of observations (Hill *et al.* 1996). Apart from the terrestrial networks, satellite-borne instruments have been recording the solar oscillations. The most important among these is the Michelson Doppler Imager (MDI) instrument (Scherrer *et al.* 1995) on board the Solar and Heliospheric Observatory (SOHO) satellite, which was launched in December 1995. With the higher spatial resolution, MDI has been able to measure solar oscillations with small associated length scales.

Solar oscillations may be regarded as a superposition of many standing waves whose frequencies are controlled by the physical properties of the medium through which they travel. The Sun can support two distinct types of wave modes: high-frequency acoustic (p-) modes for which pressure gradient provides the main restoring force and low-frequency gravity (g-) modes for which buoyancy is the dominant restoring force. These two classes of modes are separated by the fundamental (f-) modes which are essentially the surface gravity modes.

The eigenmodes of oscillations can be characterised by three quantum numbers: the angular degree, ℓ, azimuthal order, m and radial order, n. Since the oscillation amplitudes are small enough, it is possible to analyse these modes in the framework of a linearised perturbation theory. Furthermore, for an almost spherically symmetric Sun, the eigenmodes of oscillation may be expressed in terms of the spherical harmonics, $Y_\ell^m(\theta, \phi)$ to write the radial component of velocity as

$$v(r, \theta, \phi, t) = v_{n\ell}(r)Y_\ell^m(\theta, \phi)e^{-i\omega_{n\ell m}t}. \tag{3.7}$$

Here r is the radial distance from the centre, θ the colatitude, ϕ, the longitude and $\omega_{n\ell m}$, the oscillation frequency. Clearly, in the absence of rotation and magnetic field, the frequencies will be independent of the azimuthal order m, while rotation and other symmetry breaking forces can lift this degeneracy resulting in splitting of the modes for a given value of ℓ and n. The mean cyclic frequency of a given multiplet, $\nu_{n\ell}$ ($= \omega_{n\ell}/2\pi$) is determined by the spherically symmetric structure of the Sun and the frequency splittings depend on the rotation rate, magnetic field and other asphericities present in the solar interior. Extensive observations with the GONG and MDI instruments have furnished the mean frequencies of p-modes and f-modes with degree $\ell = 0$–3500 in the frequency-range, 1–10 mHz (Hill *et al.* 1996; Rhodes *et al.* 1997; Antia and Basu 1999).

The propagation characteristics of these seismic waves are affected by the physical make-up of the solar material through which they travel. Thus, a wave excited in the sub-photospheric layers by stochastic turbulence (cf., Goldreich and Keeley 1977) and propagating inwards gets refracted away from the radial direction because of the increasing sound speed encountered with rising temperature. At a certain depth the acoustic wave suffers a total internal reflection and bounces back to the surface where it is liable to get reflected because of the sharply declining density (cf., Figure 3.1). The sound waves are thus trapped within a cavity and the wave can travel inside the solar body several times establishing a standing wave pattern. The penetration depth of a given wave depends on the angle of inclination to the radial direction, which

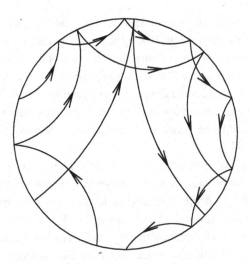

Figure 3.1. Schematic diagram showing propagation of acoustic waves corresponding to different values of ℓ. In general, the ray path will not be closed. Modes with large ℓ or small horizontal wavelength are trapped in a shallow region just below the surface while modes with low ℓ or large horizontal wavelength penetrate into deeper layers.

in turn depends on the horizontal wavelength and the frequency. Thus, oscillations with large ℓ are confined within relatively shallow cavities below the surface, while the small ℓ oscillations propagate deeper, with $\ell = 0$ modes penetrating to the centre. For a given $\ell \neq 0$, the depth of penetration increases with frequency or n. As various modes with varying ℓ, n are trapped in different regions of the solar interior, they sample thermodynamic properties of the region in which they are trapped. It is this pattern of the standing acoustic waves that provides a valuable tool to probe the internal constitution of the Sun. In fact, the diagnostic potential of solar oscillations improves with the availability of a large variety of modes for inferring the physical conditions over a sizeable fraction of solar interior. The perturbations that are detected at the photospheric level naturally encounter the surface layers. The surface layers which are not adequately understood influence the oscillation frequencies to a significant extent. It is important to filter out these surface effects properly while analysing the seismic data and deducing the structure and dynamics inside the Sun. The main source of uncertainties in outer layers is the treatment of convection as well as radiative transfer in the atmosphere and non-adiabatic effects in the sub-surface layers.

3.4   Inferences about the solar structure

The accurate helioseismic data of oscillation frequencies may be analysed in two ways: i) forward method, ii) inverse method. In the forward method, a set of solar models are constructed using the structure equations with different values of adjustable parameters. The equilibrium model is then perturbed using a linearised theory to obtain the eigenfrequencies of solar oscillations. In principle, it may be attempted to fit the theoretically computed frequencies with the accurately measured p-mode frequencies (Elsworth *et al.* 1990b). But in practice, the fit never turns out to be perfect and there are correlations between different parameters, making it difficult to determine these parameters uniquely. Nonetheless, it is possible to make some inferences from the comparison of observed and theoretically computed frequencies about the physics of the solar interior. It turns out that the thickness of the convection zone is close to 200 000 km which is larger than what was previously estimated and the helium abundance by mass Y, in the solar envelope was found to be 0.25. It was noted that an improved treatment of convection due to Canuto and Mazzitelli (1991) led to a significantly better accordance between calculated and observed p-mode frequencies (Basu and Antia 1994a).

Alternately, we can use inversion techniques to calculate the internal structure of the Sun directly from the observed frequencies. A number of inversion techniques (Gough and Thompson 1991) have, therefore, been developed to infer the internal structure of the Sun. Although these inversion techniques generally require a reference solar model to calculate the sound speed and density profiles inside the Sun, the inferred profiles are not particularly sensitive to the choice of the reference model so long as it is close to the Sun. For the purpose of inversion, the equations of adiabatic oscillations are written in the variational form (Chandrasekhar 1964) to obtain a relation between the changes in frequency and possible changes in solar structure,

$$\frac{\delta \nu_{n\ell}}{\nu_{n\ell}} = \int_0^R \mathcal{K}_{c^2,\rho}^{n\ell}(r)\frac{\delta c^2}{c^2}(r)\, dr + \int_0^R \mathcal{K}_{\rho,c^2}^{n\ell}(r)\frac{\delta \rho}{\rho}(r)\, dr + \frac{F(\nu_{n\ell})}{E_{n\ell}}, \qquad (3.8)$$

where the kernels $\mathcal{K}_{c^2,\rho}^{n\ell}(r)$ and $\mathcal{K}_{\rho,c^2}^{n\ell}(r)$ are determined by the eigenfunctions in the reference model. The perturbations $\delta \nu_{n\ell}$, δc^2 and $\delta \rho$ represent the difference between the Sun and a solar model and $E_{n\ell}$ is the mode inertia defined by

$$E_{n\ell} = \frac{4\pi \int_0^R [|\xi_r(r)|^2 + \ell(\ell+1)|\xi_h(r)|^2]\rho_0 r^2 \, dr}{M[|\xi_r(R_\odot)|^2 + \ell(\ell+1)|\xi_h(R_\odot)|^2]}, \qquad (3.9)$$

where ξ_r and ξ_h are the radial and horizontal components of the displacement eigenfunction, M the total solar mass and $\rho_0(r)$ the density profile. The last term in Eq. (3.8) accounts for the uncertainties arising from the outermost layers which are not resolved by the set of modes that may be available. For each observed mode Eq. (3.8) defines an integral equation connecting the observed frequencies to sound speed and density inside the Sun. This can be solved for the sound speed and density provided the frequencies are known. The inverse problem is in general ill-conditioned and proper treatment is required to obtain meaningful results. Several inversion techniques have been developed and tested for the purpose of obtaining reliable information about the structure (Gough and Thompson 1991).

One of the major accomplishments of the inversion techniques was the effective use of the measured solar oscillation frequencies for a reliable inference of the internal structure of the Sun (Gough et al. 1996; Kosovichev et al. 1997). Thus, the profile of the sound speed, $c = \sqrt{\Gamma_1 P/\rho}$ (where $\Gamma_1 = (\partial \ln P/\partial \ln \rho)_S$ is the adiabatic index), is now known through the bulk of the solar interior to an accuracy of better

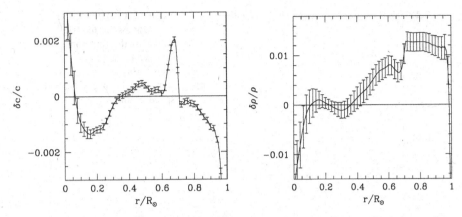

Figure 3.2. Relative difference in the sound speed and density profiles between the Sun (as inferred by seismic inversions) and the standard solar model from Christensen-Dalsgaard *et al.* (1996).

than 0.1% and the profiles of density, adiabatic index and other thermodynamic quantities are determined to somewhat lower accuracy. From the variation of sound speed beneath the convection zone it was concluded that the adopted opacities for solar modelling near the base of the convection zone, were low by about 15–20%. This was later confirmed by the more up-to-date Livermore opacity calculations (Rogers and Iglesias 1992). Similarly, it was established that helium slowly diffuses into the radiative interior (Christensen-Dalsgaard *et al.* 1993). The incorporation of revised opacity tables and diffusion of helium and heavy elements in solar interior improved the solar models significantly. Figure 3.2 shows the plots of the relative difference in sound speed and density between the Sun as inferred from helioseismic inversions and a standard solar model with gravitational settling of helium and heavy elements (Christensen-Dalsgaard *et al.* 1996). The agreement between the model and the Sun is fairly good except for a noticeable discrepancy near the base of the convection zone and a smaller discrepancy in the energy-generating core. The bump below $0.7R_\odot$ could be attributed to a sharp change in the gradient of helium abundance profile arising from diffusion in the reference model. This discrepancy occurs just below the base of the convection zone and a moderate amount of turbulent mixing (induced by say, a rotationally induced instability) in this region can alleviate this discrepant feature (Richard *et al.* 1996; Brun *et al.* 1999). This also happens to be the region where inversions for the rotation rate (see Chapter 4 of this book) show the presence of a strong shear layer, which is referred to as the tachocline (Spiegel and Zahn 1992). The shearing motion in the tachocline is probably responsible for a certain amount of mixing in this region. The dip in the relative sound speed difference around $0.2R_\odot$ is not yet well understood; it could be due to inaccurate composition profile in the solar model, possibly due to use of incorrect nuclear reaction rates. There is also a steep drop near the surface, which is likely to be due to uncertainties in the treatment

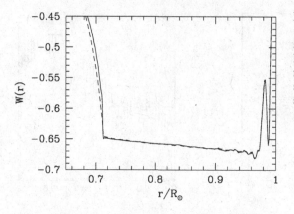

Figure 3.3. The function $W(r)$ for a solar model is shown by the continuous line, while the dashed line represents the same for the sun using the inverted sound speed profile.

of outer layers in solar models. A better treatment of these layers is found to improve the agreement. Alternately, adjusting the solar radius by about 0.03% also removes a large part of this dip. It is quite likely that the solar radius has not been estimated correctly in evolutionary models because of uncertainties plaguing the surface layers. This is owing to the fact that the surface in solar models is normally defined to be the point where the temperature equals the effective temperature. The position of this layer may not be correctly located because of uncertainties present in the structure of the outer layers.

The sound speed profile in ionisation zones is affected by the variation in the adiabatic index, Γ_1 which in turn is determined by the chemical composition. In particular, the dip in Γ_1 inside the second helium ionisation zone may be effectively used to determine the helium abundance in solar convection zone. The inverted sound speed profile can be employed to compute the quantity,

$$W(r) = \frac{r^2}{GM_\odot} \frac{dc^2}{dr} \qquad (3.10)$$

which is displayed in Figure 3.3. The peak around $r = 0.98 R_\odot$ in this curve is due to the dip in Γ_1 inside the HeII ionisation zone. This peak can be calibrated to measure the helium abundance which is found to be 0.249 ± 0.003 (Basu and Antia 1995). This value is somewhat less than what was adopted in earlier standard solar models and the discrepancy has been attributed to diffusion of helium in the solar interior, which received independent confirmation from helioseismic data (Christensen-Dalsgaard et al. 1993). Helium, being heavier than hydrogen, is expected to settle towards the centre under the influence of gravity. Although, the estimated time-scale for diffusion is larger than the age of the Sun, a part of the helium would settle in the interior causing the surface abundance to be reduced. The incorporation of gravitational settling in radiative interior, indeed, results in a significant improvement in solar models. Such a diffusion of helium and heavier elements should occur in interior of other stars as well. During the main sequence phase of stellar evolution hydrogen burning supplies the required energy to sustain the stellar luminosity; the diffusion of helium in the interior decreases the hydrogen abundance, which in turn will reduce the life-time of main sequence stars. The ages of globular clusters are determined by calibrating against theoretical calculations of stellar evolution. The inclusion of the diffusion of helium would naturally reduce the estimated age of globular clusters. This should help in resolving the age problem in the standard big bang model of cosmology.

The dip in Γ_1 inside the ionisation zone is also determined by the equation of state and the inverted sound speed in this region provides a test for the equation of state (Basu and Christensen-Dalsgaard 1997). It is found that early equations of state which were widely used in stellar evolution calculations are not good enough to model the solar interior. More sophisticated equations of state, like the MHD (Däppen *et al.* 1988) or OPAL (Rogers *et al.* 1996) equation of state, are found to yield better agreement with helioseismic data. Furthermore, the OPAL equation of state is found to be in better agreement with solar data as compared to the MHD equation of state (Basu and Antia 1995). Even these equations of state are found to be slightly discrepant in the core and this discrepancy has recently been attributed to the neglect of relativistic correction for electrons (Elliott and Kosovichev 1998).

In solar models the second derivative of temperature and hence that of the sound speed is discontinuous at the base of the convection zone, where the temperature gradient changes from adiabatic value inside the convection zone to radiative gradient below that. This discontinuity in the gradient of the function $W(r)$ (Figure 3.3) can be utilised to identify the position of the base of the convection zone (Christensen-Dalsgaard *et al.* 1991). The sound speed as well as the frequencies of p-modes are very sensitive to the depth of the convection zone and seismic inversions, therefore, enable a very accurate determination of its thickness. Using recent data the depth of the convection zone was estimated by Basu (1998) to be $(0.2865 \pm 0.0005)R_\odot$. Further, the position of the base of the convection zone in solar models depends on the opacity of solar material. We can thus estimate the opacity at the base of the convection zone (Basu and Antia 1997) and it has been found that the current opacity tables from OPAL (Iglesias and Rogers 1996) with the inferred chemical composition are consistent with helioseismic data to within an estimated error of 3%.

The convective eddies inside the convection zone are expected to penetrate beyond the theoretical local boundary, but there is no satisfactory theory to estimate the extent of overshoot. A significant overshoot can alter the stellar evolution calculations and so far the extent of penetration is treated as a parameter in stellar evolutionary calculations. Now with the availability of helioseismic data it has become possible to estimate this penetration depth below the base of the solar convection zone. The discontinuity in the derivatives of sound speed at the base of the convection zone introduces an oscillatory component (Gough 1990) in the frequencies as a function of radial order n. The amplitude of this signal is controlled by the magnitude of the discontinuity, which in turn depends on the extent of overshoot below the solar convection zone. Thus by measuring the amplitude of this oscillatory signal we can determine the extent of overshoot below the convection zone (Monteiro *et al.* 1994; Basu *et al.* 1994). The measured oscillatory signal is found to be consistent with no overshoot and on the basis of this result an upper limit of 1/20 of the local pressure scale height has been set (Basu 1997) for the overshoot distance. This is, of course, too small to affect the stellar evolution calculations significantly. It is also found that the amplitude of oscillatory signal depends on the treatment of diffusion of helium and heavy elements below the convection zone (Basu and Antia 1994b). If there is a sharp gradient in the composition profile below the base of the convection zone, the amplitude of the oscillatory signal in the frequencies is increased, but the measured amplitude from the observed frequencies is consistent with no gradient in the composition profile at the base of the convection zone. It would thus appear that the region immediately below the base of the convection zone is mixed. This inference is also confirmed, as noted earlier, by the bump in Figure 3.2, just below the base of the convection zone.

The primary inversions which have provided information about the physical quantities like the sound speed, density and adiabatic index in the solar interior are based on the equations of mechanical equilibrium. The equations of thermal equilibrium are not used because on time scales of several minutes, no significant energy exchange is expected to take place in moving elements, except in the outermost layers. The frequencies of solar oscillations are, therefore, largely unaffected by the thermal processes in the interior. However, once the sound speed and density profiles in the solar interior are deduced through primary inversions, it is possible to employ the equations of thermal equilibrium (Eqs. 3.3, 3.5) to determine the temperature and chemical composition profiles inside the Sun (Gough and Kosovichev 1990; Takata and Shibahashi 1998; Antia and Chitre 1998), provided input physics like the opacity, equation of state and nuclear energy generation rates are known. In addition, the equation of state provides a relation connecting the inverted sound speed and density with temperature and chemical composition profiles. These equations can be solved for $L(r)$, $T(r)$ and $X(r)$, provided the metal abundance $Z(r)$ is known.

In general, the computed luminosity resulting from these inferred profiles would not necessarily match the observed solar luminosity. The discrepancy between the computed and measured solar luminosity can, in fact, provide a test of input physics, and using these constraints it has been demonstrated that the nuclear reaction cross-section for the proton-proton reaction needs to be increased slightly to $(4.15 \pm 0.25) \times 10^{-25}$ MeV barns (Antia and Chitre 1998; Degl'Innocenti *et al.* 1998; Schlattl *et al.* 1999). This cross-section has a controlling influence on the rate of nuclear energy generation and neutrino fluxes, but it has never been measured in the laboratory and all estimates are based on theoretical computations. The current estimate by Adelberger *et al.* (1998) is 4.0×10^{-25} MeV barns, which is somewhat lower than the helioseismic estimate. Most of the uncertainty in the helioseismic estimate arises from uncertainty in the heavy element abundance, Z, in the solar core. Antia and Chitre (1999) have estimated the proton-proton cross-section as a function of Z to

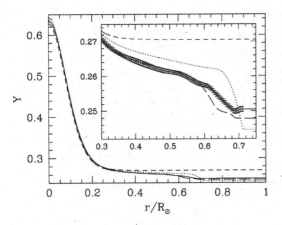

Figure 3.4. Fractional helium abundance by mass in the Sun as obtained from inversions is shown by the continuous line, while the short dashed line represents the same for a solar model without diffusion and the dotted line shows that for a model incorporating diffusion of helium and heavy elements. The long dashed line shows the helium abundance in solar model of Brun *et al.* (1999) which include mixing in the tachocline region.

find that current best estimates for Z and the pp cross-section need to be increased slightly to match helioseismic constraints. The treatment of electron screening also affects the calculation of the nuclear energy generation rate and depending on how this screening is treated the seismic estimates of the cross-section for the pp nuclear reaction can also change. All the above mentioned estimates are based on the treatment of intermediate screening by Graboske *et al.* (1973). Instead if we use the weak screening approximation (Salpeter 1954), then the required increase in the pp reaction rate is much lower, by 1% only, which is within the estimated uncertainties in theoretical values.

The inferred helium abundance profile is in good agreement with that in the standard solar model with diffusion of helium and heavier elements, except in layers just below the solar convection zone. This is the region where the solar rotation rate has a sharp gradient in radial direction (see Chapter 4 of this book). The inferred helium abundance profile, for example, shown in Figure 3.4 is essentially flat in this region. This indicates the presence of some sort of mixing process, possibly by rotationally induced instability which has not been properly accounted for. Solar models including mixing in tachocline region (Brun *et al.* 1999) are in good agreement with the helioseismically inferred composition profile. The mixing in this region can also explain the anomalously low lithium abundance in the solar envelope. It is known that the lithium abundance in the solar envelope is about a factor of 100 lower than the cosmic abundance inferred from meteorites (Vauclair 2000). Lithium can be destroyed by nuclear reactions at temperatures exceeding 2.5×10^6 K. At the base of the solar convection zone, the temperature is still not high enough to burn lithium, but if the mixing extends a little beyond the solar convection zone to a radial distance of $0.68 R_\odot$ the temperature becomes high enough to explain the low abundance of lithium. This is exactly the region where the inferred composition profile is flat, indicating the possible operation of a mixing process.

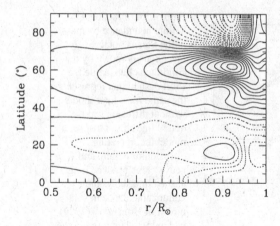

Figure 3.5. Contours of constant $\delta c^2/c^2$, the aspherical component of sound speed as a function of radius and latitude in the solar interior as obtained from inversions of temporally averaged MDI data. The continuous lines represent positive contours while dotted lines represent negative contours. The contour spacing is 2×10^{-5}.

With an allowance of up to 10% uncertainty in opacity values, the central temperature of the Sun is found to be $(15.6 \pm 0.4) \times 10^6$ K (Antia and Chitre 1995). The inferred temperature and composition profiles can be used to compute the neutrino fluxes in the seismically inferred solar models and the predicted neutrino fluxes come close to what is obtained with the current standard solar models. This suggests that the known discrepancy between the observed and predicted neutrino fluxes is likely to be due to non-standard neutrino physics. It can be shown that even if we allow for arbitrary variations in opacity or heavy element abundance in solar interior, it is not possible to construct any solar model satisfying the seismic constraint, which also matches the observed neutrino fluxes (Antia and Chitre 1997). Thus helioseismology has turned our Sun into a precision laboratory to study neutrino properties. Recent results from the Sudbury Neutrino Observatory (SNO) have confirmed (Ahmad *et al.* 2001) that the observed deficit in solar neutrinos is, indeed, due to oscillations between different neutrino species. A reliable estimate of neutrino fluxes from the Sun is required to distinguish between different possible mechanisms for oscillations between different species of neutrinos and seismic constraints have played a central role in improving these theoretical estimates of neutrino fluxes.

Apart from the spherically symmetric structure of the solar interior, it is also possible to determine helioseismically the rotation rate inside the Sun from the accurately measured rotational splittings (see Chapter 4 of this book), $D_{n\ell m} = (\nu_{n\ell m} - \nu_{n\ell -m})/2m$. It turns out that first order effects of rotation yield splittings which depend on odd powers of m, and indeed, these odd splitting coefficients have been used to determine the rotation rate as a function of depth and latitude. On the other hand, the presence of a magnetic field or other asphericities give only even order splittings and it becomes possible to separate these from the rotational effects. These even splitting coefficients allow us to study the departures from spherical symmetry inside the Sun. A study of these splitting coefficients shows that there is some aspherical perturbation located at a depth of about 50,000 km below the solar surface and around a latitude of 60° (Antia *et al.* 2000, 2001a). Figure 3.5 shows contours of constant $\delta c^2/c^2$, the aspherical component of sound speed, in the radius-latitude plane obtained by temporal average over all MDI data. The nature of this perturbation is not known and it could be due to the magnetic field concentrated in this region or the result of some aspherical perturbation to the spherically symmetric solar structure. On the other hand, near the base of the convection zone where the solar dynamo (see Chapter 6 of this book) is supposed to operate, there is no evidence for any magnetic field or asphericity. Similarly, the depth of the convection zone is also independent of latitude (Basu and Antia 2001), with a maximum limit on the variation being $0.0002 R_\odot$. This limit is two orders of magnitude smaller than the inferred latitudinal variation in the depth of tachocline (Charbonneau *et al.* 1999; Basu and Antia 2001). Thus while the tachocline is found to have a prolate shape, the convection zone boundary is spherical.

The solar oblateness has been measured to be about 10^{-5} at the solar surface (Kuhn *et al.* 1998) and there does not seem to be any evidence for temporal variation in oblateness. The oblateness is indeed consistent with what is expected from the helioseismically inferred rotation rate in the solar interior (Antia *et al.* 2000). The resulting quadrupole moment (Pijpers 1998) turns out to be $(2.18 \pm 0.06) \times 10^{-7}$, which yields a precession of the perihelion of the planet Mercury's orbit by about 0.03 arc sec/century, which is smaller than the estimated errors, thus clearly validating the general theory of relativity. Since the rotation rate varies with latitude, there are higher order components in the distortion and the estimate of the next higher order component is also consistent with observed oblateness at the surface (Antia *et al.* 2000). Thus most of the observed oblateness can be accounted for by the known rotation rate and the possible contribution due to the magnetic field should be small.

With the accumulation of helioseismic data over the past 6 years covering the rising phase of cycle 23, it has become possible to study temporal variations in the solar interior. Earlier studies had already established the temporal variation in p-mode frequencies with solar cycle (Libbrecht and Woodard 1990; Elsworth, *et al.* 1990a). It is found that p-mode frequencies shift by up to 0.4 μHz during the solar cycle and that the frequencies are larger during the phase of maximum activity. The frequency variation is found to be well correlated to solar activity (Dziembowski *et al.* 1998; Bhatnagar *et al.* 1999; Howe *et al.* 1999). Figure 3.6 shows the mean frequency shift as a function of time in the GONG data. Also shown on a suitable scale is the variation in the 10.7 cm radio flux, which is believed to be an index of solar activity. It can be seen that the two variations are reasonably well correlated. Further, the frequency shift has steep frequency dependence. Also, at the same frequency the modes which penetrate deeper and hence have larger inertia show smaller variation in frequency. This appears to suggest that the frequency variations are caused by some perturbations residing in outer layers of the Sun. In principle, it is possible to apply inversion technique to the observed frequency shifts to study the temporal variations in solar structure. But such studies have not shown any significant variation

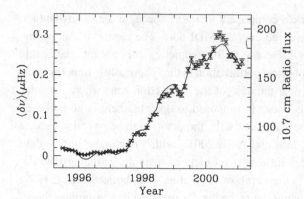

Figure 3.6. Mean frequency shift of p-modes as a function of time using GONG data is shown by the continuous line. The points with error-bars show the mean 10.7 cm radio flux on a scale marked on right hand axis.

in the interior (Basu and Antia 2000), thus confirming that the cause for most of the variations is confined to the outermost layers. In particular, no temporal variation has been detected in the depth of the convection zone which can be determined very accurately from the seismic data. Similarly, there is no signature of temporal variation in sound speed or density near the base of the convection zone, where the solar dynamo is believed to operate. It is likely that the expected variations in solar structure from dynamo action are too small to be detected helioseismically.

The frequencies of f-modes, which are surface gravity modes, are largely independent of the stratification in the solar interior and are essentially determined by the surface gravity. These frequencies which have now been measured reliably by GONG and MDI data provide an important diagnostic of the near-surface regions, as well as a precise measurement of the solar radius (Schou *et al.* 1997; Antia 1998). There is some discrepancy between different measures of solar radius which arises because different techniques effectively use a different definition of solar "surface". Even after modelling these differences there appears to be some discrepancy (Brown and Christensen-Dalsgaard 1998), which could be due to uncertainties in modelling these outer layers. Using these frequencies the solar radius can be determined to an accuracy of 1 km and thus possible changes in the solar radius of this magnitude can be determined by studying temporal variations in frequencies of f-modes. Direct observations of the Sun have given conflicting results on the variation of the solar radius with time (Delache *et al.* 1985; Wittmann *et al.* 1993; Fiala *et al.* 1994; Laclare *et al.* 1996, Noël 1997). The reported change in the measured angular radius varies from $0.1''$ to $1''$, which implies a change of 70 to 700 km in the radius. Such large changes will affect the frequencies by 0.01% to 0.1%, which are much larger than the estimated errors in these frequencies. Helioseismic data collected over the last five years show a more complicated variation in f-mode frequencies (Antia *et al.* 2001b). The observed temporal variations in f-mode frequencies can be resolved into two components: an oscillatory component with a period of 1 year and another slowly varying component which appears to be correlated to solar activity. The oscillatory component is probably an artifact of the data analysis, since its period agrees exactly with the orbital period of Earth. Both these components have a strong dependence on frequency and are therefore unlikely to arise from radius variations. Any possible change in radius should yield relative frequency variation which is independent of frequency. Detailed analysis of these frequency differences suggests that the perturbing influence is localised in the outer 1% of the solar radius. Further, by comparing the frequency shift in f-modes with those in p-modes, it appears that the observed temporal variations can be accounted for by variations in the surface magnetic field and not by perturbations in the thermal structure. To explain the observed variation in frequencies we require a varying magnetic field by about 20 G near the surface, which is consistent with observed variations over the solar cycle. Any possible variation in solar radius should be less than a few km (Dziembowski *et al.* 2001; Antia *et al.* 2001b) over the solar cycle. It can be easily shown that even a change in overall solar radius by 1 km during the solar cycle will release a large amount of energy through variation in the gravitational potential energy, which would be more than the observed solar luminosity. Thus any variation in solar radius must be confined to the rarefied outermost layers. The f-modes are trapped in the region between a depth of 4000–15000 km below the solar surface and thus the limits obtained using f-mode frequencies presumably apply to these layers.

The continuing efforts in helioseismology will hopefully reveal the nature and strength of the magnetic field present in the solar interior and will also help in ascertaining the causes that drive the cyclic magnetic activity and also in locating the seat of the solar dynamo (see Chapter 6 of this book). The global and local seismology of the Sun is clearly poised to reveal its interior to a remarkably accurate detail. With the uninterrupted accruing of the seismic data, we hope to be able to study the temporal variation of mode frequencies and amplitudes, which will indicate what changes are taking place in solar structure and dynamics. We may also learn how the Sun's magnetic field changes with the solar activity cycle and what causes the Sun's irradiance to change with the sunspot cycle.

Acknowledgements

This work utilises data obtained by the Global Oscillation Network Group (GONG) and from the Solar Oscillations Investigation/Michelson Doppler Imager on the Solar and Heliospheric Observatory (SOHO). GONG is managed by the National Solar Observatory, which is operated by AURA, Inc. under a cooperative agreement with the National Science Foundation. The data were acquired by instruments operated by the Big Bear Solar Observatory, High Altitude Observatory, Learmonth Solar Observatory, Udaipur Solar Observatory, Instituto de Astrofísico de Canarias, and Cerro Tololo Interamerican Observatory. SOHO is a project of international cooperation between ESA and NASA. SMC gratefully acknowledges support under the DAE-BRNS senior scientist scheme.

References

Adelberger, E. C. *et al.* (1998). *Rev. Mod. Phys.*, **70**, 1265–1292.

Ahmad, Q. R. *et al.* (2001). *Phys. Rev. Lett.*, **87**, 071301.

Antia, H. M. (1998). *Astron. Astrophys.*, **330**, 336.

Antia, H. M. and Basu, S. (1999). *Astrophys. J.*, **519**, 400.

Antia, H. M. and Chitre, S. M. (1995). *Astrophys. J.*, **442**, 434.

Antia, H. M. and Chitre, S. M. (1997). *Mon. Not. Roy. Astron. Soc.*, **289**, L1.

Antia, H. M. and Chitre, S. M. (1998). *Astron. Astrophys.*, **339**, 239.

Antia, H. M. and Chitre, S. M. (1999). *Astron. Astrophys.*, **347**, 1000.

Antia, H. M., Chitre, S. M., and Thompson, M. J. (2000). *Astron. Astrophys.*, **360**, 335.

Antia, H. M., Basu, S., Hill, F., Howe, R., Komm, R. W., and Schou, J. (2001a). *Mon. Not. Roy. Astron. Soc.*, **327**, 1029.

Antia, H. M., Basu, S., Pintar, J., and Schou, J. (2001b). in Proc. SOHO 10/GONG 2000 Workshop on helio- and asteroseismology at the dawn of the millennium, ESA SP-464, ed. A. Wilson (2001) p. 27.

Bahcall, J. N. and Ulrich, R. K. (1988). *Rev. Mod. Phys.*, **60**, 297.

Basu, S. (1997). *Mon. Not. Roy. Astron. Soc.*, **288**, 572.

Basu, S. (1998). *Mon. Not. Roy. Astron. Soc.*, **298**, 719.

Basu, S. and Antia, H. M. (1994a). *J. Astrophys. Astron.*, **15**, 143.

Basu, S. and Antia, H. M. (1994b). *Mon. Not. Roy. Astron. Soc.*, **269**, 1137.

Basu, S. and Antia, H. M. (1995). *Mon. Not. Roy. Astron. Soc.*, **276**, 1402.

Basu, S. and Antia, H. M. (1997). *Mon. Not. Roy. Astron. Soc.*, **287**, 189.

Basu, S. and Antia, H. M. (2000). *Solar Phys.*, **192**, 449.

Basu, S. and Antia, H. M. (2001). *Mon. Not. Roy. Astron. Soc.*, **324**, 498.

Basu, S. and Christensen-Dalsgaard J. (1997). *Astron. Astrophys.*, **322**, L5.

Basu, S., Antia, H. M., and Narasimha, D. (1994). *Mon. Not. Roy. Astron. Soc.*, **267**, 209.

Bhatnagar, A., Jain, K., and Tripathy, S. C. (1999). *Astrophys. J.*, **521**, 885.

Böhm-Vitense, E. (1958). *Z. Astrophys.*, **46**, 108.

Brown, T. M. and Christensen-Dalsgaard, J. (1998). *Astrophys. J.*, **500**, L195.

Brun, A. S., Turck-Chièze, S., and Zahn, J. P. (1999). *Astrophys. J.*, **525**, 1032.

Canuto, V. M. and Mazzitelli, I. (1991). *Astrophys. J.*, **370**, 295.

Chandrasekhar, S. (1964). *Astrophys. J.*, **139**, 664.

Christensen-Dalsgaard, J. and Däppen, W. (1992). *Astron. Astrophys. Rev.*, **3**, 267.

Charbonneau P., Christensen-Dalsgaard J., Henning R., Larsen R. M., Schou J., Thompson M. J. and Tomczyk S. (1999). *Astrophys. J.*, **527**, 445.

Christensen-Dalsgaard, J., Gough, D. O., and Thompson, M. J. (1991). *Astrophys. J.*, **378**, 413.

Christensen-Dalsgaard, J., Proffitt, C. R., and Thompson, M. J. (1993). *Astrophys. J.*, **403**, L75.

Christensen-Dalsgaard, J. *et al.* (1996) *Science*, **272**, 1286.

Cox, J. P. and Giuli, R. T. (1968). *Principles of Stellar Structure*, Gordon and Breach, New York.

Däppen, W., Mihalas, D., Hummer, D. G., and Mihalas, B. W. (1988). *Astrophys. J.*, **332**, 261.

Degl'Innocenti S., Fiorentini G., and Ricci B. (1998). *Phys. Lett.*, **B416**, 365.

Delache P., Laclare F. and Sadsaoud H. (1985). *Nature*, **317**, 416.

Deubner, F.-L. (1975). *Astron. Astrophys.*, **44**, 371.

Dziembowski, W. A., Goode, P. R., DiMauro, M. P., Kosovichev, A. G., and Schou, J. (1998). *Astrophys. J.*, **509**, 456.

Dziembowski, W. A., Goode, P. R., and Schou, J. (2001). *Astrophys. J.*, **553**, 897.

Eggleton, P. P., Faulkner, J., and Flannery, B. P. (1973). *Astron. Astrophys.*, **23**, 325.

Elliott, J. R. and Kosovichev, A. G. (1998). *Astrophys. J.*, **500**, L199.

Elsworth, Y., Howe, R., Isaak, G. R., McLeod, C. P., and New, R. (1990a). *Nature*, **345**, 322.

Elsworth, Y., Howe, R., Isaak, G. R., McLeod, C. P., and New, R. (1990b). *Nature*, **347**, 536.

Fiala A. D., Dunham D. W., and Sofia S. (1994). *Solar Phys.*, **152**, 97.

Goldreich, P. and Keeley, D. A. (1977). *Astrophys. J.*, **212**, 243.

Gough, D. O. (1990). Comments on helioseismic inference, *Progress of seismology of the sun and stars, Lecture Notes in Physics*, vol. **367**, 283, ed. Osaki, Y. and Shibahashi, H., Springer, Berlin.

Gough, D. O. and Kosovichev, A. G. (1990). *in Proc. IAU Colloquium No 121, Inside the Sun*, p. 327, ed. Berthomieu G. and Cribier M., Kluwer, Dordrecht.

Gough, D. O. and Thompson, M. J. (1991). The inversion problem, *Solar interior and atmosphere*, ed. Cox, A. N., Livingston, W. C. and Matthews, M., p. 519, Space Science Series, University of Arizona Press.

Gough, D. O. *et al.* (1996). *Science*, **272**, 1296.

Graboske, H. C., Dewitt, H. E., Grossman, A. S., and Cooper, M. S. (1973). *Astrophys. J.*, **181**, 457.

Guzik, J. A. and Cox, A. N. (1993). *Astrophys. J.*, **411**, 394.

Harvey, J. W. *et al.* (1996). *Science*, **272**, 1284.

Hill, F. *et al.* (1996). *Science*, **272**, 1292.

Howe, R., Komm, R., and Hill, F. (1999). *Astrophys. J.*, **524**, 1084.

Iglesias, C. A. and Rogers, F. J. (1996). *Astrophys. J.*, **464**, 943.

Kosovichev, A. G. *et al.* (1997). *Solar Phys.*, **170**, 43.

Kuhn, J. R., Bush, R. I., Scheick, X., and Scherrer, P. (1998). *Nature*, **392**, 155.

Laclare F., Delmas C., Coin J. P., and Irbah A. (1996). *Solar Phys.*, **166**, 211.

Libbrecht, K. G. and Woodard, M. F. (1990). *Nature*, **345**, 779.

Leibacher, J. and Stein, R. F. (1971). *Astrophys. Lett.*, **7**, 191.

Leighton, R. B., Noyes, R. W., and Simon, G. W. (1962). *Astrophys. J.*, **135**, 474.

Monteiro, M. J. P. F. G., Christensen-Dalsgaard, J., and Thompson, M. J. (1994). *Astron. Astrophys.*, **283**, 247.

Noël F. (1997). *Astron. Astrophys.*, **325**, 825.

Pijpers, F. P. (1998). *Mon. Not. Roy. Astron. Soc.*, **297**, L76.

Rhodes, E. J., Jr., Kosovichev, A. G., Schou, J., Scherrer, P. H., and Reiter, J. (1997). *Solar Phys.*, **175**, 287.

Richard O., Vauclair S., Charbonnel C., and Dziembowski W. A. (1996). *Astron. Astrophys.*, **312**, 1000.

Rogers, F. J. and Iglesias, C. A. (1992). *Astrophys. J. Suppl.*, **79**, 507.

Rogers, F. J., Swenson, F. J., and Iglesias, C. A. (1996). *Astrophys. J.*, **456**, 902.

Salpeter, E. E. (1954). *Aust. J. Phys.*, **7**, 373.

Scherrer, P. H. *et al.* (1995). *Solar Phys.*, **162**, 129.

Schlattl, H., Bonanno, A., and Paternó, L. (1999). *Phys. Rev.*, **D60**, 113002.

Schou, J., Kosovichev, A. G., Goode, P. R., and Dziembowski, W. A. (1997). *Astrophys. J.*, **489**, L197.

Schwarzschild, M. (1958). *Structure and Evolution of Stars*. Princeton University Press.

Spiegel E. A. and Zahn J.-P. (1992). *Astron. Astrophys.*, **265**, 106.

Takata, M. and Shibahashi, H. (1998). *Astrophys. J.*, **504**, 1035.

Ulrich, R. K. (1970). *Astrophys. J.*, **162**, 993.

Vauclair, S. (2000). *J. Astrophys. Astron.*, **21**, 323.

Wittmann A. D., Alge E. and Bianda M. (1993). *Solar Phys.*, **145**, 205.

4

Rotation of the solar interior

J. Christensen-Dalsgaard

Teoretisk Astrofysik Center, Danmarks Grundforskningsfond, and

Institut for Fysik og Astronomi, Aarhus Universitet, DK-8000 Aarhus C, Denmark

M.J. Thompson

Space and Atmospheric Physics Group, The Blackett Laboratory, Imperial College,

London SW7 2BZ, UK

4.1 Introduction

Solar rotation has been known at least since the early seventeenth century when, with the newly invented telescope, Fabricius, Galileo and Scheiner observed the motion of sunspots across the solar disk (for a brief review, see Charbonneau *et al.* 1999). The rotation of the Sun and other stars originates from the contracting interstellar gas clouds from which stars are born; these clouds share the rotation of the Galaxy. As the clouds contract, they rotate more rapidly, as a result of the conservation of angular momentum and the reduction in the moment of inertia with contraction (for a discussion of star formation, see Lada and Shu 1990). Although the details of star formation within the contracting clouds are uncertain and involve mass loss and interaction with disks around the star which will transport angular momentum from one part of the cloud to another, it is plausible that newly formed star should be spinning quite rapidly. This is indeed observed: the rotation of the stellar surface causes a broadening of the lines in the star's spectrum, owing to the Doppler effect, and from measurements of this effect it is inferred that many young stars rotate at near the break-up speed, where the centrifugal force at the equator almost equals gravity.

Stars tend to slow down when they get older. At least for stars of roughly solar type, the observations show that the rotation rate decreases with increasing age (Skumanich 1972). This is thought to take place through angular-momentum loss in a wind, magnetically coupled to the outer parts of the Sun (e.g. Mestel 1968; Mestel and Spruit 1987). The extent to which the slowdown affects the deep interior of the Sun then depends on the efficiency of the coupling between the inner and outer parts. In fact, simple models of the dynamics of the solar interior tend to predict that the core

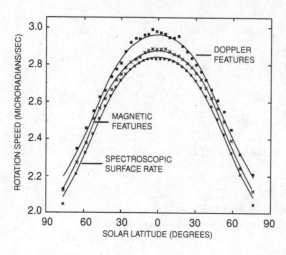

Figure 4.1. Near-surface solar rotation rates as determined from surface spectroscopic Doppler-velocity measurements, tracking magnetic features, and tracking Doppler features resulting from large-scale convective flow patterns. A rotation period of 36 days corresponds to an angular speed of 2.02 microrad/s, and a period of 25 days corresponds to 2.91 microrad/s. (Adapted from Snodgrass and Ulrich 1990.)

of the Sun is rotating up to fifty times as rapidly as the surface (e.g. Pinsonneault *et al.* 1989). Such a rapidly rotating solar core could have serious consequences for the tests of Einstein's theory of general relativity based on observations of planetary motion: a rapidly rotating core would flatten the Sun and hence perturb the gravitational field around it. Even a subtle effect of this nature, difficult to see directly on the Sun's turbulent surface, might be significant.

Very detailed observations have been carried out of the solar surface rotation by tracking the motion of surface features such as sunspots and, more recently, by Doppler-velocity measurements. It was firmly established by the nineteenth century, by careful tracking of sunspots at different latitudes on the Sun's surface, that the Sun is not rotating as a solid body: at the equator the rotation period is around 25 days, but it increases gradually towards the poles where the period is estimated to be in excess of 36 days. Figure 4.1 shows the near-surface solar rotation determined from surface Doppler measurements, as well as from tracking magnetic features and large-scale convective flow patterns, as a function of solar latitude. The origin of this *differential rotation* is almost certainly linked to the otherwise dynamic nature of the outer parts of the Sun. In the outer 29% of the Sun's radius (or 200 Mm), energy is transported by *convection*, in rising elements of warm gas and sinking elements of colder gas: this region is called the *convection zone* (for an overview of solar structure, see Christensen-Dalsgaard *et al.* 1996). The convection can be seen directly using high-resolution observations of the solar surface, in the *granulation*, with brighter areas of warm gas just arrived at the surface, surrounded by colder lanes of sinking gas. The gas motions also transport angular momentum, and hence provide a link between rotation in different parts of the convection zone. Furthermore, convection is affected by rotation, which may introduce anisotropy in the angular momentum transport. Indeed, it is likely that this transport is responsible for the differential rotation, although the details are far from understood.

Similarly complex dynamical interactions are also found in the giant gaseous planets (Jupiter, Saturn, Uranus and Neptune) which, like the Sun, are vigorously convecting as they rotate (see Ingersoll 1990). Here the interaction probably gives rise to the banded structures immediately visible on Jupiter, and more faintly on Saturn. Even closer to home, the Earth's atmosphere and oceans are rotating fluid systems and exhibit, among other things, large-scale circulations and meandering jets such as the jet stream. In all these systems, rotation plays a significant role in the observed dynamical behaviour.

4.2   Helioseismic probes of the solar interior

In recent years, the observation that the Sun is oscillating simultaneously in many small-amplitude global resonant modes has provided a new diagnostic of the solar interior. The frequencies of these global modes depend on conditions inside the Sun (Gough and Toomre 1991; see also Chapter 3, of this book), and so by measuring these frequencies we are able to make deductions about the state of the interior. This field is known as *helioseismology*. The observed oscillations are sometimes called five-minute oscillations, because they have periods in the vicinity of five minutes. The modes are distinguished not only by their different frequencies, but also by their different patterns on the surface of the Sun. Specifically, the behaviour of a mode is characterized by a spherical harmonic

$$Y_l^m(\theta, \phi) = c_{lm} P_l^m(\cos\theta) \exp(im\phi) \tag{4.1}$$

as function of co-latitude θ and longitude ϕ; here P_l^m is a Legendre function, and c_{lm} is a normalization constant. The spherical harmonics are characterized by two integer numbers, their *degree l* and their *azimuthal order m*; a few examples are illustrated in Figure 4.2. In addition, a mode is characterized by its radial order n which, approximately, is given as the number of nodes in the radial direction. The dependence on time t of the oscillation can be written as $\exp(-i\omega t)$, where ω is the angular frequency. Different modes are sensitive to different regions of the Sun, depending on their frequency, degree and azimuthal order (see also Chapter 3). In particular, in the radial direction the modes are essentially confined outside an inner *turning point* at the distance $r = r_t$ from the centre, where r_t satisfies

$$\frac{c(r_t)}{r_t} = \frac{\omega}{l + 1/2}, \tag{4.2}$$

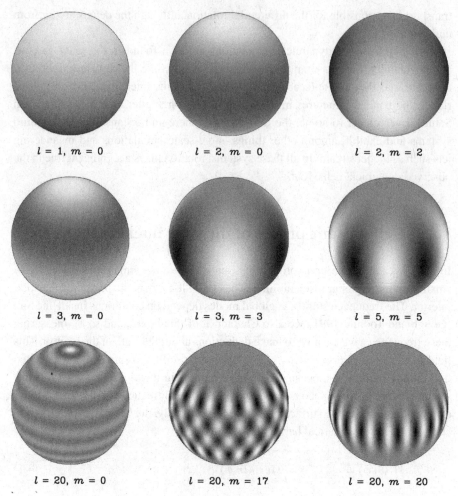

$l = 1, m = 0$ $l = 2, m = 0$ $l = 2, m = 2$

$l = 3, m = 0$ $l = 3, m = 3$ $l = 5, m = 5$

$l = 20, m = 0$ $l = 20, m = 17$ $l = 20, m = 20$

Figure 4.2. Examples of spherical harmonic patterns, for different values of the degree l and order m. For clarity the polar axis has been inclined 30° relative to the page.

c being the sound speed; thus low-degree modes extend over much of the Sun while high-degree modes are confined close to the solar surface. By exploiting the different sensitivities of the modes, helioseismology is able to make inferences about localized conditions inside the Sun.

4.2.1   Rotational effects on the oscillation frequencies

One of the factors that affect the mode frequencies is the Sun's rotation. The dominant effect of rotation on the oscillation frequencies is quite simple: the oscillation patterns illustrated in Figure 4.2 actually correspond to waves running around the equator.

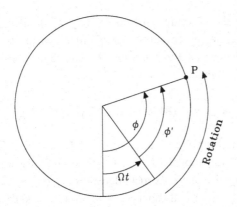

Figure 4.3. Geometry of rotational splitting, in a star rotating with angular velocity Ω. The point **P** has longitude ϕ' in the system rotating with the star and longitude $\phi = \phi' + \Omega t$ in the inertial system.

Patterns travelling in the same direction as the rotation of the Sun would appear to rotate a little faster, patterns rotating in the opposite direction a little more slowly. When observed at a given position on the Sun the oscillations in the former case would have slightly higher frequency, and in the latter case slightly lower frequency, than if the Sun had not been rotating. The frequency difference between these two cases therefore provides a measure of the rotation rate of the Sun.

This simple description can be made more precise by noting that, as a function of longitude and time t, the oscillations behave as $\cos(\omega t - m\phi)$ (apart from an arbitrary phase). We now consider a star that is rotating uniformly, with angular velocity Ω, and introduce a coordinate system rotating with the star, with longitude ϕ'; the longitude ϕ in an inertial frame is related to ϕ' by $\phi' = \phi - \Omega t$ (cf. Figure 4.3). We furthermore consider an oscillation with frequency ω_0 in the rotating frame, and hence varying as $\cos(m\phi' - \omega_0 t)$. Consequently, the oscillation as observed from the inertial frame depends on ϕ and t as

$$\cos(m\phi - m\Omega t - \omega_0 t) \equiv \cos(m\phi - \omega_m t) ,$$

where

$$\omega_m = \omega_0 + m\Omega . \tag{4.3}$$

Thus the frequencies are split according to m, the separation between adjacent values of m being simply the angular velocity; this is obviously just the result of the advection of the wave pattern with rotation.

In reality other effects must be taken into account to describe the frequency shifts caused by rotation, which are often referred to as the *rotational frequency splitting*. The *Coriolis force* affects the dynamics of the oscillations and hence their frequencies, although it turns out that for the modes observed in the five-minute region this effect is modest; owing to the slow rotation of the Sun centrifugal effects are essentially negligible and in any case have a different functional dependence on m (see Section 4.2.2). However, the variation of angular velocity $\Omega(r, \theta)$ with position in the Sun must be taken into account[†]. Each mode feels an average angular velocity, where the average is determined by the mode's frequency, degree and azimuthal order (Hansen, Cox and van Horn 1977; Gough 1981). The precise form of this spatial average is described by a weight function, or *kernel*, such that

$$\omega_{nlm} = \omega_{nl0} + m \int_0^R \int_0^\pi K_{nlm}(r, \theta) \Omega(r, \theta) r \, dr \, d\theta \, , \tag{4.4}$$

where R is the total radius of the Sun; the kernels K_{nlm} can be calculated from the eigenfunctions for a non-rotating model (Schou, Christensen-Dalsgaard and Thompson 1994). It might be noted that the kernels depend only on m^2, so that the rotational splitting $\omega_{nlm} - \omega_{nl0}$ is an odd function of m. Also, the kernels are symmetrical around the equator; as a result, the rotational splitting is only sensitive to the component of Ω which is similarly symmetrical. In the special case where $\Omega = \Omega(r)$ is independent of θ, the integral in Eq. (4.4) is independent of m and hence the rotational splitting is simply proportional to m; note that this is the same linear dependence on m as for the simple effect of advection (cf. Eq. 4.3).

These weight functions vary from mode to mode. As already indicated in Figure 4.2, modes with $m = l$ are concentrated near the equator, increasingly so with increasing l, whereas modes of lower azimuthal order extend to higher latitudes. Thus modes with $m = l$ feel an average of the rotation near the equatorial plane, whereas modes of lower azimuthal order sense the average rotation over a wider range of latitudes. In a similar manner, the high-degree five-minute modes (i.e., with large values of l) sense only conditions near the surface of the Sun, while modes of low degree feel conditions averaged over much of the solar interior (cf. Eq. 4.2).

These properties can be illustrated by a few examples of weight functions, as shown in Figure 4.4. The observed modes include some that penetrate essentially to the solar centre, others that are trapped very near the surface, and the whole range of intermediate penetration depths, with a similar variation in latitudinal extent. Thus the observed frequency splittings provide a similarly wide range of averages of the internal rotation.

4.2.2   Data on rotational splitting

The rotational splittings must of course be measured from precise observations of the Sun's global oscillations. Several instruments and observational networks are engaged in this work (see Section 4.3). For practical considerations (including the effects of mode realization arising from the finite lifetime of the modes), the observers commonly do not measure the individual rotational splittings for each m value but

---

[†] Specifically, $\Omega(r, \theta)$ refers to the azimuthal component of the azimuthally averaged flow field within the Sun.

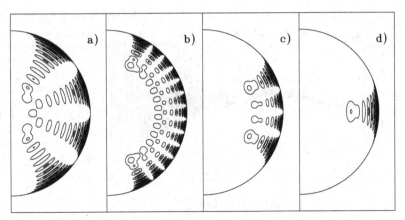

Figure 4.4. Contour plot of kernel weight functions determining the sensitivity of different modes to the solar internal rotation. All modes have frequencies near 2 mHz; their degree l and azimuthal order m are, from left to right: $(l, m) = (5, 2)$, $(20, 8)$, $(20, 17)$, and $(20, 20)$. Note that the kernels in panels (a) and (b) have roughly the same latitudinal extent, because $m/(l + 1/2)$ is almost the same for these modes, but the lower-l mode senses the deeper regions of the Sun; also that for fixed l the modes are more confined to low latitudes as m increases.

instead fit to the observational data an expression of the form

$$\nu_{nlm} = \nu_{nl0} + \sum_{j=1}^{j_{max}} a_j(n, l)\, \mathcal{P}_j^{(l)}(m)\,. \tag{4.5}$$

Here then the inferred quantities which convey information about the rotation are the so-called a coefficients $a_j(n, l)$, and the $\mathcal{P}_j^{(l)}$ are polynomials of degree j, suitably chosen to obtained statistically independent coefficients (cf. Schou *et al.* 1994). The odd a coefficients (i.e., those with odd values of j) contain information about the rotation through the advection and Coriolis effects on the modes, and these are used to infer the Sun's internal rotation.

Figure 4.5 shows the first two odd a coefficients obtained from observations with the SOI-MDI experiment on the SOHO spacecraft. To provide some indication of the variation of rotation with depth, the data are shown against $\nu/(l + 1/2)$ ($\nu = \omega/2\pi$ being the cyclic frequency), which according to Eq. (4.2) determines the location of the turning point, $r = r_t$, of the mode. The turning-point radius is indicated as the upper abscissas. The increase in a_1 with increasing $\nu/(l + 1/2)$, i.e., with increasing depth of penetration of the modes, indicates that the rotation rate increases with depth in the outer $5 - 10\%$ of the solar radius. Also, it should be noticed that a_3 decreases towards zero for those modes that penetrate below $r = 0.8R$, as is indeed the case for the higher a coefficients also; thus the rotational splitting tends to become linear in m, suggesting that rotation tends towards latitude independence in the inner parts of the Sun (cf. Section 4.2.1). These crude inferences are confirmed by the more detailed analysis presented in Section 4.3.

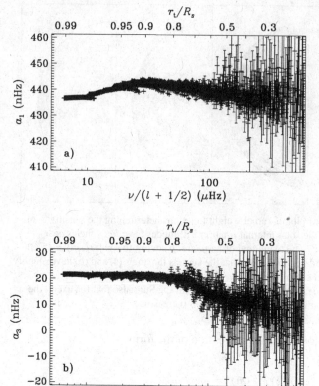

Figure 4.5. Observed a coefficients a_1 and a_3 from the expansion given in Eq. (4.5) of frequency splittings from SOI-MDI observations. The error bars correspond to one standard deviation. The upper abscissas indicate the turning-point radius, defined by Eq. (4.2).

The even a coefficients contain information about aspects of the Sun which affect modes of azimuthal orders $\pm m$ equally: these include centrifugal distortion and other departures of the Sun from spherical symmetry. The even coefficients are generally small in value and are not used for determining the form of the rotation, though they can be used to make inferences about the Sun's internal asphericity; this is discussed in Chapter 3 of this book.

4.2.3   Inversion for solar internal rotation

The wealth of data on rotational splitting allows the determination of the detailed variation of rotation with position in the solar interior. Modes of high degree, trapped near the surface, provide measures of the rotation of the superficial layers of the Sun. Having determined that, its effect on the somewhat more deeply penetrating modes can be eliminated, leaving just a measure of rotation at slightly greater depths. In this way, information about rotation in the Sun can be 'peeled' layer by layer, much as one could an onion, in a way that allows us to obtain a complete image of solar internal rotation.

It is fairly evident that this process gets harder, the deeper one attempts to probe, since fewer and fewer modes penetrate to the required depth; furthermore, the effect of rotation decreases because of the smaller size of the region involved. Thus the rotation of the solar core is difficult to determine. Similarly, all modes are affected by the equatorial rotation while only modes of low m extend to the vicinity of the poles, and the polar regions have relatively little effect on the oscillations, complicating the determination of the high-latitude rotation. However, as we shall see the quality of current data is such that the angular velocity can be determined quite near the poles, at least in the outer parts of the convection zone.

The analysis of the oscillation data to infer properties of the solar interior is often characterized as *inversion* (Gough and Thompson 1991). To illustrate aspects of these procedures, we write Eq. (4.4) as

$$\Delta_i = \int K_i(\mathbf{x})\Omega(\mathbf{x})d\mathbf{x} , \qquad (4.6)$$

where the label i stands for the modes (n, l, m) and $\mathbf{x}$ replaces the coordinates (r, θ); also, $\Delta_i = m^{-1}(\omega_{nlm} - \omega_{nl0})$ are the observed data. Likewise, kernels similar to the $K_i(\mathbf{x})$ can be derived if the data Δ_i are in the form of a coefficients, as in Eq. (4.5). The goal of the analysis is obviously to infer the properties of $\Omega(\mathbf{x})$ from the data, taking into account also the observational errors. In most cases considered so far, the analysis corresponds either implicitly or explicitly to making linear combinations of the data; thus to infer Ω at a point $\mathbf{x}_0 = (r_0, \theta_0)$, say, coefficients $c_i(\mathbf{x}_0)$ are determined such that, as far as possible,

$$\Omega(\mathbf{x}_0) \simeq \bar{\Omega}(\mathbf{x}_0) = \sum_i c_i(\mathbf{x}_0)\Delta_i , \qquad (4.7)$$

where $\bar{\Omega}$ is the inferred approximation to the true angular velocity. From Eq. (4.6) it then follows that

$$\bar{\Omega}(\mathbf{x}_0) = \int \mathcal{K}(\mathbf{x}_0, \mathbf{x})\Omega(\mathbf{x}) \, d\mathbf{x} , \qquad (4.8)$$

where the *averaging kernels* $\mathcal{K}$ are given by

$$\mathcal{K}(\mathbf{x}_0, \mathbf{x}) = \sum_i c_i(\mathbf{x}_0)K_i(\mathbf{x}) . \qquad (4.9)$$

Also, if the standard errors $\sigma(\Delta_i)$ of the observations are known, the error in the inferred angular velocity $\bar{\Omega}(\mathbf{x}_0)$ can be found from Eq. (4.7).

It is evident from Eq. (4.8) that the coefficients $c_i(\mathbf{x}_0)$ should be determined such that $\mathcal{K}(\mathbf{x}_0, \mathbf{x})$ is as far as possible localized near $\mathbf{x} = \mathbf{x}_0$. The extent of $\mathcal{K}$ provides a measure of the resolution of the inversion. At the same time, it must be ensured that the error on $\bar{\Omega}(\mathbf{x}_0)$ is reasonable. In general, there is a trade-off between error and resolution, determined by one or more parameters of the procedure.

In one commonly used inversion procedure, the so-called regularized least-squares (or RLS) procedure, the solution is parametrized and the parameters are determined through a least-squares fit of the data to the right-hand side of Eq. (4.6). To ensure that the solution is well-behaved and the errors are of reasonable magnitude, the fit is regularized by limiting also, for example, a measure of the square of the second derivative of the solution, thus penalizing rapid variations. From the results of the fit the coefficients $c_i(x_0)$, and hence the averaging kernels, can be determined. In a second class of procedures, the optimally-localized averages (or OLA) procedures, the $c_i(x_0)$ are determined such as to obtain a localized averaging kernel $\mathcal{K}(x_0, x)$, while at the same time limiting the errors. Details of these procedures and their results were given by Schou *et al.* (1998).

As already remarked, the weight functions (Figure 4.4) are symmetrical around the solar equator, and so we can infer only the similarly symmetric component of rotation. This must be kept in mind in the following, when interpreting the results. We note that this restriction can be avoided by applying local helioseismic techniques to the data: such techniques are described in Chapter 5 of this book.

4.3   The solar internal rotation

Early helioseismic data on rotational splittings provided information only about the modes with $m \simeq \pm l$; as a result, they were sensitive mainly to rotation near the equator. Duvall *et al.* (1984) showed that there was relatively little variation of rotation with depth; in particular, there were no significant indications of a rapidly rotating core. A few years later initial data on the dependence of the splitting on m were obtained (Brown 1985; Libbrecht 1988, 1989). Strikingly, they indicated that the surface latitudinal differential rotation persisted through the convection zone, whereas there was little indication of variation with latitude in the rotation beneath the convection zone (e.g. Brown and Morrow 1987; Brown *et al.* 1989; Christensen-Dalsgaard and Schou 1988; see also Figure 4.6).

In the last few years the amount and quality of helioseismic data on solar rotation have increased dramatically, as a result of several ground-based and space-based experiments (Duvall 1995). The LOWL instrument of the High Altitude Observatory has provided high-quality data on modes of low and intermediate degree over the past more than five years. The BiSON and IRIS networks, observing low-degree modes in Doppler velocity integrated over the solar disk, have yielded increasingly tight constraints on the rotation of the solar core, while the GONG six-station network is setting a new standard for ground-based helioseismology (Harvey *et al.* 1996). Further, the SOI-MDI experiment on SOHO (Scherrer *et al.* 1995) has yielded a wealth of data on modes of degree up to 300, allowing a detailed analysis of the properties of rotation in the convection zone. The results we present below are the combined knowledge that has emerged from these observational efforts (e.g. Tomczyk, Schou

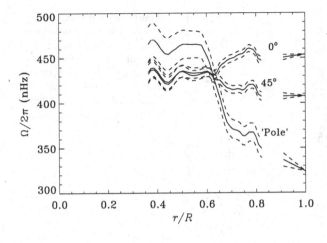

Figure 4.6. Rotation rate $\Omega/2\pi$ inferred from inversion of a coefficients from BBSO (Libbrecht 1989), targeted at the equator, latitude 45° and the pole. The dashed lines indicate the $1 - \sigma$ error limits. The inversion was only possible between $0.4R$ and $0.8R$; for clarity, the results have been connected with the directly observed surface rates. (Adapted from Christensen-Dalsgaard and Schou 1988.)

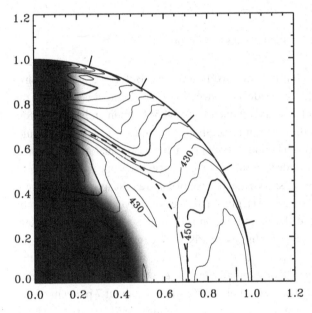

Figure 4.7. Rotation of the solar envelope inferred from observations by SOI-MDI (Schou *et al.* 1998). The equator is at the horizontal axis and the pole is at the vertical axis, both axes being labelled by fractional radius. Some contours are labelled in nHz, and, for clarity, selected contours are shown by bolder curves. The dashed circle indicates the base of the convection zone and the tick marks at the edge of the outer circle are at latitudes 15°, 30°, 45°, 60°, and 75°. The shaded area indicates the region in the Sun where no reliable inference can be made with the current data.

and Thompson 1995; Thompson *et al.* 1996; Corbard *et al.* 1997, 1999; Di Mauro and Dziembowski 1998; Schou *et al.* 1998). Some of these results are illustrated in Figures 4.7 and 4.8. These are the result of a so-called regularized least squares inversion applied to SOI-MDI data (Schou *et al.* 1998). Figure 4.7 shows the inferred rotation as a contour plot, in the region where the results are judged to be reliable; Figure 4.8 shows the radial cuts through the same solution, at a few selected latitudes.

4.3.1   Rotation of the solar convection zone

In discussing what we now know about the rotation inside the Sun, we shall start from the near-surface layers and work towards the centre. As we have already discussed,

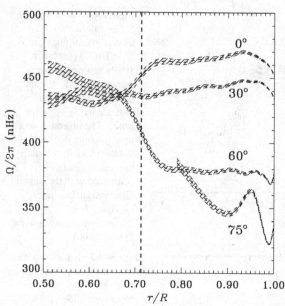

Figure 4.8. Rotation of the solar envelope as a function of radius, at the latitudes indicated, inferred from observations by SOI-MDI (Schou *et al.* 1998). The heavy dashed line marks the base of the convection zone.

the outer 29 per cent of the Sun is convectively unstable. Before helioseismology, models predicted that the rotation inside the convection zone would organize itself on cylinders aligned with the rotation axis (Glatzmaier 1985; Gilman and Miller 1986; see also Section 4.4). Thus the rotation at depth at, say, equatorial latitudes would match the surface rotation at high latitudes, rather than the faster equatorial rotation at the surface, and so at given latitude the rotation in the convection zone would decrease with depth. Helioseismology has shown that this is not so: to a first approximation it is more accurate to say that the rotation at a given latitude is nearly constant with depth, or to put it another way the differential rotation seen at the surface imprints itself through the convection zone. This finding is clearly visible in Figures 4.7 and 4.8. In detail, the situation is more complicated. At low latitudes, immediately beneath the solar surface the rotation rate actually initially *increases* with depth. The equatorial rotation reaches a maximum at a depth of about 50 Mm (i.e., about 7 per cent of the way in from the surface to the centre of the Sun): at this point, the rotation rate is about 5 per cent higher than it is at the surface. This is consistent with a variety of surface measurements of rotation. Tracking sunspots tends to give a slightly higher rotation rate than that obtained by making direct spectroscopic measurements of the velocity of the surface (Korzennik *et al.* 1990; see also Figure 4.1). Probably the reason is that the sunspots extend to some depth below the surface, and so are dragged along at a rate that is more similar to the subsurface rotation a few per cent beneath the surface which helioseismology has revealed.

Figure 4.9 shows averaging kernels (cf. Eq. 4.9) corresponding to the solution in Figure 4.7, at a few locations inside the Sun. These give an indication of the resolution attained by the inversion. Note that the radial resolution is finer for points high in the

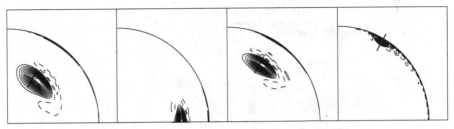

Figure 4.9. Averaging kernels for the inversion solution shown in Figure 4.7, at selected radii and latitudes in the Sun, which are (from left to right) as follows: $0.540R$, $60°$; $0.692R$, $0°$; $0.692R$, $60°$; $0.952R$, $60°$. The corresponding locations are indicated with crosses. Positive contours are shows as solid lines, negative contours as dashed lines. (Adapted from Schou *et al.* 1998.)

convection zone and poorer in the radiative interior. This is largely a reflection of the local vertical wavenumber of the waves at the various depths.

The rotational velocity at the surface of the Sun is about 2 km per second, dropping off rather smoothly towards higher latitudes. However, it has now been found that superimposed on this are bands of faster and slower rotation, a few metres per second higher or lower than the mean flow (Kosovichev and Schou 1997; Schou 1999). The origin of this behaviour, illustrated in Figure 4.10, is not understood, but it is reminiscent of the more pronounced banded flow patterns seen on Jupiter and Saturn. Evidence for such bands had been obtained previously from direct Doppler measurements on the solar surface (Howard and LaBonte 1980). However, the seismic inferences have shown that they extend to a depth probably exceeding 40 Mm beneath the surface (Howe *et al.* 2000a). Moreover, these bands migrate from high latitudes towards the equator over the solar cycle.

It has been customary to represent the directly measured surface rotation rate in terms of a simple low-order expansion in $\sin \psi$, where ψ is latitude on the Sun. This in fact quite successfully captured the observed behaviour; however, since the solar rotation axis is close to the plane of the sky, direct measurements of rotation near the poles are difficult and uncertain. Strikingly, the helioseismic results have shown a marked departure from this behaviour, at latitudes above about $60°$: Relative to the simple fit, the actual rotation rate decreases quite markedly there. The origin or significance of this behaviour is not yet understood. There is also evidence, hinted at in Figure 4.7, of a more complex behaviour of rotation at high latitudes. Some analyses have shown a 'jet', i.e., a localized region of more rapid rotation, at a latitude around $75°$ and a depth of about 35 Mm beneath the solar surface (Schou *et al.* 1998). Also, evidence has been found that the rotation rate shows substantial variations in time at high latitudes, over time scales of order months. It is probably fair to say that the significance of these results is still somewhat uncertain, however. Also, it should be kept in mind, as mentioned above, that the results provide an average of

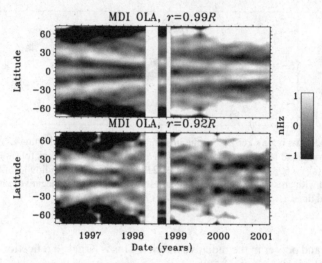

Figure 4.10. The evolution with time of the fine structure in the near-surface solar rotation, based on observations from the SOI-MDI instrument on the SOHO spacecraft, after subtraction of the time-averaged rotation rate. The result is represented as a function of time (horizontal axis) and latitude (vertical axis), the grey scale at the right giving the scale in nHz; 1 nHz corresponds to a speed of around 4 m/s at the equator. The banded structure, apparently converging towards the equator as time goes by, should be noted. The vertical white bands correspond to time intervals when the spacecraft was temporarily inactive. (Adapted from Howe *et al.* 2001.)

rotation in the Northern and Southern hemispheres and, evidently, an average over the observing period of at least 2 – 3 months. Thus the interpretation of the inferred rotation rates in terms of the actual dynamics of the solar convection zone is not straightforward.

4.3.2 The tachocline

At the base of the convection zone, a remarkable transition occurs: the variation of rotation rate with latitude disappears, so that the region beneath the convection zone rotates essentially rigidly, at a rate corresponding to the surface rate at mid-latitudes (see Figure 4.12). The region over which the transition occurs is very narrow, no more than a few per cent of the total radius of the Sun (e.g. Kosovichev 1996, Charbonneau *et al.* 1999). This layer has been called the *tachocline* (Spiegel and Zahn 1992). Why the differential rotation does not persist beneath the convection zone is not yet known, but it is possible that a large-scale weak magnetic field permeates the inner region and enforces nearly rigid rotation by dragging the gas along at a common rate (Gough and McIntyre 1998). Such a field is quite possible as a relic from the original collapsing gas cloud from which the Sun condensed.

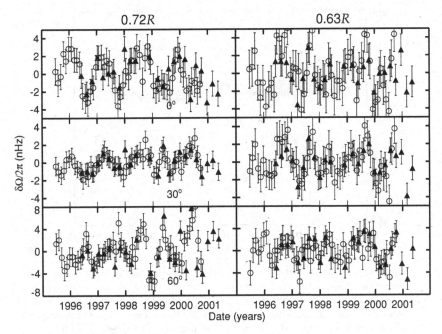

Figure 4.11. Deviation from the mean rotation rate inferred from inversions at various locations near the base of the convection zone, as a function of time. Open circles represent results from the GONG network, and filled triangles are from the SOI-MDI experiment. (From Howe *et al.* 2001.)

The discovery of the tachocline, and of the form of the rotation in the convection zone, has led to an adjustment of our theories of the *solar dynamo* (see Chapter 6 of this book). One idea is that the dynamo action consists of two components: a twisting of the magnetic field by the motion of convective elements, and a shearing out of the field by differential rotation. Prior to the helioseismic findings, the simulations of rotation implied that the radial gradient of differential rotation in the convection zone could provide the second ingredient, so it was thought that the dynamo action occurred in that region. Now, however, the tachocline with its very substantial radial gradient seems a more likely location for the dynamo action producing the large-scale magnetic field (Gilman, Morrow and DeLuca 1989).

If the magnetic field is built up in the tachocline over the course of the solar cycle, one might expect to see variations over time in the tachocline properties, including perhaps the rotation rate there. Variations in the rotation rate have in fact been found in the vicinity of the tachocline, but the timescale was unexpected. Rather than the 11-year timescale of the solar cycle, Howe *et al.* (2000b) have reported quasi-periodic variations in the equatorial region just above and below the tachocline with a period of 1.3 years (Figure 4.11). The oscillations are revealed by subtracting out the temporal mean rotation rate at each location and looking at the residuals as a function of time.

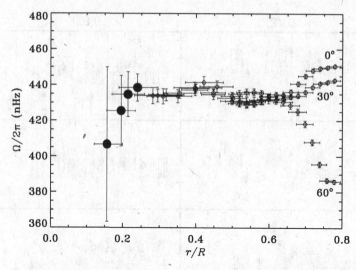

Figure 4.12. The inferred rotation as a function of fractional radius inside the Sun at three solar latitudes: the equator, 30° and 60°; the vertical axis shows the rotation frequency in nHz. The vertical error bars indicate the statistical uncertainty on the rotation rate (±1 standard deviation), wheras the horizontal bars provide a measure of the radial resolution of the inversion. Note that the result becomes much more uncertain in the deep interior, where furthermore the different latitudes cannot be separated. The observations used to infer the rotation were from the LOWL instrument and the BiSON network (Chaplin *et al.* 1999).

The equatorial variations at radius $0.72R$ show the 1.3-year oscillation most strongly, and the variations at radius $0.63R$ exhibit the same period but are in antiphase. This implies that the radiative interior also partakes in the temporal variation, and suggests perhaps a back-and-forth exchange of angular momentum between the two locations. The signal is much weaker at 30° latitude, and the variations at higher latitudes are possibly not significant. These findings are exciting because they may represent the first direct observation of variability at the seat of the solar dynamo.

4.3.3   The radiative interior

Even deeper in the Sun, in the radiative interior, the helioseismic results on the rotation are more uncertain due to the fact that so few of the observed five-minute modes (only the low-degree modes) have any sensitivity to this region. Indeed, the results have been somewhat contradictory (for a review, see Eff-Darwich and Korzennik 1998), some indicating rotation faster than the surface rate and others indicating rotation slower than or comparable to the rotation rate at the base of the convection zone; an example is illustrated in Figure 4.12. However, down to within 15 per cent of the solar radius from the centre, which is the deepest point at which present observations permit localized inferences to be made, all the modern results agree that the rotation rate is not more than a factor two different from the surface rate: thus early models which predicted that the whole of the nuclear-burning core was rotating much faster are firmly ruled out. Again, this finding would be consistent with a magnetic field linking the core to the bulk of the radiative interior.

4.4   Modelling solar rotation

Although helioseismology has provided us with a remarkably detailed view of solar internal rotation, the theoretical understanding of the inferred behaviour is still incomplete. In the convection zone, the problem is to model the complex combined dynamics of rotation and convection, the latter occurring on scales from probably less than a few hundred kilometres to the scale of the entire convection zone and time scales from minutes to years. Viscous dissipation is estimated to occur on even smaller spatial scales, of order 0.1 km or less. Capturing this range of scales is entirely outside the possibility of current numerical simulations; thus simplifications are required. Detailed simulations of near-surface convection, on a scale of a few Mm, have been remarkably successful in reproducing the observed properties of the granulation (Stein and Nordlund 1989, 1998) but are evidently not directly relevant to the question of rotation. Simulations of the entire convection zone are necessarily restricted to rather large scales and hence cannot capture the near-surface details. Such simulations, therefore, typically exclude the outer 30 Mm of the convection zone. Early examples of such simulations by Gilman and Glatzmaier, of fairly limited resolution, showed a tendency for rotation to organize itself on cylinders (Glatzmaier 1985; Gilman and Miller 1986): the rotation rate depended primarily on the distance to the rotation axis. The convection itself in these simulations was dominated by so-called *banana cells* – long, thin, large-scale convection cells oriented in the north-south direction – as seen also, for example, in laboratory convection as observed in the GFFC Spacelab experiment (Hart *et al.* 1985).

Rotation on cylinders is predicted for simple systems by the *Taylor-Proudman theorem*, which may be derived as follows. The velocity $\mathbf{u}$ in an inviscid fluid in a gravitational potential Φ satisfies an equation of motion

$$\frac{\partial \mathbf{u}}{\partial t} + \mathbf{u} \cdot \nabla \mathbf{u} = -\frac{1}{\rho}\nabla p - \nabla \Phi \,, \tag{4.10}$$

where p and ρ are pressure and density respectively, and t is time. Suppose that the velocity is purely rotational and independent of time, so in cylindrical coordinates (s, ϕ, z) we write $\mathbf{u} = s\Omega(s, z)\mathbf{e}_\phi$, where s is the distance from the rotation axis, z distance along the axis, and $\mathbf{e}_\phi$ is a unit vector in the azimuthal direction. Taking the curl of Eq. (4.10) gives

$$-s\frac{\partial \Omega^2}{\partial z}\mathbf{e}_\phi = \frac{1}{\rho^2}\nabla \rho \times \nabla p \,. \tag{4.11}$$

In the solar convection zone the fluid is essentially isentropic (uniform specific entropy) and so the pressure can be regarded as a function of the density alone: hence the right-hand side of Eq. (4.11) is zero, and thus Ω depends only on s, not on z, i.e., the rotation rate is uniform on cylindrical surfaces.

If the Taylor-Proudman theorem applied in the Sun, the rotation rate on a given cylinder would obviously be observable where the cylinder intersected the surface; thus the observed decrease of rotation rate with increasing latitude would correspond to a similar decrease of rotation rate with depth at, say, the equator. The actual solar rotation, shown in Figure 4.7, is obviously very different from this behaviour and from the simulation results mentioned above. The overall variation of rotation within the convection zone is evidently predominantly with latitude, with little variation in the radial direction except in the tachocline. Given the necessary simplifications of the calculations, their failure to model solar rotation is perhaps not surprising. In particular, the effects of smaller-scale turbulence (beneath the smallest scale resolved in the simulation) is typically represented as some form of viscosity; it was suggested by Gough (1976), and later by others, that the effect of rotation on the small-scale motion might render this turbulent viscosity non-isotropic, with important effects on the transport of angular momentum within the convection zone. Of course, such effects were not included in our derivation of the Taylor-Proudman theorem, where the inviscid fluid equations were used and the flow was assumed to be described just by the (large-scale) rotation. In contrast, simple models of convection-zone dynamics, with parametrized anisotropic viscosity, have in fact had some success in reproducing the helioseismically inferred rotation rate (Pidatella *et al.* 1986).

Recent advances in computing power have led to improved numerical simulations (Miesch *et al.* 2000; Brun and Toomre 2002), which come closer to representing turbulent convective flow regimes such as exist in the Sun's convection zone. Figure 6 shows results from two such simulations by Miesch, Elliott and Toomre. The simulations can yield a range of differential rotation profiles, depending on the conditions imposed at the top and bottom boundaries of the simulation region, and on the parameter values adopted for the problem. Since it is not obvious what are the most appropriate boundary conditions and parameter values to choose, it is necessary to explore various possibilities and study the different responses. Simulation B has rotation contours at mid-latitudes which are nearly radial, as in the Sun (compare Figure 4.7), but the contrast in rotation rate between low and high latitudes is not as great as is observed in the Sun (about 70 nHz, rather than 130 nHz). In case A, the latitudinal variation of the Sun's rotation is better reproduced, but the mid-latitude contours do not look quite as similar to those in Figure 4.7. Nonetheless these results are encouraging indications that we may be close to reproducing theoretically the gross features of the solar rotation inferred by helioseismology. There is still though much work ahead, both observational and theoretical, in getting a detailed understanding of the Sun's rotation and with that, we hope, a better understanding of the solar activity cycle and of large-scale rotating fluid systems on planets and stars.

4.5   Final remarks

The advent of helioseismology has revolutionized our knowledge of the Sun's internal rotation, showing that the dynamics of the solar interior are quite different from the pre-helioseismic model simulations. The rotation within the convection zone is now well established except perhaps at high latitudes, and the more turbulent simulations appear to be getting close to matching observation. The existence of the tachocline and the near-uniform rotation in much of the deeper interior are also well established, and suggest that a magnetic field enforces essentially rigid rotation in at least the outer part of the Sun's radiative interior. The rotation of the very deepest part of the Sun is still somewhat uncertain but a very fast-spinning core is ruled out unless it is very small. The temporal variations of the near-surface banded flows are firmly established but their origin is still incompletely understood. The less certain aspects of the helioseismic inferences are the temporal variations around the tachocline region and the high-latitude jet: we must wait to see whether these are confirmed by future observations.

In addition to the global helioseismic investigations, the newer field of local helioseismology (described more fully in Chapter 5 of this book) is providing knowledge of the dynamics of the near-surface layers of the Sun. These techniques, which by virtue of their local character can also provide information about how flows and structure vary with longitude and in the northern and southern hemispheres, reveal not only the rotational flow field, but also meridional (i.e., North-South) flows. Meridional flows are likely to play a significant role in understanding the angular-momentum budget in the convective envelope. Recent findings with one such local helioseismic technique (Haber *et al.* 2001) indicate fascinating changes in the structure of the meridional circulation patterns in one hemisphere of the Sun from year to year; also, they confirm the migrating banded flows in the near-surface rotation but indicate that they can be markedly asymmetric about the equator. These near-surface findings, combined with the deeper inferences of the global methods, continue to provide an intriguing insight into the dynamics of the solar interior.

Acknowledgements

We are grateful to Prof. R. K. Ulrich for providing Figure 4.1, to Dr. R. Howe for providing Figures 4.10 and 4.11, and to Dr. M. S. Miesch for providing Figure 4.13. We are happy to acknowledge the financial support of the UK Particle Physics

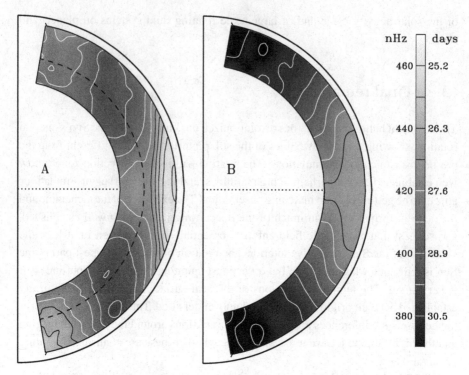

Figure 4.13. The results of two simulations of convection-zone rotation by Miesch, Elliott and Toomre (Miesch 2000). The two simulations use different boundary conditions and parameter values, and illustrate some of the range of possible responses of the differential rotation to the form of the convection. Also note that Simulation A has a higher resolution and includes penetration into a stable region beneath the convection zone, whereas the convective motions in Simulation B are more laminar and there is no penetration beneath the convection zone.

and Astronomy Research Council, and the Danish National Research Foundation through its establishment of the Theoretical Astrophysics Center. Our thanks also go to Prof. Juri Toomre and Dr. Michael Knölker for hospitality at JILA and HAO respectively, during the time when much of this chapter was written.

References

Brown, T. M. (1985). *Nature*, **317**, 591.

Brown, T. M. and Morrow, C. A. (1987). *Astrophys. J.*, **314**, L21.

Brown, T. M., Christensen-Dalsgaard, J., Dziembowski, W. A., Goode, P., Gough, D. O., and Morrow, C. A. (1989). *Astrophys. J.*, **343**, 526.

Brun, A. S. and Toomre, J. (2002). *Astrophys. J.*, **570**, 865.

Chaplin, W. J., Christensen-Dalsgaard, J., Elsworth, Y., Howe, R., Isaak, G. R., Larsen, R. M., New, R., Schou, J., Thompson, M. J., and Tomczyk, S. (1999). *Mon. Not. R. Astr. Soc.*, **308**, 405.

Charbonneau, P., Christensen-Dalsgaard, J., Henning, R., Larsen, R. M., Schou, J., Thompson, M. J., and Tomczyk, S. (1999). *Astrophys. J.* **527**, 445.

Christensen-Dalsgaard, J. and Schou, J. (1988). In *Seismology of the Sun and Sun-like Stars*, ed. V. Domingo and E. J. Rolfe, ESA SP-286 (ESA Publications Division, Noordwijk, The Netherlands), 149.

Christensen-Dalsgaard, J., Däppen, W., Ajukov, S. V., Anderson, E. R., Antia, H. M., Basu, S., Baturin, V. A., Berthomieu, G., Chaboyer, B., Chitre, S. M., Cox, A. N., Demarque, P., Donatowicz, J., Dziembowski, W. A., Gabriel, M., Gough, D. O., Guenther, D. B., Guzik, J. A., Harvey, J. W., Hill, F., Houdek, G., Iglesias, C. A., Kosovichev, A. G., Leibacher, J. W., Morel, P., Proffitt, C. R., Provost, J., Reiter, J., Rhodes Jr., E. J., Rogers, F. J., Roxburgh, I. W., Thompson, M. J., and Ulrich, R. K. (1996). *Science*, **272**, 1286.

Corbard, T., Berthomieu, G., Morel, P., Provost, J., Schou, J., and Tomczyk, S. (1997). *Astron. Astrophys.*, **324**, 298.

Corbard, T., Blanc-Féraud, L., Berthomieu, G., and Provost, J. (1999). *Astron. Astrophys.*, **344**, 696.

Di Mauro, M. P. and Dziembowski, W. A. (1998). *Mem. Soc. Astron. Ital.*, **69**, 559.

Duvall, T. L. (1995). In *Proc. Fourth SOHO Workshop: Helioseismology*, eds. J. T. Hoeksema, V. Domingo, B. Fleck, and B. Battrick, ESA SP-376, vol. 1 (ESA Publications Division, Noordwijk, The Netherlands), 107.

Duvall, T. L., Dziembowski, W. A., Goode, P. R., Gough, D. O., Harvey, J. W., and Leibacher, J. W. (1984). *Nature*, **310**, 22.

Eff-Darwich, A. and Korzennik, S. G. (1998). In *Structure and dynamics of the interior of the Sun and Sun-like stars; Proc.*

SOHO 6/GONG 98 Workshop, ed. S. G. Korzennik and A. Wilson, ESA SP-418 (ESA Publications Division, Noordwijk, The Netherlands), 685.

Gilman, P. A. and Miller, J. (1986). *Astrophys. J. Suppl.*, **61**, 585.

Gilman, P. A., Morrow, C. A., and DeLuca, E. E. (1989). *Astrophys. J.*, **338**, 528.

Glatzmaier, G. (1985). *Astrophys. J.*, **291**, 300.

Gough, D. O. (1976). In *Problems of stellar convection, IAU Colloq. No. 38, Lecture Notes in Physics*, vol. **71**, ed. E. Spiegel and J.-P. Zahn (Springer-Verlag, Berlin), 15.

Gough, D. O. (1981). *Mon. Not. R. Astr. Soc.*, **196**, 731.

Gough, D. O. and McIntyre, M. E. (1998). *Nature*, **394**, 755.

Gough, D. O. and Thompson, M. J. (1991). In *Solar interior and atmosphere*, ed. A. N. Cox, W. C. Livingston and M. Matthews, Space Science Series (University of Arizona Press, Tucson, AZ), 519.

Gough, D. O. and Toomre, J. (1991). *Ann. Rev. Astron. Astrophys.*, **29**, 627.

Haber, D. A., Hindman, B. W., Toomre, J., Bogart, R. S., and Hill, F. (2001). In *Helio- and Asteroseismology at the Dawn of the Millennium: Proc. SOHO 10 / GONG 2000 Workshop*, ed. A. Wilson, ESA SP-464 (ESA Publications Division, Noordwijk, The Netherlands), 213.

Hansen, C. J., Cox, J. P., and van Horn, H. M. (1977). *Astrophys. J.*, **217**, 151.

Hart, J. E., Glatzmaier, G. A. and Toomre, J. (1986). *J. Fluid Mech.*, **173**, 519.

Harvey, J. W., Hill, F., Hubbard, R. P., Kennedy, J. R., Leibacher, J. W., Pintar, J. A., Gilman, P. A., Noyes, R. W., Title, A. M., Toomre, J., Ulrich, R. K., Bhatnagar, A., Kennewell, J. A., Marquette, W.,

Partrón, J., Saá, O., and Yasukawa, E. (1996). *Science*, **272**, 1284.

Howard, R. and LaBonte, B. J. (1980). *Astrophys. J.*, **239**, L33.

Howe, R., Christensen-Dalsgaard, J., Hill, F., Komm, R. W., Larsen, R. M., Schou, J., Thompson, M. J., and Toomre, J. (2000a). *Astrophys. J.*, **533**, L163.

Howe, R., Christensen-Dalsgaard, J., Hill, F., Komm, R. W., Larsen, R. M., Schou, J., Thompson, M. J., and Toomre, J. (2000b). *Science*, **287**, 2456.

Howe, R., Christensen-Dalsgaard, J., Hill, F., Komm, R.W., Larsen, R. M., Schou, J., Thompson, M.J., and Toomre, J. (2001). In *Helio- and Asteroseismology at the Dawn of the Millennium: Proc. SOHO 10 / GONG 2000 Workshop*, ed. A. Wilson, ESA SP-464 (ESA Publications Division, Noordwijk, The Netherlands), 19.

Ingersoll, A. P. (1990). *Science*, **248**, 308.

Korzennik, S. G., Cacciani, A., Rhodes, E. J., and Ulrich, R. K. (1990). In *Progress of seismology of the sun and stars*, *Lecture Notes in Physics*, vol. **367**, eds. Y. Osaki and H. Shibahashi (Springer-Verlag, Berlin), 341.

Kosovichev, A. G. (1996). *Astrophys. J.*, **469**, L61.

Kosovichev, A. G. and Schou, J. (1997). *Astrophys. J.*, **482**, L207.

Lada, C. J. and Shu, F. H. (1990). *Science*, **248**, 564.

Libbrecht, K. G. (1988). In *Seismology of the Sun and Sun-like Stars*, ed. V. Domingo and E. J. Rolfe, ESA SP-286, ESA Publications Division, Noordwijk, The Netherlands, 131.

Libbrecht, K. G. (1989). *Astrophys. J.*, **336**, 1092.

Mestel, L. (1968). *Mon. Not. R. Astr. Soc.*, **138**, 359 – 391.

Mestel, L. and Spruit, H. C. (1987). *Mon. Not. R. Astr. Soc.*, **226**, 57 – 66.

Miesch, M. S. (2000). *Solar Phys.*, **192**, 59.

Miesch, M. S., Elliott, J. R., Toomre, J., Clune, T. L., Glatzmaier, G. A., and Gilman, P. A. (2000). *Astrophys. J.*, **532**, 593.

Pidatella, R. M., Stix, M., Belvedere, G., and Paterno, L. (1986). *Astron. Astrophys.*, **156**, 22.

Pinsonneault, M. H., Kawaler, S. D., Sofia, S., and Demarque, P. (1989). *Astrophys. J.*, **338**, 424.

Scherrer, P. H., Bogart, R. S., Bush, R. I., Hoeksema, J. T., Kosovichev, A. G., Schou, J., Rosenberg, W., Springer, L., Tarbell, T. D., Title, A., Wolfson, C. J., Zayer, I., and the MDI engineering team (1995). *Solar Phys.*, **162**, 129.

Schou, J., Antia, H. M., Basu, S., Bogart, R. S., Bush, R. I., Chitre, S. M., Christensen-Dalsgaard, J., Di Mauro, M. P., Dziembowski, W. A., Eff-Darwich, A., Gough, D. O., Haber, D. A., Hoeksema, J. T., Howe, R., Korzennik, S. G., Kosovichev, A. G., Larsen, R. M., Pijpers, F. P., Scherrer, P. H., Sekii, T., Tarbell, T. D., Title, A. M., Thompson, M. J., and Toomre, J. (1998). *Astrophys. J.*, **505**, 390.

Schou, J. (1999). *Astrophys. J.*, **523**, L181.

Schou, J., Christensen-Dalsgaard, J., and Thompson, M. J. (1994). *Astrophys. J.*, **433**, 389.

Skumanich, A. (1972). *Astrophys. J.*, **171**, 565.

Snodgrass, H. B. and Ulrich, R. K. (1990). *Astrophys. J.*, **351**, 309.

Spiegel, E. A. and Zahn, J.-P. (1992). *Astron. Astrophys.*, **265**, 106.

Stein, R. F. and Nordlund, Å. (1989). *Astrophys. J.*, **342**, L95.

Stein, R. F. and Nordlund, Å. (1998). *Astrophys. J.*, **499**, 914.

Thompson, M. J., Toomre, J., Anderson, E. R., Antia, H. M., Berthomieu, G., Burtonclay, D., Chitre, S. M., Christensen-Dalsgaard, J., Corbard, T., DeRosa, M., Genovese, C. R., Gough, D. O., Haber, D. A., Harvey, J. W., Hill, F., Howe, R., Korzennik, S. G., Kosovichev, A. G., Leibacher, J. W., Pijpers, F. P., Provost, J., Rhodes Jr., E. J., Schou, J., Sekii, T., Stark, P. B., and Wilson, P. R. (1996). *Science*, **272**, 1300.

Tomczyk, S., Schou, J., and Thompson, M. J. (1995). *Astrophys. J.*, **448**, L57.

5

Helioseismic tomography

A.G. Kosovichev

W.W. Hansen Experimental Physics Laboratory,
Stanford University, CA 95305-4085, USA

5.1   Introduction

Helioseismic tomography is a promising new method for probing 3-D structures and flows beneath the solar surface. This is important for studying the birth of active regions in the Sun's interior and for understanding the relation between the internal dynamics of active regions and chromospheric and coronal activity. In this method, the time for waves to travel along subsurface ray paths is determined from the temporal cross correlation of signals at two separated surface points. By measuring the times for many pairs of points from Dopplergrams covering the visible hemisphere, a tremendous quantity of information about the state of the solar interior is derived.

Solar oscillations are excited by turbulent convection in the upper convective boundary layer of the Sun. Predominantly excited waves are acoustic (p) and surface gravity waves (f) with oscillation periods of 3-10 min in a wide range of wavenumbers. The combined amplitude of the oscillations is about 200 m s^{-1}. The oscillations are observed by measuring the displacement of the solar surface through the Doppler shift of a solar absorption line formed in the lower part of the solar atmosphere.

There are two basic approaches to infer solar properties from the oscillation data. The first approach, sometimes called 'global' helioseismology is to study the resonant properties of the interior by determining the oscillation frequencies of solar normal modes (e.g. Rhodes *et al.* 1997). The second approach, 'local-area' helioseismology, is to investigate local properties of solar oscillations and interior. Helioseismic tomography belongs to the second approach. It is based on measurements of wave travel times between different points on the surface (Duvall *et al.* 1997). Because of the stochastic nature of solar oscillations substantial spatial and temporal averaging of data is required to accurately measure the frequencies and travel times. These two approaches are complementary: global helioseismology is mainly used to infer large-scale properties through the whole Sun, whereas solar tomography has been useful for determining local properties of convective and magnetic structures in the subsurface layers. However, important cross-checks have been made for the solar differential rotation and meridional circulation (Giles, 1999).

In this chapter we discuss the basic ideas and procedures for interpretation and inversion of the time-distance data. The first step is in establishing relations between the observed travel-time variations and the internal properties of the Sun (variations of the sound speed, flow velocity and magnetic field). The second step is inversion of these relations which are typically linear integral equations, to infer the internal properties. We present results on the internal structure of supergranulation, meridional circulation, sunspots and emerging active regions. An active region which emerged on the solar disk in January 1998, was studied from SOHO/MDI for eight days, both before and after its emergence at the surface. The tomographic images show a complicated structure of the emerging region in the interior, and suggest that the emerging flux ropes travel very quickly through the depth range of our observations. The sound-speed and flow maps beneath a large sunspot shed light into sunspot formation and stability. We present also initial results on detecting active regions on the far side of the Sun using a technique of 'acoustic holography' which is similar to the time-distance approach.

5.2   Method of helioseismic tomography

Solar acoustic waves are excited by turbulent convection near the solar surface and propagate through the interior with the speed of sound. Because the sound speed increases with depth the waves are refracted and reappear on the surface at some distance from the source. The wave propagation is illustrated in Figure 5.1. Waves excited at point A will reappear at the surface points B, C, D, E, F, and others after propagating along the ray paths indicated by the curves connecting these points.

The basic idea of helioseismic tomography is to measure the acoustic travel time between different points on the solar surface, and then to use these measurements for inferring variations of the structure and flow velocities in the interior along the wave paths connecting the surface points. This idea is similar to the seismology of the Earth. However, unlike in Earth, the solar waves are generated stochastically by numerous acoustic sources in the subsurface layer of turbulent convection.

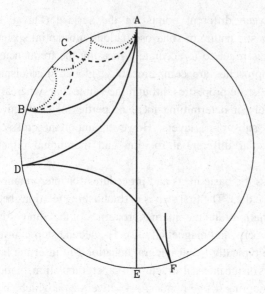

Figure 5.1. A cross-section diagram through the solar interior showing a sample of wave paths inside the Sun.

Therefore, the wave travel time is determined from the cross-covariance function, $\Psi(\tau, \Delta)$, of the oscillation signal, $f(t, r)$, between different points on the solar surface (Duvall *et al.* 1993):

$$\Psi(\tau, \Delta) = \int_0^T f(t, r_1) f^*(t + \tau, r_2) dt, \tag{5.1}$$

where Δ is the horizontal distance between the points with coordinates r_1 and r_2, τ is the delay time, and T is the total time of the observations. Because of the stochastic nature of excitation of the oscillations, function Ψ must be averaged over some areas on the solar surface to achieve a signal-to-noise ratio sufficient for measuring travel times τ. The oscillation signal, $f(t, r)$, is usually the Doppler shift or intensity of a spectral line. A typical cross-covariance function shown in Figure 5.2a displays a set of ridges which correspond to the first (the lowest ridge), the second (the second ridge from the bottom), the third and so on bounces of packets of acoustic waves from the surface.

The origin of the multiple bounces is illustrated in Figure 5.1. Waves originated at point A may reach point B directly (solid curve) forming the first-bounce ridge, or after one bounce at point C (dashed curve) forming the second-bounce ridge, or after two bounces (dotted curve) - the third-bounce ridge and so on. Because the sound speed is higher in the deeper layers the direct waves arrive first, followed by the second-bounce and higher-bounce waves.

The cross-covariance function represents a solar 'seismogram'. Figure 5.3 shows the cross-covariance signal as a function of time for a distance of 30 degrees. It consists of three wave packets corresponding to the first, second and third bounces. Ideally, the seismogram should be inverted to infer the structure and flows using a wave theory. However, in practice, as in terrestrial seismology (Aki and Richards, 1980)

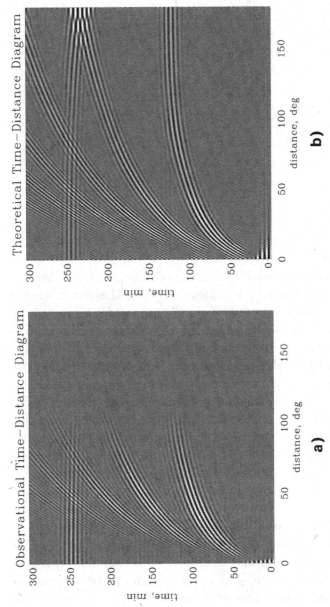

Figure 5.2. The observational (a) and theoretical (b) cross-covariance functions as a function of distance on the solar surface, Δ, and the delay time, τ. The lowest set of ridges ('first bounce') corresponds to waves propagated to the distance, Δ, without additional reflections from the solar surface. The second from the bottom ridge ('second bounce') is produced by the waves arriving to the same distance after one reflection from the surface, and the third ridge ('third bounce') results from the waves arriving after two bounces from the surface. The backward ridge at τ ≈ 250 min is a continuation of the second-bounce ridge due to the choice of the angular distance range from 0 to 180 degrees (that is, the counterclockwise distance ADF in Figure 5.1 is substituted with the clockwise distance AF). Because of foreshortening close to the solar limb the observational cross-covariance function covers only 110 degrees of distance.

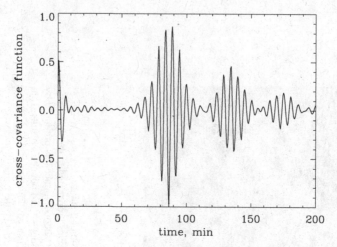

Figure 5.3. The observed cross-covariance signal as a function of time at the distance of 30 degrees.

different approximations are employed, the most simple and powerful of which is the geometrical acoustic (ray) approximation.

Generally, the observed solar oscillation signal corresponds to displacement or pressure perturbation, and can be represented in terms of normal modes, or standing waves. Therefore, the cross-covariance function can be expressed in terms of normal modes, and then represented as a superposition of traveling wave packets (Kosovichev and Duvall, 1997). An example of the theoretical cross-covariance function of p modes of the standard solar model is shown in Figure 5.2b. This model reproduces the observational cross-covariance function very well in the observed range of distances, from 0 to 90 degrees. The theoretical model was calculated for larger distances, including points on the far side of the Sun, which is not accessible for measurements. A backward propagating ridge originating from the second-bounce ridge at 180 degrees is a geometrical effect due to the choice of the range of the angular distance from 0 to 180 degrees. In the theoretical diagram (Figure 5.2b) one can notice a weak backward ridge between 30 and 70 degrees and at 120 min. This ridge is due to reflection from the solar core. However, this signal has not been detected in observations because it is very weak.

By grouping the normal modes in narrow ranges of the angular phase velocity, $v = \omega_{nl}/L$, where ω_{nl} is the oscillation frequency of a normal mode of angular degree l and radial order n, $L = \sqrt{l(l+1)}$, and applying the method of stationary phase, the cross-covariance function can be approximately represented in the form (Kosovichev and Duvall, 1997):

$$\Psi(\tau, \Delta) \propto \sum_{\delta v} \cos\left[\omega_0\left(\tau - \frac{\Delta}{v}\right)\right] \exp\left[-\frac{\delta\omega^2}{4}\left(\tau - \frac{\Delta}{u}\right)^2\right], \qquad (5.2)$$

where δv is a narrow interval of the phase speed, $u \equiv (\partial\omega/\partial k_h)$ is the horizontal component of the group velocity, $k_h = L/R$ is the angular component of the wave vector, R is the solar radius, ω_0 is the central frequency of a Gaussian frequency filter applied to the data, and $\delta\omega$ is the characteristic width of this filter. Therefore, the phase and group travel times are measured by fitting individual terms of eq. (5.2) to the observed cross-covariance function using a least-squares technique.

This technique measures both phase (Δ/v) and group (Δ/u) travel times of the p-mode wave packets. The previous time-distance measurements provided either group (Jefferies *et al.* 1994), or phase travel times (Duvall *et al.* 1996). It was found that the noise level in the phase-time measurements is substantially lower than in the group-time measurements. Variations of the local travel times at different points on the surface relative to the travel times averaged over the observed area are used to infer variations of the internal structure and flow velocities using a perturbation theory. Because of the stochastic nature of solar oscillation it is generally required to average the cross-correlation for a particular range of distances both over some spatial areas and in time in order to accumulate a good signal-to-noise ratio. Two typical schemes of the spatial averaging are shown in Figure 5.4. For the so-called 'surface-focusing' scheme (Figure 5.4a) the measured travel times are mostly sensitive to the near surface condition at the central point where the ray paths are focused. However, by measuring the travel times for several distances and applying an inversion procedure it is possible to infer the distribution of the variations of the wave speed and flow velocities with depth (Kosovichev and Duvall, 1997). The averaging also can be done in such a way that the 'focus' point is located beneath the surface (Duvall, 1995).

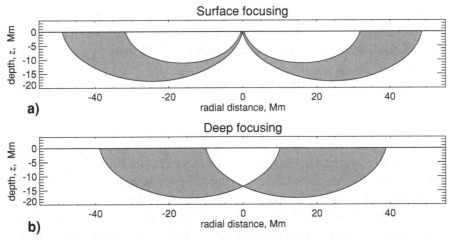

Figure 5.4. The regions of ray propagation (shaded areas) as a function of depth, z, and the radial distance, Δ, from a point on the surface for two observing schemes: 'surface focusing' (a) and 'deep focusing' (b). The rays are also averaged over a circular regions on the surface, forming three-dimensional figures of revolution.

An example of the 'deep-focusing' scheme is shown in Figure 5.4b. In this case the travel times are more sensitive to deep structures but still inversions are required for correct interpretation. First results show that deep-focusing allows us to see more features in the interior.

The phase travel times (or phase shifts) are also employed in two other methods, 'acoustic holography' (Braun and Lindsey, 2000; Lindsey and Braun, 2000a) and 'acoustic imaging' (Chou, 2000; Chou and Duvall, 2000; Chou and Dai, 2001). These methods employ the deep-focusing scheme but have different from the time-distance technique averaging procedures for the cross-covariance function. Woodard (2002) has developed a new approach which employs the cross-covariance function directly without determining the travel times. His method is similar to 'wave-form tomography' in geophysics (e.g. Pratt, 1999).

5.3   The ray approximation

In the ray approximation, the travel times are sensitive only to the perturbations along the ray paths given by Hamilton's equations. The variations of the travel time obey Fermat's Principle (e.g. Gough, 1993):

$$\delta\tau = \frac{1}{\omega}\int_{\Gamma}\delta k dr, \tag{5.3}$$

where δk is the perturbation of the wave vector due to the structural inhomogeneities and flows along the unperturbed ray path, Γ. Using the dispersion relation for acoustic waves in the convection zone the travel-time variations can be expressed in terms of the sound-speed, magnetic field strength and flow velocity (Kosovichev and Duvall, 1997).

The effects of flows and structural perturbations are separated from each other by taking the difference and the mean of the reciprocal travel times:

$$\delta\tau_{\text{diff}} \approx -2\int_{\Gamma}\frac{(\mathrm{nU})}{c^2}ds; \tag{5.4}$$

$$\delta\tau_{\text{mean}} \approx -\int_{\Gamma}\frac{\delta c}{c}Sds, \tag{5.5}$$

where c is the adiabatic sound speed, n is a unit vector tangent to the ray, $S = k/\omega$ is the phase slowness. Magnetic field causes anisotropy of the mean travel times, which allows us to separate, in principle, the magnetic effects from the variations of the sound speed (or temperature). So far, only a combined effect of the magnetic fields and temperature variations has been measured reliably. The development of a more accurate theory for the travel times, based on the Born approximation is currently under way (Birch and Kosovichev, 2000, 2001; Gizon and Birch, 2002).

5.4   The Born approximation

Generally, a solution to the wave equation

$$\mathcal{L}f + \omega^2 f = F_\omega(r_s) \tag{5.6}$$

with a point source function $F_\omega(r_s)$ describing solar oscillations with amplitude f can be expressed in terms of the Green's function, $G_\omega(r, r_s)$:

$$f(r, r_s, t) = \int_0^\infty G_\omega(r, r_s)F_\omega(r_s)e^{-i\omega t}d\omega. \tag{5.7}$$

Using this equation and Eq. (5.1), the cross-covariance function $\Psi(\Delta, \tau)$ can be calculated for randomly distributed point sources. This calculation can be easily done for a spherically symmetrical model (Kosovichev and Duvall, 1997). For this model the cross-covariance function between two points is equivalent to a signal from a point source located at one point and received at the other point. The idea that the cross-correlation function would mimic an explosion type signal was also developed in geophysics and known as 'Claerbout's Conjecture' (see Rickett and Claerbout, 2000): "By cross-correlating noise traces recorded at two locations on the surface, we can construct the wavefield that would be recorded at one of the locations if there was a source at the other". This significantly simplifies the interpretation of the time-distance measurements and allows us to take into account wave effects in the theory of travel-time variations. The simplest approach is to assume that the non-spherical perturbations are weak and apply the standard Born approximation.

For a point-source case the unperturbed Green's function, G_0 satisfies the following equation:

$$\mathcal{L}_0 G_0 + \omega^2 G_0 = \delta(r - r_s), \tag{5.8}$$

where $\mathcal{L}_0$ is the unperturbed wave operator. The corresponding equation for the perturbation to the Green's function, G_1, caused by perturbation operator $\mathcal{L}_1$ is:

$$(\mathcal{L}_0 + \mathcal{L}_1)(G_0 + G_1) + \omega^2(G_0 + G_1) = \delta(r - r_s). \tag{5.9}$$

Taking the difference between these two equations we obtain:

$$\mathcal{L}_0 G_1 + \omega^2 G_1 = \mathcal{L}_1(G_0 + G_1). \tag{5.10}$$

This is known as the Lippman-Swinger equation. The Born approximation is taking by neglecting the G_1 term in the right-hand side of this equation. Then, the solution to the approximate equation can be easily expressed in terms of the unperturbed Green's function and the perturbation described by $\mathcal{L}_1$. The unperturbed Green's function, $\mathcal{G}_0$, can be represented in terms of the normal mode eigenfunctions of a spherically symmetrical solar model. Consequently, the kernels, $K_T(r, \Delta)$, for the travel-time variation:

$$\delta\tau(\Delta) = \int_V K_T(r, \Delta)\frac{\delta w}{w}dr \tag{5.11}$$

are represented in terms of the mode eigenfunctions similarly to the kernels for normal mode frequency variations (e.g. Kosovichev, 1999). This procedure is described in more detail by Marquering *et al.* (1999), and an example of its application to the Sun is given by Birch and Kosovichev (2000). A general procedure for randomly distributed sources has been proposed by Gizon and Birch (2002). It reveals some important characteristics of the kernel functions, that are missing in the single-source kernels.

One unexpected feature of the single-source travel-time kernels calculated in the Born approximation is that these kernels have zero value along the ray path (called 'banana-doughnut kernels' by Marquering *et al.* 1999; Dahlen *et al.* 2000). Examples of the Born kernels for the first and the second bounces are shown in Figure 5.5. The kernels are mostly sensitive to perturbations within the first Fresnel zone. The first-bounce kernels for relatively large distances can be reasonably approximated by a simple semi-analytical formula (Jensen *et al.* 2000). However, for short distances and multiple bounces the kernels have to be calculated numerically.

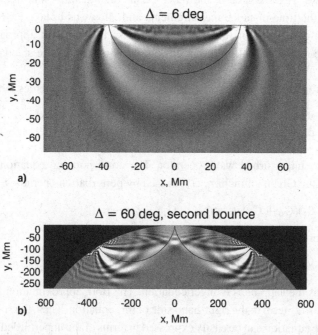

Figure 5.5. Travel-time sensitivity kernels in the first Born approximation for sound-speed variations as a function of the horizontal, x, and vertical, y, coordinates for: a) the first-bounce signal for distance $\Delta = 6$ degrees, b) the second-bounce signal for $\Delta = 60$ degrees. The solid curves show the corresponding ray paths at frequency $\nu = 3$ mHz.

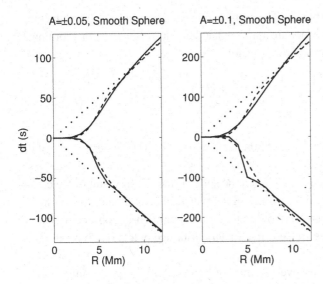

Figure 5.6. Tests of the ray and Born approximations: travel times for smooth spheres as functions of sphere radius at half maximum. The solid lines are the numerical results. The dashed curves are the Born approximation travel times and the dotted lines are the first order ray approximation. The left panel shows the two perturbations of the relative amplitude, $A = \pm 0.05$. The right panel is for the cases $A = \pm 0.1$. (Birch *et al.* 2001).

Figure 5.6 shows the test results for both the ray and Born approximations for a simple model of a smooth sphere in an uniform medium by comparing with precise numerical results (Birch *et al.* 2001; see also Hung *et al.* 2001). These results show that for typical perturbations in the solar interior the Born approximation is sufficiently accurate, while the ray approximation significantly overestimates the travel times for perturbations smaller than the size of the first Fresnel zone. That means that the inversion results based on the ray theory may significantly underestimate the strength of the small-scale perturbations.

5.5   Inversion methods

Typically, we measure times for acoustic waves to travel between points on the solar surface and surrounding quadrants symmetrical relative to the North, South, East and West directions. In each quadrant, the travel times are averaged over narrow ranges of travel distance Δ. Then, the times for northward-directed waves are subtracted from the times for south-directed waves to yield the time, $\tau_{\text{diff}}^{\text{NS}}$, which predominantly measures north-south motions. Similarly, the time difference, $\tau_{\text{diff}}^{\text{EW}}$, between westward- and eastward directed waves yields a measure of east-ward motion. The time, $\tau_{\text{diff}}^{\text{oi}}$, between outward- and inward-directed waves, averaged over the full annulus, is mainly sensitive to vertical motion and the horizontal divergence. The time, τ_{mean}, which measures sound-speed perturbations is also averaged over the full annulus (for more details see Duvall *et al.* 1997, and Kosovichev and Duvall, 1997).

The next step is to determine the variations of the sound speed and flow velocity from the observed travel times using equations (5.4) and (5.5). It is assumed that the convective structures and flows do not change during the observations and can be represented by a discrete model. In this model, the 3-D region of wave propagation is divided into rectangular blocks. The perturbations of the sound speed and the three components of the flow velocity are approximated by linear functions of coordinates within each block, e.g. for the flow velocity

$$U(x, y, z) = \sum_{i,j,k} U_{ijk} \left[1 - \frac{|x - x_i|}{x_{i+1} - x_i} \right] \left[1 - \frac{|y - y_j|}{y_{j+1} - y_j} \right] \left[1 - \frac{|z - z_k|}{z_{k+1} - z_k} \right],$$

(5.12)

where x_i, y_j, z_k are the coordinates of the rectangular grid, U_{ijk} are the values of the velocity in the grid points, and $x_i \leq x \leq x_{i+1}$, $y_j \leq y \leq y_{j+1}$, and $z_k \leq z \leq z_{k+1}$.

The travel time measured at a point on the solar surface is the result of the cumulative effects of the perturbations in each of the traversed rays of the 3D ray systems. This pattern is then translated for different surface points in the observed area, so that overall the travel times are sensitive to all subsurface points in the depth range covered by ray paths.

We average the equations over the ray systems corresponding to the different radial distance intervals of the data, using approximately the same number of ray paths as in the observational procedure. As a result, we obtain two systems of linear equations that relate the data to the sound speed variation and to the flow velocity, e.g. for the velocity field,

$$\delta \tau_{\text{diff}; \lambda \mu \nu} = \sum_{ijk} A_{\lambda \mu \nu}^{ijk} \cdot U_{ijk},$$

(5.13)

where vector-matrix A maps the structure properties into the observed travel time variations, and indices λ and μ label the location of the central point of a ray system on the surface, and index ν labels the annulus. These equations are solved by a regularized least-squares technique using the LSQR algorithm (Paige and Saunders, 1982). Jensen $et\ al.$ (1997) suggested to speed up the inversion by doing most of the calculation in the Fourier domain.

The results of test inversions of Kosovichev and Duvall (1997) and Zhao $et\ al.$ (2001) demonstrate an accurate reconstruction of sound-speed variations and the horizontal components of subsurface flows. However, vertical flows in deep layers are not resolved because of the predominantly horizontal propagation of the rays in these layers. The vertical velocities are also systematically underestimated by 10-20% in the upper layers. Similarly, the sound-speed variations are underestimated in the bottom layers. These limitations of the solar tomography should be taken into account in interpretation of the inversion results.

5.6   Diagnostics of supergranulation

Helioseismic tomography has been successfully used to infer local properties of large-scale zonal and meridional flows (Giles $et\ al.$ 1997; Giles and Duvall, 1998),

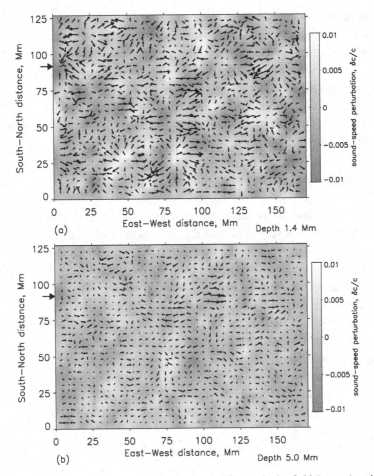

Figure 5.7. The supergranulation horizontal flow velocity field (arrows) and the sound-speed perturbation (grey-scale background) at the depths of 1.4 Mm (a) and 5.0 Mm (b), as inferred from the SOHO/MDI high-resolution data of 27 January 1996. (Kosovichev and Duval, 1997).

convective flows and structures (Duvall *et al.* 1997; Kosovichev and Duvall *et al.* 1997), structure and dynamics of active regions (Kosovichev, 1996), flows around sunspots (Duvall *et al.* 1996). Here we present some results of tomographic inversion for large-scale convective cells ('supergranulation').

The data used were for 8.5 hours on 27 January, 1996 from the high resolution mode of the MDI instrument. The results of inversion of these data are shown in Figure 5.7. It has been found that, in the upper layers, 2-3 Mm deep, the horizontal flow is organized in supergranular cells, with outflows from the center of the supergranules. The characteristic size of the cells is 20-30 Mm. Comparing with MDI magnetograms, it was found that the cell boundaries coincide with the areas of enhanced magnetic field. These results are consistent with the observations of supergranulation on the solar surface. However, in the layers deeper than 5 Mm, the supergranulation pattern disappears. This suggests that supergranulation is only 5 Mm deep. An alternative interpretation suggesting a depth of 8 Mm was presented by Duvall (1998). The vertical flows correlate with the supergranular pattern in the upper layers. Typically, there are upflows in the 'hotter' areas where the sound speed is higher than average, and downflows in the 'colder' areas. A surprising result of this investigation is that the downflows frequently occur in the middle of supergranules. This is not understood and requires further investigation.

5.7   Large-scale flows

In general, convective flows must play a major role in the mechanisms of solar activity. Therefore, it is very important for both understanding the physics of solar activity and forecasting to study various aspects of the internal dynamics. Duvall and Gizon (2000) have developed a diagnostic of near-surface layers by applying the time-distance approach to surface gravity waves (f modes). These waves are much less sensitive to temperature variations, and therefore the mean travel-differences for these waves are mostly due to magnetic field. The f-mode waves also provide very accurate measurements for horizontal mass flows just beneath the solar surface, in the upper layer 1–2 Mm thick, and allow us to detect both large-scale and small-scale flow patterns associated with solar activity (Gizon *et al.* 2000). Figure 5.8 shows large-scale flow patterns during Carrington Rotation 1948 (April 4, 1999 – May 1, 1999). These results reveal converging large-scale flows associated with active regions, with a velocity of $\simeq 50$ m/s.

These results are obtained from the MDI full-disk Dynamics data which are available typically for two months each year. With longer continuous observing runs it will be possible to study the interaction between the large-scale magnetic fields and flows, and determine if, by observing the flow patterns, one can predict future solar activity. The f-mode method is very promising and will receive further development.

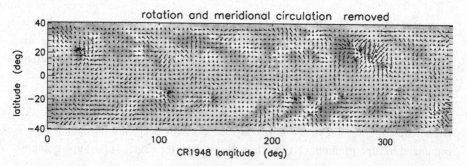

Figure 5.8. Large-scale near-surface flows associated with active regions, obtained from surface gravity waves (f-modes). The background color map shows magnetic field. (Gizon *et al.* 2001).

5.8   Meridional circulation

Meridional flows from the equator towards the north and south poles have been observed on the solar surface in direct Doppler-shift measurements (e.g. Duvall, 1979). The MDI observations by Giles *et al.* (1997) have provided the first evidence that such flows persist to great depths, and, thus, possibly play an important role in the 11-year solar cycle. The poleward flow can transport the magnetic remnants of sunspots generated at low latitudes to higher latitudes and, therefore, contribute to the cyclic polar field reversal.

Giles *et al.* (1997) applied a version of the "deep-focusing" scheme and a regularized least-squares inversion technique to the MDI data to determine zonal and meridional flows in the convection zone (Figure 5.9). The spatial resolution drastically decreases with depth. Figure 5.9b shows a sample of averaging kernels which characterize the averaging properties of the inversion results and provide a measure of the spatial resolution.

Some results of this investigation are presented in Figure 5.10 which shows the rotation rate and the speed of the meridional circulation as a function latitude at 6 different depths. The symmetrical with respect to the equator component of the rotation rate has been determined also by standard methods of "global" helioseismology, from rotational splitting of normal mode frequencies of solar oscillations. The results are compared in Figure 5.10a, and demonstrate good agreement. This serves as a test of the time-distance approach and provides confidence in the results.

The meridional circulation inside a star has been determined for the first time by Giles *et al.* (1997). It has a maximum speed of 20 m/s in the subsurface layers (Figure 5.10b). The flow is directed from the equator to the poles. The speed becomes smaller with depth. However, the return flow has not been detected even close to the bottom of the convection zone. However, the bottom has not been reached. Such measurements require very long ($\sim$ 1 year) time series because deep signals are much weaker than the surface or near-surface signals. The travel-time variations for the deep flows are of the order of 1 sec. So far the deep inferences have been made for 2D flows. However, this study suggests the possibility of full 3D imaging of the convection zone. Of course, the prime goal is to reduce the integration time.

5.9   Emerging active regions

One of the important tasks of heliotomography is diagnostics of emerging active regions in the interior. For space weather predictions it would be very important to detect active regions before they emerge. However, this task has proven to be very difficult. Here we present the results for an emerging active region observed in January, 1998. This active region (NOAA 8131) was a high-latitude region of the new solar cycle which began in 1997.

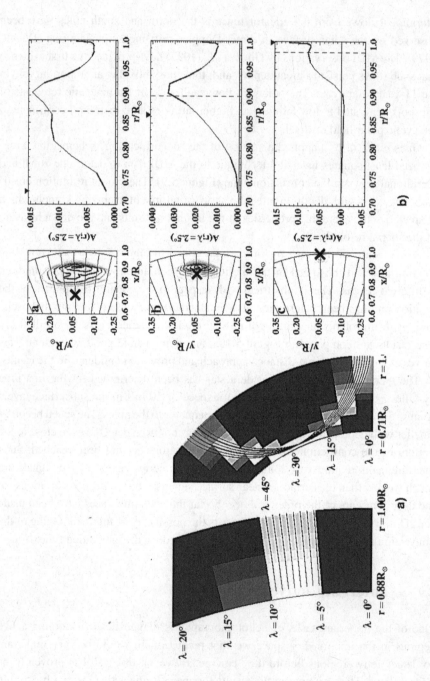

Figures 5.11a-f show the sound speed variations in a vertical cross-section in the region of the emerging flux and in a horizontal plane at a depth of 18 Mm, for several 8-hour intervals. The perturbations of the magnetosonic speed shown in these figures are associated with the magnetic field and temperature variations in the emerging magnetic structures. It is clear that the 8-hour integration time which yields a good signal-to-noise ratio is not sufficiently short for detecting emerging flux before it appears on the surface. Only reducing the integration time to 2 hours we were able to detect a strong perturbation at the bottom of our observing region (Kosovichev and Duvall, 2000). From the investigation of emerging active regions we conclude that it is necessary to probe much deeper layers of the solar convection zone because the emerging flux propagates very rapidly in the top 20 Mm. The estimated emergence speed is approximately 1.3 km/s. This is higher than predicted by theories. The high emergence speed makes the problem of detection of emerging active regions in the solar interior very challenging.

The typical amplitude of the sound-speed variation in the perturbation is about 0.5 km/s. This may correspond to a magnetic field strength of 500 G at the bottom of the box, or a temperature variation of 300 K. After the emergence we observed a gradual increase of the perturbation in the subsurface layers, and the formation of sunspots. The observed development of the active region suggests that the sunspots were formed as result of the concentration of magnetic flux close to the surface.

5.10   Structure and dynamics of sunspots

The high-resolution data from SOHO/MDI has allowed to investigate structures and flows beneath sunspots in some detail (Kosovichev *et al.* 2000; Zhao *et al.* 2001). Figure 5.12 shows an example of the internal structure of a large sunspot observed on June 17, 1998. An image of the spot taken in the continuum is shown at the top. The sound-speed perturbations under the sunspot are much stronger than these of

Figure 5.9. a) The sensitivity kernels for the zonal velocity for latitude $\lambda = 7.5°$, distance $\Delta = 18.1°$ (left) and for the meridional velocity for $\lambda = 24.4°$, $\Delta = 44.1°$ (right). The light colors show the sensitivity to horizontal flows, with dark areas being those with no sensitivity. The black curves depict ray paths for the individual measurements which were averaged together to get the single kernel shown here; 52 rays were used for the zonal velocity kernel, and 39 rays - for the meridional velocity kernel. b) Selected averaging kernels for the meridional velocity inversion. The left column shows contour plots of the kernels. The contour spacing is 5% of the maximum amplitude of each kernel; dashed contours indicate a negative amplitude. The nominal locations of the three kernels, in each case marked with ×, are: (i) 0.72 R, (ii) 0.87 R and (iii) 1 R; $\lambda = 2.5°$. In the right column, a cross section of each kernel is plotted, as a function of radius, at the central latitude. The arrow above each plot marks the nominal radius of the kernel; the dashed line marks the location of the centroid. (Giles 1999).

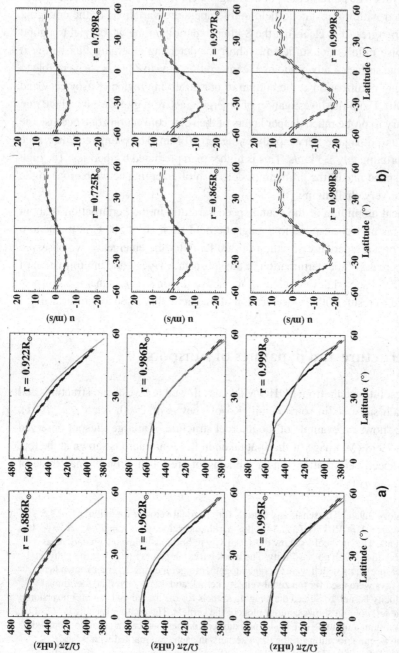

Figure 5.10. a) Comparison of the time-distance and normal mode methods for determining the solar rotation. The angular frequency is plotted versus latitude for six different choices of radial coordinate. The dotted curve is the result of an inversion of frequency splittings. The solid curve is the symmetric component of the time-distance result from this work; the dashed lines are the formal errors from the inversion. b) Inversion results for the meridional circulation from 792 days of the MDI Structure Program, showing the meridional velocity as a function of latitude λ, for six different depths. Positive velocities are northward. (Giles 1999).

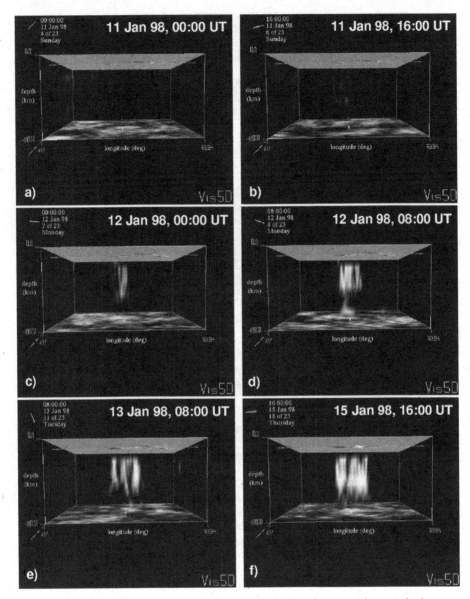

Figure 5.11. The sound-speed perturbations in the solar convection zone in the emerging active region of January 1998 obtained with 8-hour time resolution. The horizontal size of the box is approximately 38 degrees (460 Mm), the vertical size is 18 Mm. The panels on the top are MDI magnetograms showing the surface magnetic field of positive (white) and negative polarities (black). The perturbations of the sound speed (light features) are approximately in the range from -1 to +1 km/s.

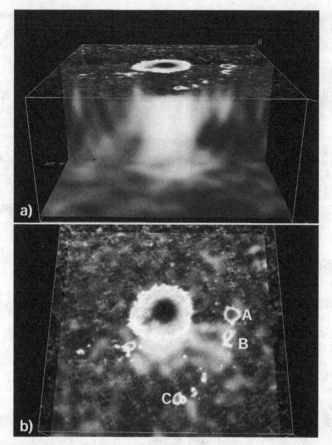

Figure 5.12. The sound-speed perturbation in a large sunspot observed on June 20, 1998, are shown as vertical and horizontal cuts. The horizontal size of the box is 13 degrees (158 Mm), the depth is 24 Mm. The positive variations of the sound speed are shown in light gray, and the negative variations (just beneath the sunspot) in dark. The upper semitransparent panel is the surface intensity image (dark color shows umbra, and light color shows penumbra). In the lower panel the horizontal sound-speed plane is located at the depth of 4 Mm, and shows long narrow structures ('fingers') connecting the main sunspot structure with surrounding pores A and B of the same magnetic polarity as the spot. Pores of the opposite polarity (e.g. C) are not connected to the spot.

the emerging flux, and reach more than 3 km/s. It is interesting that beneath the spot the perturbation is negative in the subsurface layers and becomes positive in the deeper interior. One can suggest that the negative perturbations beneath the spot are, probably, due to the lower temperature. It follows that magnetic inhibition of convection that makes sunspots cooler is most effective within the top 2-3 Mm of the convection zone. The strong positive perturbation below suggests that the deep sunspot structure is hotter than the surrounding plasma. However, the effects of temperature and magnetic field have not been separated in these inversions. Separating these effects is an important problem of solar tomography. These data also show at a depth of 4 Mm connections to the spot of small pores, A and B, which have the same magnetic polarity as the main spot. Pore C of the opposite polarity is not connected. This suggests that sunspots represent a tree-like structure in the upper convection zone.

Figure 5.13 shows flow maps beneath the sunspot. In Figure 5.13a which shows results for the first layer corresponding to an average of depth of 0–3 Mm, one can clearly identify a ring of strong downflows around the sunspot, with relatively weaker downflows inside the ring. Converging flows at the sunspot center can also be seen in this graph. Figure 5.13b shows the flows in the layer, corresponding to a depth of 6–9 Mm. The sunspot region is occupied by a ring of upflows with almost zero velocity at the center. Outside this region, the results are much noisier, and no clear upflows or downflows can be identified. At the same time, strong outflows from the sunspot center can be seen. Clearly, the motion to the South-East direction is much stronger.

Figures 5.13c-d show two vertical cut graphs, one with East-West direction, the other North-South direction, through the center of the sunspot. in between two layers by use of linear interpolation. Converging and downward flows can be seen in both diagrams right below the sunspot region from 1.5 Mm to about 5 Mm. Below that, the horizontal outflows seem to dominate in this region, though relatively weaker upflows also appear. Below a depth of ∼10 Mm, the flows seem not to be concentrated in the region vertically below the sunspot. This can be seen more clearly in the East-West cut (Figure 5.13c). It is intriguing that an upflow toward East dominates in the region from 10 Mm to 18 Mm. In South-North cut (Figure 5.13d), this pattern is not so clear but still can be seen, with the upflow toward South direction stronger than toward the North.

The inversion result shows converging flows in the top layer of the sunspot region which are opposite to the well-known diverging Evershed flow observed spectroscopically on the solar surface. Gizon *et al.* (2000) have studied the horizontal flows in sunspots using the f-mode diagnostics. The velocity amplitude of the Evershed flow inferred from the f-mode is systematically lower than that at the surface. This suggests that the Evershed flow is a shallow phenomenon. This can explain the apparent disagreement with the results obtained from acoustic modes.

The observational evidence of converging flows beneath the spot supports a cluster model of sunspots suggested by Parker (1979). According to this model the field of a sunspot divides into many separate tubes within the first 1-2 Mm below the surface, and that a converging downward flow beneath the sunspot holds the separate tubes in a loose cluster.

5.11   Imaging the far side of the Sun

A new approach for predicting new active regions on the far side of the Sun has been recently demonstrated by Lindsey and Braun (2000). They used a holographic

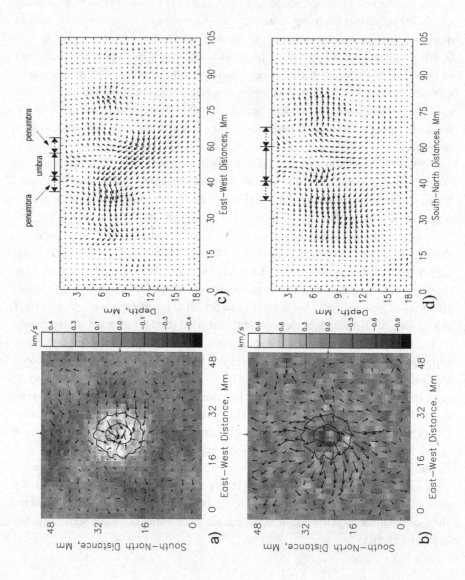

technique to image the far side of the Sun. This allows early detection of new active regions which a few days later will rotate to the front side and affect the Earth's space environment. The time-distance technique also can provide similar measurements of the travel-time perturbations on the far side (Duvall and Kosovichev, 2001). The holographic results are illustrated in Figure 5.14. The measurement scheme shown in Figure 5.14a is essentially a deep-focusing scheme with the focus point chosen on the far-side surface. Figure 5.14b shows a map of the travel time variations on the far side. The travel time becomes shorter by $\simeq 6$ sec when there is an active region in the focus. The far-side signal was recovered from 48-hour time series of medium resolution images.

Braun and Lindsey (2001b) have also developed an asymmetrical scheme with one-bounce wave paths on one side and two-bounce paths on the other side, which allows to map the whole far-side hemisphere of the Sun. The far-side images provide important information for long-term space weather forecasts.

5.12   Conclusion

Helioseismic tomography provides unique information about three-dimensional structures and flows associated with magnetic field and turbulent convection in the solar interior. This method is at the very beginning of its development. In this chapter, we have reviewed some basic principles of this technique based on the geometrical ray and Born approximations, and presented some initial inversion results. Developing helioseismic tomography is one of the most challenging and exciting problems of solar physics.

The initial results are very promising. They reveal interesting dynamics of super-granulation, meridional circulation, emerging active regions and the formation of sunspots in the upper convection zone. Further studies of the Sun's interior by solar tomography will shed more light on the mechanisms of solar activity.

Figure 5.13. The mass flows in a sunspot area at the depth of 0–3 Mm (a) and depth of 6–9 Mm (b). The arrows show the magnitude and directions of the horizontal flows, and the background gray-scale map shows the vertical flows (light color indicates downflow). The contours at the center correspond to the umbra and penumbra boundaries. The longest arrow represents 1.0 km/s in panel a), and 1.6 km/s in panel b). The arrows outside the frame indicate where the vertical cuts shown in panels c) and d) are made. The vertical cuts are made through the sunspot center with the cut direction of East-West (panel c, with East on the left side) and South-North (panel d, with South on the left side). The range covered by the line arrows indicate the area of umbra, and the range covered by the dotted arrow indicate the area of penumbra. The longest arrow represents 1.4 km/s. (Zhao *et al.* 2001)

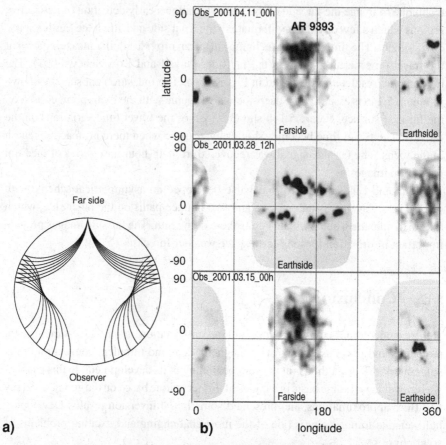

Figure 5.14. a) A 'deep-focusing' scheme with the acoustic ray focus point on the far side of the Sun. b) Carrington maps of the far-side travel-time variations (top and bottom panels) and the Earth-side magnetic field (middle panel) for March 15 (top), March 28 (middle) and April 11 (bottom), 2000. Active region NOAA 9393 located at Carrington longitude 150 degrees was on the far side on March 15 and April 11, and is visible in the travel-time maps. On March 28 this region was on the Earth side, and is shown on a low-resolution magnetic map (middle panel).

References

Aki, K. and Richards, P. (1980). *Quantitative Seismology. Theory and Methods*, (San Francisco, Freeman).

Birch, A.C. and Kosovichev, A.G. (2000). *Solar Phys.*, **192**, 193.

Birch, A.C. and Kosovichev, A.G. (2001). Sensitivity of travel times to sound speed perturbations, in *Recent Insights into the Physics of the Sun and Heliosphere*, ed. P. Brekke, B. Fleck, and J.B. Gurman (The Astronomical Society of the Pacific, San Francisco), p. 180.

Birch, A. C., Kosovichev, A. G., Price, G. H., and Schlottmann, R. B. (2001). *Astrophys. J.*, **561**, L229.

Braun, D.C. and Lindsey, C. (2000). *Solar Phys.*, **192**, 307.

Braun, D.C., and Lindsey, C. (2001). *Astrophys. J.*, **560**, L189.

Chou, D.-Y. (2000). *Solar Phys.*, **192**, 241.

Chou, D.-Y. and Duvall, T.L., Jr. (2000). *Astrophys. J.*, **533**, 568.

Chou, D.-Y. and Dai, D.-C. (2001). *Astrophys. J.*, **559**, L175.

Dahlen, F. A., Hung, S.-H., and Nolet, G. (2000). *Geophysical J. Int.*, **141**, 157.

Duvall, T. L., Jr. (1979). *Solar Phys.*, **63**, 3.

Duvall, T.L., Jr., Jefferies, S.M., Harvey, J. W., and Pomerantz, M. A. (1993). *Nature*, **362**, 430.

Duvall, T. L., Jr. (1995). in *GONG '94: Helio- and Astero-Seismology from the Earth and Space*, ed. R.K. Ulrich, E.J. Rhodes, Jr., W.Dappen, ASP Conf. Series, **76**, (Astronomical Society of the Pacific, San Francisco), p. 93.

Duvall, T. L., Jr., D'Silva, S., Jefferies, S. M., Harvey, J. W., and Schou, J. (1996). *Nature*, **379**, 235.

Duvall,T.L., Jr., Kosovichev, A.G., Scherrer, P.H., Bogart, R.S., Bush, R.I., De Forest, C., Hoeksema, J.T., Schou, J., Saba, J.L.R., Tarbell, T.D., Title, A.M., Wolfson, C.J., and Milford, P.N. (1997). *Solar Phys.*, **170**, 63.

Duvall, T.L., Jr. (1998). in *Structure and Dynamics of the Interior of the Sun and Sun-like Stars*, Proc. SOHO-6/GONG-98 Workshop, ESA SP-418, ed. S.G. Korzennik and A. Wilson, (European Space Agency, Noordwijk), p. 581.

Duvall T.L. and Gizon L. (2000). *Solar Phys.*, **192**, 177.

Duvall T.L. and Kosovichev, A.G. (2001). in *Recent Insights into the Physics of the Sun and Heliosphere*, ed. P. Brekke, B. Fleck, and J.B. Gurman (The Astronomical Society of the Pacific, San Francisco).

Giles, P. M., Duvall, T. L., Jr, Scherrer, P. H., and Bogart, R.S. (1997). *Nature*, **390**, 52.

Giles, P.M. and Duvall, T.L., Jr. (1998). in *New Eyes to See Inside the Sun and Stars. Pushing the Limits of Helio- and Asteroseismology with New Observations from the Ground and from Space*, Proc. IAU Symposium 185, ed. F.-L. Deubner, J. Christensen-Dalsgaard, and D. Kurtz (Kluwer Academic Publishers, Netherlands), 149.

Giles, P.M. (1999). Time-Distance Measurements of Large-Scale Flows in the Solar Convection Zone, Stanford University, D. Phil. thesis.

Gizon, L., Duvall, T.L., Jr., and Larsen, R.M. (2000). *Journal of Astrophysics and Astronomy*, **21**, 339.

Gizon L., Duvall T.L., Jr., and Larsen R.M. (2001). in *Recent Insights into the Physics of the Sun and Heliosphere*, ed. P. Brekke, B. Fleck, and J.B. Gurman (The Astronomical Society of the Pacific, San Francisco), 189–191.

Gizon, L. and Birch, A.C. (2002). *Astrophys. J.*, submitted.

Gough, D. O. (1993). in: *Astrophysical Fluid Dynamics*, ed. J.-P. Zahn and J. Zinn-Justin, (Elsevier Science Publ.), p. 339.

Hung, S.-H., Dahlen, F. A., and Nolet, G. (2001). *Geophysical J. Int.*, **146**, 289.

Jefferies, S. M., Osaki, Y., Shibahashi, H. , Duvall, T. L., Jr., Harvey, J. W., and Pomerantz, M. A. (1994). *Astrophys. J.*, **434**, 795.

Jensen, J. M., Jacobsen, B. H., and Christensen-Dalsgaard, J. (1997). in *Proceedings, Interdisciplinary Inversion*

Workshop 5 (Aarhus University, Aarhus), p. 57.

Jensen, J.M., Jacobsen, B.H., and Christensen-Dalsgaard, J. (2000). *Solar Phys.*, **192**, 231.

Kosovichev, A.G. (1996). *Astrophys. J.*, **461**, L55.

Kosovichev, A. G. and Duvall, T. L., Jr. (1997). in: *SCORe'96 : Solar Convection and Oscillations and their Relationship*, ed. F.P. Pijpers, J. Christensen-Dalsgaard, and C.S. Rosenthal, (Kluwer Academic Publishers, Netherlands) p. 241.

Kosovichev, A.G. (1999). *J. Comput. Appl. Math.* **109**, 1.

Kosovichev, A.G. and Duvall, T.L., Jr. (1999). *Current Science*, **77**, 1467.

Kosovichev, A. G., Duvall, T. L., and Scherrer, P. H. (2000). *Solar Phys.*, **192**, 159.

Lindsey, C. and Braun, D.C. (2000a). *Solar Phys.*, **192**, 261.

Lindsey, C. and Braun, D.C. (2000b). *Science*, **287**, 1799.

Marquering, H., Dahlen, F. A., and Nolet, G. (1999). *Geophysical J. Int.*, **137**, 805.

Paige, C. C. and Saunders, M. A. (1982). *ACM Trans. Math. Software*, **8**, 43.

Parker, E. N. (1979). *Astrophys. J.*, **230**, 905.

Pratt, R.G. (1999). *Geophysics*, **64**, 902.

Rickett, J. E., and Claerbout, J. F. (2000). *Solar Phys.*, **192**, 203.

Rhodes, E. J., Jr., Kosovichev, A. G., Schou, J.; Scherrer, P. H., and Reiter, J. (1997). *Solar Phys.*, **175**, 287.

Woodard, M.F. (2002). *Astrophys. J.*, in press.

Zhao, Junwei, Kosovichev, A.G., and Duvall, T.L., Jr, (2001). *Astrophys. J.*, **557**, 384.

6

The solar dynamo as a model of the solar cycle

A.R. Choudhuri

Dept. of Physics, IISc, Bangalore-560012, India

6.1   Introduction

In elementary textbooks on stellar structure, a star is usually modeled as a spherically symmetric, non-rotating, non-magnetic object. It is mainly the magnetic field which makes our Sun much more intriguing than such a textbook star. Several other chapters in this volume should convince the reader of this. It comes, therefore, as no surprise that one of the central problems in solar physics is to understand the origin of the Sun's magnetic field. The solar dynamo theory attempts to address this problem. The basic idea of this theory is that the solar magnetic fields are generated and maintained by complicated nonlinear interactions between the solar plasma and magnetic fields. As we shall see, there are still many difficulties with this theory and we are still far from having a completely satisfactory explanation of why the Sun's magnetic field behaves the way it does. However, no alternate theory of the origin of solar magnetism has so far been able to explain even a fraction of what dynamo theory has explained. Some of us, therefore, are still struggling to put the solar dynamo theory on firmer footing, with the fond hope that we are probably approximately on the correct path.

Dynamo theory is based on the principles of magnetohydrodynamics (abbreviated as MHD), in which hydrodynamics equations are combined with Maxwell's electrodynamics equations. Comprehensive introductions to MHD can be found in the books by Alfvén and Fälthammar (1963), Cowling (1976), Parker (1979), Priest (1982) and Choudhuri (1998). Some books devoted primarily to dynamo theory are by Moffatt (1978), Krause and Rädler (1980) and Zeldovich et al. (1983). We also refer to the review articles on the solar dynamo by Ruzmaikin (1985), Gilman (1986), Hoyng (1990), Brandenburg and Tuominen (1990), Schmitt (1993)

103

and Choudhuri (1999). The present chapter is an updated version of the last mentioned review.

Within the limited scope of one chapter, it is not possible to present a technical subject like the solar dynamo theory in such a fashion that a reader without any previous knowledge of the subject learns it fully and then becomes an expert on it! After summarizing the relevant observations in §6.2, we write just enough about the basics in §6.3–4 to give a rough idea of the subject. Then in §6.5–7 we discuss some of the important issues from current research frontiers. While surveying current research, we shall mainly focus our attention on discussing how successful the solar dynamo theory has been as a model for the solar cycle. To keep the length of the chapter reasonable, we shall refrain from discussing various other issues connected with the foundations of dynamo theory. This chapter is primarily meant for readers who are not experts on dynamo theory and does not attempt to review all aspects of the solar dynamo problem. It is hoped that a reader who studies this chapter in conjunction with Chapter 16 of Choudhuri (1998) should be in a position to read research papers on this subject.

6.2   Relevant observations

The next chapter by Stenflo will discuss the spectral techniques to detect and measure magnetic fields at the solar surface. Hale (1908) was the first to discover the evidence of Zeeman effect in sunspot spectra and made the momentous announcement that sunspots are regions of strong magnetic fields. This is the first time that somebody found conclusive evidence of large-scale magnetic fields outside the Earth's environment. The typical magnetic field of a large sunspot is about 3000 G.

Even before it was realized that sunspots are seats of solar magnetism, several persons have been studying the occurrences of sunspots. Schwabe (1844) noted that the number of sunspots seen on the solar surface increases and decreases with a period of about 11 years. Now we believe that the Sun has a cycle with twice that period, i.e. 22 years. Since the Sun's magnetic field changes its direction after 11 years, it takes 22 years for the magnetic field to come back to its initial configuration. Carrington (1858) found that sunspots seemed to appear at lower and lower latitudes with the progress of the solar cycle. In other words, most of the sunspots in the early phase of a solar cycle are seen between 30° and 40°. As the cycle advances, new sunspots are found at increasingly lower latitudes. Then a fresh half-cycle begins with sunspots appearing again at high latitudes. Individual sunspots live from a few days to a few weeks.

After finding magnetic fields in sunspots, Hale and his coworkers made another significant discovery (Hale et al., 1919). They found that often two large sunspots are seen side by side and they invariably have opposite polarities. The line joining the centres of such a bipolar sunspot pair is, on an average, nearly parallel to the solar

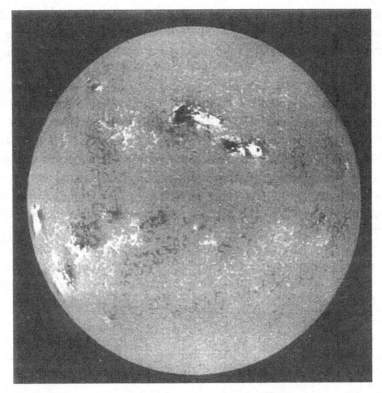

Figure 6.1. A magnetogram image of the full solar disk. The regions with positive and negative magnetic polarities are respectively shown in white and black, with grey indicating regions where the magnetic field is weak. Courtesy: Karen Harvey.

equator. Hale's coworker Joy, however, noted that there is a systematic tilt of this line with respect to the equator and that this tilt increases with latitude (Hale *et al.*, 1919). This result is usually known as *Joy's law*. The tilts, however, show a considerable amount of scatter around the mean given by Joy's law. The sunspot in the forward direction of rotation is called the leading spot and the other the following spot. The tilt is such that the leading spot is usually found nearer the equator than the following spot. It was also noted that the sunspot pairs have opposite polarities in the two hemispheres. In other words, if the leading sunspot in the northern hemisphere has positive polarity, then the leading sunspot in the southern hemisphere has negative polarity. This is clearly seen in Figure 6.1, which is a magnetic map of the Sun's disk obtained with a magnetogram. The regions of positive and negative polarities are shown in white and black respectively. The polarities of the bipolar sunspots in any hemisphere get reversed from one half-cycle of 11 years to the next half-cycle.

After the development of the magnetograph by Babcock and Babcock (1955), it became possible to study the much weaker magnetic field near the poles of the Sun. This magnetic field is of the order of 10 G and reverses its direction at the time of solar maximum (i.e. when the number of sunspots seen on the solar surface is maximum) (Babcock, 1959). This shows that this weak, diffuse field of the Sun is somehow coupled to the much stronger magnetic field of the sunspots and is a part of the same solar cycle. Low-resolution magnetograms show the evidence of weak magnetic field even in lower latitudes. The true nature of this field is not very clear. It has been found (Stenflo, 1973) that the magnetic field on the solar surface outside sunspots often exists in the form of fibril flux tubes of diameter of the order of 300 km with field strength of about 2000 G (large sunspots have sizes larger than 10,000 km). One is not completely sure if the field found in the low-resolution magnetograms is truly a diffuse field or a smearing out of the contributions made by fibril flux tubes. Keeping this caveat in mind, we should refer to the field outside sunspots as seen in magnetograms as the 'diffuse' field. It was found that there were large unipolar patches of this diffuse field on the solar surface which migrated polewards (Bumba and Howard, 1965). Even when averaged over longitude, one finds predominantly one polarity in a belt of latitude which drifts polewards (Howard and LaBonte, 1981; Wang *et al.*, 1989a). The reversal of polar field presumably takes place when sufficient field of opposite polarity has been brought near the poles. This opposite polarity field often arises out of large impulses of activity.

Figure 6.2 shows the distribution of both sunspots and the weak, diffuse field in a plot of latitude vs. time. The various shades of gray indicate values of longitude-averaged diffuse field, whereas the latitudes where sunspots were seen at a particular

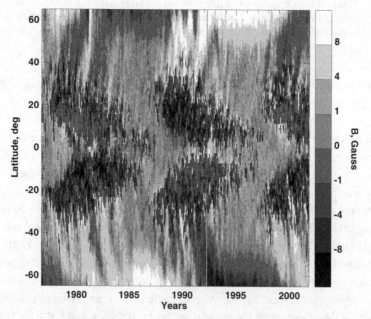

Figure 6.2. Shades of gray showing the latitude-time distribution of longitudinally averaged weak, diffuse magnetic field (B is in Gauss) with a 'butterfly diagram' of sunspots superimposed on it. Courtesy: Alexander Kosovichev.

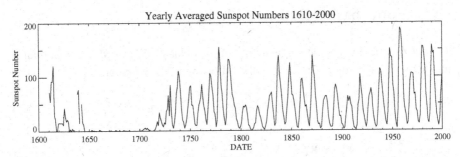

Figure 6.3. The yearly averaged sunspot number plotted against the year for the period 1610–2000. Courtesy: David Hathaway.

time are marked in black. The sunspot distribution in a time-latitude plot is often referred to as a *butterfly diagram*, since the pattern (the regions marked in black in Figure 6.2) reminds one of butterflies. Such butterfly diagrams were first plotted by Maunder (1904). Historically, most of the dynamo models concentrated on explaining the distribution of sunspots and ignored the diffuse field. Only during the last few years, it has been realized that the diffuse fields give us important clues about the dynamo process and they should be included in a full self-consistent theory. The aim of such a theory should be to explain diagrams like Figure 6.2 (i.e. not just the butterfly diagram).

We have provided above a summary of the various regular features in the Sun's activity cycle. One finds lots of irregularities and fluctuations superposed on the underlying regular behaviour, as can be seen in Figure 6.2. These irregularities are more clearly visible in Figure 6.3, where the number of sunspots seen on the solar surface is plotted against time. Galileo was one of the first persons in Europe to study sunspots at the beginning of the 17th century. After Galileo's work, sunspots were almost not seen for nearly a century! Such a grand minimum has not occurred again in the last 300 years.

It may be noted that all the observations discussed above pertain to the Sun's surface. We have no direct information about the magnetic field underneath the Sun's surface. The new science of helioseismology, however, has provided us lots of information about the velocity field underneath the solar surface. For an account of this subject, the readers may turn to Chapter 4 by Christensen-Dalsgaard and Thompson. We shall have occasions to refer to some of the helioseismic findings in our discussion later. It is to be noted that heat is transported by convection in the outer layers of the Sun from about $0.7 R_\odot$ to $R_\odot$ (where $R_\odot$ is the solar radius). This region is called the convection zone, within which the plasma is in a turbulent state. The job of a theorist now is to construct a detailed model of the physical processes in this turbulent plasma such that all the surface observations of magnetic fields are properly explained – a fairly daunting problem, of which the full solution is still a distant dream.

6.3   Some basic MHD considerations

The velocity field $\mathbf{v}$ and the magnetic field $\mathbf{B}$ in a plasma (regarded as a continuum) interact with each other according to the following MHD equations:

$$\frac{\partial \mathbf{v}}{\partial t} + (\mathbf{v} \cdot \nabla)\mathbf{v} = -\frac{1}{\rho}\nabla\left(p + \frac{B^2}{2\mu}\right) + \frac{(\mathbf{B} \cdot \nabla)\mathbf{B}}{\mu\rho} + \mathbf{g} + \nu\,\nabla^2\mathbf{v}, \tag{6.1}$$

$$\frac{\partial \mathbf{B}}{\partial t} = \nabla \times (\mathbf{v} \times \mathbf{B}) + \lambda\,\nabla^2\mathbf{B}. \tag{6.2}$$

Here ρ is density, p is pressure, $\mathbf{g}$ is gravitational field, ν is kinematic viscosity and

$$\lambda = \frac{1}{\mu\sigma} \tag{6.3}$$

is magnetic diffusivity (σ is electrical conductivity). Equation (6.1) is essentially the Navier–Stokes equation, to which magnetic forces have been added. It is clear from (6.1) that the magnetic field has two effects: (i) it gives rise to an additional pressure $B^2/2\mu$; and (ii) the other magnetic term $(\mathbf{B} \cdot \nabla)\mathbf{B}/\mu\rho$ is of the nature of a tension along magnetic field lines.

Equation (6.2) is known as the *induction equation* and is the key equation in MHD. If V, B and L are the typical values of velocity, magnetic field and length scale, then the two terms on the R.H.S. of (6.2) are of order VB/L and $\lambda B/L^2$. The ratio of these two terms is a dimensionless number, known as the magnetic Reynolds number, given by

$$R_m = \frac{VB/L}{\lambda B/L^2} = \frac{VL}{\lambda}. \tag{6.4}$$

Since R_m goes as L, it is expected to be much larger in astrophysical situations than it is in the laboratory. In fact, usually one finds that $R_m \gg 1$ in astrophysical systems and $R_m \ll 1$ in laboratory-size objects. Hence the behaviours of magnetic fields are very different in laboratory plasmas and astrophysical plasmas. For example, it is not possible to have a laboratory analogue of the Sun's self-sustaining magnetic field. If $R_m \gg 1$ in an astrophysical system, then the diffusion term in (6.2) is negligible compared to the term preceding it. In such a situation, it can be shown that the magnetic field is frozen in the plasma and moves with it. This result was first recognized by Alfvén (1942) and is often referred to as *Alfvén's theorem of flux-freezing*.

It is known that the Sun does not rotate like a solid body. The angular velocity at the equator is about 20% faster than that at the poles. Because of the flux freezing, this differential rotation would stretch out any magnetic field line in the toroidal direction (i.e. the ϕ direction with respect to the Sun's rotation axis). This is indicated in Figure 6.4. We, therefore, expect that the magnetic field inside the Sun may be predominantly in the toroidal direction.

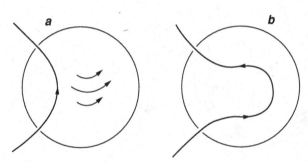

Figure 6.4. The production of a strong toroidal magnetic field underneath the Sun's surface. **a.** An initial poloidal field line. **b.** A sketch of the field line after it has been stretched by the faster rotation near the equatorial region.

We already mentioned in §6.2 that energy is transported by convection in the layers underneath the Sun's surface. To understand why the magnetic field remains concentrated in structures like sunspots instead of spreading out more evenly, we need to study the interaction of the magnetic field with the convection in the plasma. This subject is known as magnetoconvection. The linear theory of convection in the presence of a vertical magnetic field was studied by Chandrasekhar (1952). The nonlinear evolution of the system, however, can only be found from numerical simulations pioneered by Weiss (1981). It was found that space gets separated into two kinds of regions. In certain regions, magnetic field is excluded and vigorous convection takes place. In other regions, magnetic field gets concentrated, and the tension of magnetic field lines suppresses convection in those regions. Sunspots are presumably such regions where magnetic field is piled up by surrounding convection. Since heat transport is inhibited there due to the suppression of convection, sunspots look darker than the surrounding regions. We refer Thomas and Weiss (1992) for a review of our understanding of sunspots.

Although we have no direct information about the state of the magnetic field under the Sun's surface, it is expected that the interactions with convection would keep the magnetic field concentrated in bundles of field lines throughout the solar convection zone. Such a concentrated bundle of magnetic field lines is called a flux tube. In the regions of strong differential rotation, therefore, we may have the magnetic field in the form of flux tubes aligned in the toroidal direction. If a part of such a flux tube rises up and pierces the solar surface as shown in Figure 6.5b, then we expect to have two sunspots with opposite polarities at the same latitude. But how can a configuration like Figure 6.5b arise? The answer to this question was provided by Parker (1955a) through his idea of magnetic buoyancy. We have seen in (6.1) that a pressure $B^2/2\mu$ is associated with a magnetic field. If p_{in} and p_{out} are the gas pressures inside and outside a flux tube, then we need to have

$$p_{out} = p_{in} + \frac{B^2}{2\mu} \tag{6.5}$$

to maintain pressure balance across the surface of a flux tube. Hence

$$p_{in} \leq p_{out}, \tag{6.6}$$

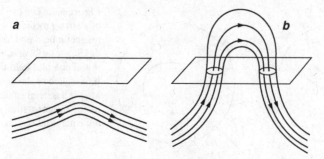

Figure 6.5. Magnetic buoyancy of a flux tube. **a.** A nearly horizontal flux tube under the solar surface. **b.** The flux tube after its upper part has risen through the solar surface.

which often, though not always, implies that the density inside the flux tube is less than the surrounding density. If this happens in a part of the flux tube, then that part becomes buoyant and rises against the gravitational field to produce the configuration of Figure 6.5b starting from Figure 6.5a.

A look at Figure 6.4 now ought to convince the reader that the sub-surface toroidal field in the two hemispheres should have opposite polarity. If this toroidal field rises due to magnetic buoyancy to produce the bipolar sunspot pairs, then we expect the bipolar sunspots to have opposite polarities in the two hemispheres as seen in Figure 6.1. We thus see that combining the ideas of flux freezing, magnetoconvection and magnetic buoyancy, we can understand many aspects of the bipolar sunspot pairs. We now turn our attention to the central problem – dynamo generation of the magnetic field.

6.4   The turbulent dynamo and mean field MHD

We now address the question whether it is possible for motions inside the plasma to sustain a magnetic field. Ideally, one would like to solve (6.1) and (6.2) to understand how velocity and magnetic fields interact with each other. Solving these two equations simultaneously in any non-trivial situation is an extremely challenging job. In the early years of dynamo research, one would typically assume a velocity field to be given and then solve (6.2) to find if this velocity field would sustain a magnetic field. This problem is known as the *kinematic dynamo problem*. One of the first important steps was a negative theorem due to Cowling (1934), which established that an axisymmetric solution is not possible for the kinematic dynamo problem. One is, therefore, forced to look for more complicated, non-axisymmetric solutions.

A major breakthrough occurred when Parker (1955b) realized that turbulent motions inside the solar convection zone (which are by nature non-axisymmetric) may be able to sustain the magnetic field. We have indicated in Figure 6.4 how a magnetic field line in the poloidal plane may be stretched by the differential rotation

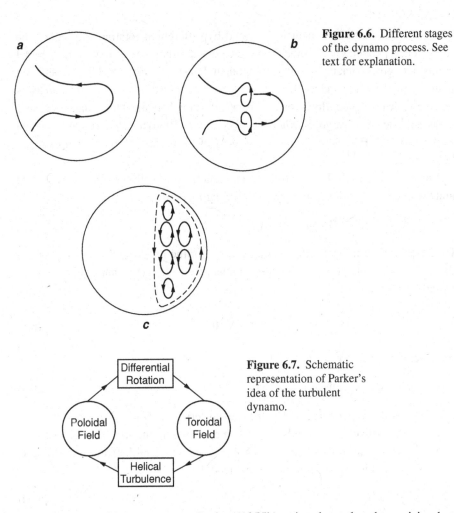

Figure 6.6. Different stages of the dynamo process. See text for explanation.

Figure 6.7. Schematic representation of Parker's idea of the turbulent dynamo.

to produce a toroidal component. Parker (1955b) pointed out that the uprising hot plasma blobs in the convection zone would rotate, as they rise, because of the Coriolis force of solar rotation (just like cyclones in the Earth's atmosphere) and such helically moving plasma blobs would twist the toroidal field shown in Figure 6.6a to produce magnetic loops in the poloidal plane as shown in Figure 6.6b. Keeping in mind that the toroidal field has opposite directions in the two hemispheres and helical motions of convective turbulence should also have opposite helicities in the two hemispheres, we conclude that the poloidal loops in both hemispheres should have the same sense as indicated in Figure 6.6c. Although we are in a high magnetic Reynolds number situation and the magnetic field is nearly frozen in the plasma (i.e. molecular resistivity is negligible), we expect the turbulent mixing to give rise to an effective diffusion and the poloidal loops in Figure 6.6c should eventually coalesce to give the large-scale poloidal field as sketched by the broken line in Figure 6.6c.

Figure 6.7 captures the basic idea of Parker's turbulent dynamo. The poloidal and toroidal components of the magnetic field feed each other through a closed loop. The poloidal component is stretched by differential rotation to produce the toroidal component. On the other hand, the helical turbulence acting on the toroidal component gives back the poloidal component. Parker (1955b) developed a heuristic mathematical formalism based on these ideas and showed by mathematical analysis that these ideas worked. However, a more systemic mathematical formulation of these ideas had to await a few years, when Steenbeck *et al.* (1966) developed what is known as mean field MHD. Some of the basic ideas of mean field MHD are summarized below.

Since we have to deal with a turbulent situation, let us split both the velocity field and the magnetic field into average and fluctuating parts, i.e.

$$\mathbf{v} = \bar{\mathbf{v}} + \mathbf{v}', \quad \mathbf{B} = \bar{\mathbf{B}} + \mathbf{B}'. \tag{6.7}$$

Here the overline indicates the average and the prime indicates the departure from the average. On substituting (6.7) in the induction equation (6.2) and averaging term by term, we obtain

$$\frac{\partial \bar{\mathbf{B}}}{\partial t} = \nabla \times (\bar{\mathbf{v}} \times \bar{\mathbf{B}}) + \nabla \times \mathcal{E} + \lambda \nabla^2 \bar{\mathbf{B}} \tag{6.8}$$

on remembering that $\overline{\mathbf{v}'} = \overline{\mathbf{B}'} = 0$. Here

$$\mathcal{E} = \overline{\mathbf{v}' \times \mathbf{B}'} \tag{6.9}$$

is known as the *mean e.m.f.* and is the crucial term for dynamo action. This term can be perturbatively evaluated by a scheme known as the *first order smoothing approximation* (see, for example, §16.5 of Choudhuri, 1998). If the turbulence is isotropic, then this approximation scheme leads to

$$\mathcal{E} = \alpha \bar{\mathbf{B}} - \beta \nabla \times \bar{\mathbf{B}}, \tag{6.10}$$

where

$$\alpha = -\frac{1}{3} \overline{\mathbf{v}'.(\nabla \times \mathbf{v}')} \, \tau \tag{6.11}$$

and

$$\beta = \frac{1}{3} \overline{\mathbf{v}'.\mathbf{v}'} \, \tau. \tag{6.12}$$

Here τ is the correlation time of turbulence. On substituting (6.10) in (6.8), we get

$$\frac{\partial \bar{\mathbf{B}}}{\partial t} = \nabla \times (\bar{\mathbf{v}} \times \bar{\mathbf{B}}) + \nabla \times (\alpha \bar{\mathbf{B}}) + (\lambda + \beta)\nabla^2 \bar{\mathbf{B}} \tag{6.13}$$

It should be clear from this that β is the turbulent diffusion. This is usually much larger than the molecular diffusion λ so that λ can be neglected in (6.13). It follows from (6.11) that α is a measure of average helical motion in the fluid. It is this coefficient which describes the production of the poloidal component from the toroidal component by helical turbulence. This term would go to zero if turbulence has no net average helicity (which would happen in a non-rotating frame).

Equation (6.13) is known as the *dynamo equation* and has to be solved to understand the generation of magnetic field by the dynamo process. A variant of this equation was first derived by rather intuitive arguments in the classic paper of Parker (1955b). The mean field MHD developed by Steenbeck *et al.* (1966) put this equation on a firmer footing. In the kinematic dynamo approach, one has to specify a velocity field $\bar{\mathbf{v}}$ and then solve (13). Using spherical polar coordinates with respect to the rotation axis of the Sun, we can write

$$\bar{\mathbf{v}} = \Omega(r, \theta) r \sin\theta \, \mathbf{e}_\phi + \bar{\mathbf{v}}_P, \tag{6.14}$$

where $\Omega(r, \theta)$ is the angular velocity in the interior of the Sun and $\bar{\mathbf{v}}_P$ is some possible average flow in the meridional plane. Until a few years ago, almost all the calculations of the kinematic dynamo problem were done by taking $\bar{\mathbf{v}}_P = 0$. If this is the case, then one has to specify some reasonable $\Omega(r, \theta)$ and $\alpha(r, \theta)$ before proceeding to solve the dynamo equation (6.13). In the 1970s almost an industry grew up presenting solutions of the dynamo equation for different specifications of Ω and α.

The first pioneering solution in rectangular geometry was obtained by Parker (1955b) himself. He showed that periodic and propagating wave solutions of the dynamo equation are possible. Presumably this offers an explanation for the solar cycle. Sunspots migrate from higher to lower latitudes with the solar cycle because sunspots are produced (by magnetic buoyancy) where the crest of the propagating dynamo wave lies. Parker (1955b) found that the parameters α and Ω have to satisfy the following condition in the northern hemisphere to make the dynamo wave propagate in the equatorward direction (so as to explain the butterfly diagram of sunspots):

$$\alpha \frac{d\Omega}{dr} \leq 0. \tag{6.15}$$

Steenbeck and Krause (1969) were the first to solve the dynamo equation in a spherical geometry appropriate for the Sun and produced the first theoretical butterfly diagram of the distribution of sunspots in time-latitude. Then many dynamo solutions were worked out by Roberts (1972), Köhler (1973), Yoshimura (1975), Stix (1976) and others. One might have felt complacent about the varieties of butterfly diagrams produced by these authors. However, it has to be admitted that many basic physics questions remained unanswered. Since nothing was known at that time about the conditions in the interior of the Sun, different authors were choosing different α and Ω subject only to the condition (6.15), and thereby were trying to fit the observational data better. Eventually it appeared that it was becoming a game in which you could get solutions according to your wishes by tuning your free parameters suitably. Further progress in solar dynamo theory became possible only by asking fundamental questions about the basic physics in the interior of the Sun, rather than by blindly solving the dynamo equation. These efforts will be described in the next two sections. It may be noted that all the authors of this period focused their attention on explaining the equatorward propagation of sunspots, by assuming that sunspots were produced in the regions where the toroidal component had the peak value. No serious attempt was made to connect the behaviour of the weak, diffuse magnetic field with the dynamo process or to explain the poleward migration of this field, although Köhler (1973) and Yoshimura (1975) presented some models that show a polar branch, i.e. a region near the poles where the dynamo wave propagates poleward.

6.5   Dynamo in the overshoot layer?

Where does the solar dynamo work? Since one needs convective turbulence to drive the dynamo, it used to be tacitly assumed in the 1970s that the dynamo works in the solar convection zone and the different researchers of that period used to take α and Ω non-zero in certain regions of the convection zone. This approach had to be questioned when Parker (1975) started looking at the effect of magnetic buoyancy on the solar dynamo. Magnetic buoyancy is particularly destabilizing in the interior of the convection zone, where convective instability and magnetic buoyancy reinforce each other. On the other hand, if a region is stable against convection, then magnetic buoyancy can be partially suppressed there (see, for example, Parker, 1979, §8.8). Calculations of buoyant rise by Parker (1975) showed that any magnetic field in the convection zone would be removed from there by magnetic buoyancy fairly quickly. Hence it is difficult to make the dynamo work in the convection zone, since the magnetic field has to be stored in the dynamo region for a sufficient time to allow for dynamo amplification.

It is expected that there is a thin overshoot layer (probably with a thickness of the order of 10^4 km) just below the bottom of the convection zone. This is a layer which is convectively stable according to a local stability analysis, but convective motions are induced there due to convective plumes from the overlying unstable layers overshooting and penetrating there. Several authors (Spiegel and Weiss, 1980; van Ballegooijen, 1982) pointed out that this layer is a suitable location for the operation of the dynamo. Although there would be enough turbulent motions in this layer to drive the dynamo, magnetic buoyancy would be suppressed by the stable temperature gradient there. This idea turned out to be a really prophetic theoretical guess, since helioseismology observations a few years later indeed discovered a region of strong differential rotation at the bottom of the solar convection zone (see Chapter 4 by Christensen-Dalsgaard and Thompson). So it is certainly expected that a strong toroidal magnetic field should be generated just below the bottom of the convection zone due to this strong differential rotation. It may be noted that there have been other ideas as well for suppressing magnetic buoyancy at the bottom of the convection zone. Parker (1987) suggested 'thermal shadows', whereas van Ballegooijen and Choudhuri (1988) showed that an equatorward meridional circulation at the base of the convection zone can help in suppressing magnetic buoyancy there.

For about a decade starting from the mid-1980s, most researchers in this field believed that the whole dynamo process in the Sun, as summarized in Figure 6.7, takes place in the overshoot layer. Properties of such a dynamo operating in the overshoot layer were studied by several authors (DeLuca and Gilman, 1986; Gilman et al., 1989; Schmitt and Schüssler, 1989; Choudhuri, 1990; Belvedere et al., 1990; Rüdiger and Brandenburg, 1995). If the dynamo operates in the overshoot layer, then some new questions arise. Previously when the solar dynamo was supposed to work in the convection zone, the sunspots seen on the solar surface could be regarded as direct signatures of the dynamo process. One could assume that sunspots appeared wherever the dynamo produced strong toroidal fields just underneath the surface. On the other hand, if the dynamo works at the bottom of the convection zone, the whole depth of the convection zone separates the region where the magnetic fields are generated and the solar surface where sunspots are seen. In order to understand the relation between the solar dynamo and sunspots, one then has to study how the magnetic fields generated at the bottom of the convection zone rise through the convection zone to produce sunspots.

The best way to study this is to treat it as an initial-value problem. First an initial configuration with some magnetic flux at the bottom of the convection zone is specified, and then its subsequent evolution is studied numerically. The evolution depends on the strength of magnetic buoyancy, which is in turn determined by the value of the magnetic field. If the dynamo is driven by turbulence, then one would expect an equipartition of energy between the dynamo-generated magnetic field and the fluid kinetic energy, i.e.

$$\frac{B^2}{2\mu} \approx \frac{1}{2}\rho v^2. \tag{6.16}$$

This suggests $B \approx 10^4$ G on the basis of standard models of the convection zone. Because of the strong differential rotation, we expect the magnetic field at the bottom of the convection zone to be mainly in the toroidal direction. One, therefore, has to take a toroidal magnetic flux tube going around the rotation axis as the initial configuration. The evolution of such magnetic flux tubes due to magnetic buoyancy was first studied by Choudhuri and Gilman (1987) and Choudhuri (1989). It was found that the Coriolis force due to the Sun's rotation plays a much more important role in this problem than what anybody suspected before. If the initial magnetic field is taken to have a strength around 10^4 G, then the flux tubes move parallel to the rotation axis and emerge at very high latitudes rather than at latitudes where sunspots are seen. Only if the initial magnetic field is taken as strong as 10^5 G, then magnetic buoyancy is strong enough to overpower the Coriolis force and the magnetic flux tubes can rise radially to emerge at low latitudes.

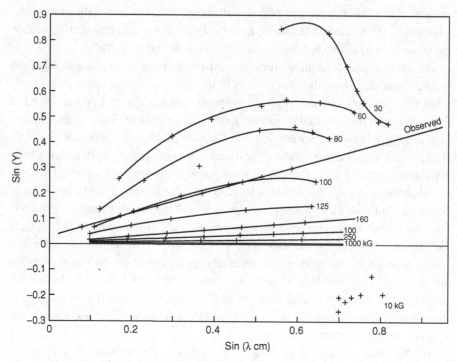

Figure 6.8. Plots of sin(tilt) against sin(latitude) theoretically obtained for different initial values of magnetic field indicated in kG. The observational data indicated by the straight line fits the theoretical curve for initial magnetic field 100 kG (i.e. 10^5 G). Reproduced from D'Silva and Choudhuri (1993).

D'Silva and Choudhuri (1993) extended these calculations to look at the tilts of emerging bipolar regions at the surface. These tilts are also produced by the action of the Coriolis force on the rising flux tube. Figure 6.8 taken from D'Silva and Choudhuri (1993) shows the observational tilt vs. latitude plot of bipolar sunspots (i.e. Joy's law) along with the theoretical plots obtained by assuming different values of the initial magnetic field. It is clearly seen that theory fits observations only if the initial magnetic field is about 10^5 G. Apart from providing the first quantitative explanation of Joy's law nearly three-quarters of a century after its discovery, these calculations put the first stringent limit on the value of the toroidal magnetic field at the bottom of the convection zone. Several other authors (Fan *et al.*, 1993; D'Silva and Howard, 1993; Schüssler *et al.*, 1994; Caligari *et al.*, 1995) soon performed similar calculations and confirmed the result. The evidence is now mounting that the magnetic field at the bottom of the convection zone is indeed much stronger than the equipartition value given by (6.16) (see Schüssler, 1993, for a review of this topic). We already mentioned that the tilts of active regions have a large amount of scatter around the mean given by Joy's law. Longcope and Fisher (1996) showed that this scatter can arise due to the turbulent buffeting of the flux tubes when they rise through the convection zone. It is found that active regions emerging with initial tilts inconsistent with Joy's law change their orientations in the next few days to bring the tilts close to values given by Joy's law (Howard, 1996). Although this observation at first seems to cast a doubt whether Joy's law is produced by the Coriolis force, Longcope and Choudhuri (2002) have shown how this observation can be fitted within the current theoretical framework.

If the magnetic field is much stronger than the equipartition value, then it would be impossible for the helical turbulence (entering the mathematical theory through the α term defined in (6.11)) to twist the magnetic field lines. The dynamo process as envisaged in Figure 6.7 is, therefore, not possible. We need a very different type of dynamo model. Schmitt (1987) and Feriz-Mas *et al.* (1994) proposed that the buoyant instability of the strong magnetic field itself may lead to a magnetic configuration which was previously thought to be created by helical turbulence. Parker (1993) suggested an 'interface dynamo' in which the helical turbulence acts in a region above the bottom of the convection zone. This idea has been further explored by Charbonneau and MacGregor (1997) and Markiel and Thomas (1999). In the next section, we discuss what we regard as the most promising approach to build a model of the solar dynamo that can account for the very strong toroidal magnetic field at the bottom of the convection zone.

6.6   The Babcock–Leighton approach and advective dynamo models

We saw in the previous two sections that one of the crucial ingredients in turbulent dynamo theory is the role of helical turbulence in generating the poloidal component from the toroidal component, which is mathematically modeled through mean field MHD. This approach to the dynamo problem will be called the Parker–Steenbeck–Krause–Rädler or the PSKR approach. In this approach, the dynamo is supposed to operate within a region where convective turbulence exists and no special attention is paid to phenomena taking place at the solar surface. Several decades ago, Babcock (1961) and Leighton (1969) developed a somewhat different approach, which we call the Babcock–Leighton or the BL approach. Even in this approach, the toroidal magnetic field is believed to be produced by the differential rotation of the Sun. For the production of the poloidal component, however, a totally different scenario is invoked. The strong toroidal component leads to bipolar sunspots due to magnetic buoyancy. We have noted that these bipolar sunspots have a tilt with respect to latitudinal lines (i.e. Joy's law). Therefore, when these bipolar sunspots eventually decay, the magnetic flux gets spread around in such a way that the flux at the higher latitude has more contribution from the polarity of the sunspot which was at the higher latitude. In this way, a poloidal component arises.

Compared to the PSKR approach, the BL approach was heuristic and semi-qualitative. The mean field MHD is, no doubt, based on some assumptions and approximations, and it is not clear whether these hold in the conditions prevailing in the Sun's interior. However, for an ideal system satisfying these assumptions and approximations, the mean field MHD is a rigorous mathematical theory. No quantitative mathematical theory of comparable sophistication was developed for the BL approach. Nearly all the self-consistent dynamo calculations in the 1970s and 1980s, therefore, followed the PSKR approach. The BL idea that the poloidal field at the solar surface is produced from the tilted active regions was developed further by a group in NRL (DeVore et al., 1984; Sheeley et al., 1985; Wang et al., 1989a, 1989b), who were studying the spread of magnetic flux from the decay of sunspots. There is evidence of a general meridional flow with an amplitude of about a few m s^{-1} near the Sun's surface proceeding from the equator to the pole. Techniques of helioseismology are now being used to probe the nature of the meridional flow underneath the solar surface (Giles et al., 1997; Braun and Fan, 1998). The poloidal magnetic field produced from the decay of tilted bipolar sunspots is advected poleward by this meridional circulation. As we have already pointed out in §6.2, the weak diffuse magnetic field on the solar surface migrates towards the pole, in contrast to sunspots which migrate equatorward. One presumably has to identify the weak diffuse field as the poloidal component of the Sun's magnetic field, whereas sunspots formed from the much stronger toroidal component. The main aim of the NRL group was to model the evolution of the weak diffuse field, assuming that this was entirely coming from the decay of bipolar sunspots. No attempt was made to address the full dynamo problem. They even made the drastically simple assumption that the magnetic field is a scalar residing on the solar surface and the appropriate partial differential equation was solved only on this two-dimensional surface.

Dikpati and Choudhuri (1994, 1995) and Choudhuri and Dikpati (1999) attempted to make a vectorial model of the evolution of the weak diffuse field and to connect it to the dynamo problem. Since the meridional flow at the surface is poleward, there must be an equatorward flow in the lower regions of the convection zone, rising near the equator. If the dynamo operated at the base of the convection zone, then, in accordance with the ideas prevalent a few years ago, the poloidal component produced by this dynamo would be brought to the surface by the meridional circulation. This can be an additional source of the weak diffuse field at the surface, apart the contributions coming from the decay of sunspots. Figure 6.9 from Choudhuri and Dikpati (1999) shows a theoretical time-latitude distribution of the weak diffuse field on the surface, obtained by *assuming* a dynamo wave at the bottom of the convection zone *as given*. In other words, to produce this figure – which should be compared with the observational Figure 6.2 – the dynamo problem was not solved self-consistently. The aim now should be to develop a self-consistent model of the dynamo, which should be able to explain the behaviours of both the sunspots (i.e. the toroidal component) and the weak diffuse field (i.e. the poloidal component).

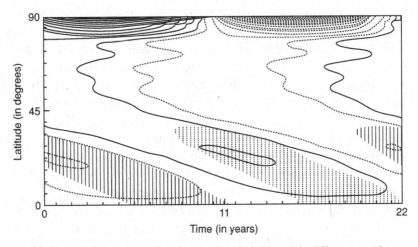

Figure 6.9. A theoretical time-latitude distribution of the weak, diffuse magnetic field on the solar surface, with 'half-butterfly diagrams' obtained from a running dynamo wave assumed given at the base of the convection zone. Reproduced from Choudhuri and Dikpati (1999).

With helioseismology establishing the existence of strong differential rotation at the base of the convection zone, there is little doubt that the toroidal magnetic field is produced there and has a magnitude of the order of 10^5 G – a result pinned down by the simulations of buoyant flux rise. It will, however, be impossible for helical turbulence to twist such strong magnetic fields, and it seems improbable that the generation of the poloidal field from the toroidal field by helical turbulence – as envisaged in the PSKR approach – takes place at the base of the convection zone. For the generation of the poloidal field, we then invoke the BL idea that it is produced by the decay of bipolar sunspots at the surface. The meridional circulation can then carry this field poleward, to be eventually brought to the bottom of the convection zone where it is stretched by the differential rotation to produce the strong toroidal field. If a mean field formulation is made of the process of poloidal field generation near the surface by invoking an α coefficient concentrated near the solar surface, then such a model of the dynamo will incorporate the best features of both the PSKR and the BL approaches. On the one hand, detailed quantitative calculations will be possible, as in the PSKR approach. On the other hand, the surface phenomena emphasized in the BL approach are integrated in the dynamo problem. The meridional circulation plays an important role in such a model so that a suitable form of $\overline{\mathbf{v}}_P$ in (6.14) is to be specified. Since advection by meridional circulation is so crucial in this model, we can call it an advection-dominated or advective dynamo model. It is hoped that this advective dynamo model will account for both the equatorward migration of the strong toroidal field at the base of the convection zone and the poleward drift of the poloidal field at the surface.

This advective dynamo model has one other attractive feature. Researchers in 1970s built kinematic models of the solar dynamo by arbitrarily specifying $\alpha(r, \theta)$ and $\Omega(r, \theta)$. These were regarded as free parameters to be tuned suitably so as to give solutions with desired characteristics. In the advective models, these important ingredients to the dynamo process are directly based on observations. Helioseismology has given us $\Omega(r, \theta)$, with its shear concentrated at the base of the convection zone. The α coefficient also arises out of the observed decay of bipolar sunspots on the surface. Previously it used to be even debated whether α is positive or negative. Researchers used to fudge α such that the inequality (6.15) was satisfied. The direction of tilt of bipolar sunspots on the surface, however, clearly indicates that α arising out of their decay has to be positive in the northern hemisphere – a point made by Stix (1976) long ago. Once these key ingredients are fixed directly by observation, we no longer have the freedom to fudge them according to our wishes, which researchers in 1970s could do. This leads to one problem. Helioseismology shows that $d\Omega/dr$ is positive in lower latitudes where sunspots are seen. If α is also positive in the northern hemisphere, then clearly inequality (6.15) is not satisfied and dynamo waves are expected to propagate poleward!

The first calculations on the advective model were reported by Choudhuri *et al.* (1995). Dynamo waves were indeed found to propagate poleward, *if meridional circulation was switched off.* The toroidal field and the poloidal field are respectively produced in layers near the base of the convection zone and near the solar surface. When meridional circulation is switched off, any field can reach from one layer to the other layer by diffusion with a time scale of L^2/β, where L is the separation between the layers (i.e. the thickness of the convection zone). If the meridional flow has a typical velocity of the order V, then it takes time V/L for the meridional circulation to carry some quantity between the two layers. When the time scale V/L is shorter than the diffusion time scale L^2/β, the problem is dominated by meridional circulation, and Choudhuri *et al.* (1995) found that the strong toroidal component at the bottom of the convection zone actually propagates equatorward, overriding the inequality (6.15). Thus the inequality (6.15), which was regarded as inviolable for four decades since Parker (1955b) derived it, is found not to hold in the presence of a meridional circulation having a time scale shorter than diffusion time, thereby opening up the possibility of constructing new type of advection-dominated models of the solar dynamo.

Further calculations on the advective dynamo model have been reported in a series of papers by Durney (1995, 1996, 1997), who followed Leighton (1969) closely in approximating the decaying bipolar sunspots by two rings of opposite polarity and did not use the α-effect. It has been shown by Nandy and Choudhuri (2001) that this method gives results qualitatively similar to results obtained by using the α-effect. Advective dynamo models with solar-like differential rotation have been presented by Dikpati and Charbonneau (1999) and Küker *et al.* (2001). It should be emphasized that all these studies are still of rather exploratory nature. They demonstrate the viability of the advective dynamo models and study their different characteristics. The nature of dynamo waves in the presence of solar-like rotation becomes extremely complicated (Belvedere *et al.*, 2000) and we are still far from building a sufficiently realistic model, putting in all the details, that would account for the observational data presented in Figure 6.2. The advection-dominated dynamo models with solar-like rotation tend to produce a strong toroidal field at high latitudes, whereas we know that sunspots are seen at low latitudes. Although these models show the equatorward migration of the strong toroidal field and the poleward migration of the weak poloidal field, we are still unable to produce a theoretical plot which fully resembles Figure 6.2 in detail (Nandy and Choudhuri, in preparation). The theoretical Figure 6.9, which is a reasonable match to Figure 6.2, was obtained by taking a dynamo wave at low latitudes for granted (i.e. by not solving the dynamo equation self-consistently). Although the advective dynamo models seem promising, we need to understand many aspects of them before we can think of arriving at a standard model of the solar dynamo. It may be mentioned that nearly all the recent kinematic models assume axisymmetry to make the problem two-dimensional rather than three-dimensional. Some early efforts of constructing non-axisymmetric kinematic dynamo models have been summarized by Ruzmaikin (1985).

6.7   Miscellaneous ill–understood issues

We have primarily discussed those aspects of kinematic dynamo models which directly pertain to the matching of theory with observational data. It should, however, be kept in mind that many fundamental issues of dynamo theory are still very ill-understood. Until we have a better understanding of these issues, the kinematic models can, at best, be considered superficial attempts at a very deep physics problem. The reader may look up the IAU Symposium volume on *The Cosmic Dynamo* (Krause *et al.*, 1993). for several articles dealing with these fundamental issues. Here we make only very brief comments on some of these issues.

We have seen in §6.4 that the turbulent dynamo theory is developed by averaging over turbulent fluctuations. The existence of magnetic flux concentrations clearly indicates that the fluctuations are much larger than the average values (often by orders of magnitude). Does a mean field theory make sense in such a situation? Can we trust the perturbative procedures like the first order smoothing approximation? Hoyng (1988) raised some questions regarding the interpretation of the averaged quantities. The dynamo equation (6.13) admits of several possible modes in spherical geometry. The preferred mode seems to be the mode with dipole symmetry, in which the toroidal component is oppositely directed in the two hemispheres. This mode approximately corresponds to the observational data. However, Stenflo and Vogel (1986) pointed out that one hemisphere of the Sun often has more sunspots than the other, indicating that there may be a superposition with higher modes having different symmetry. Analyzing the statistics of sunspot data for several decades, Gokhale and Javaraiah (1990) claimed to have found evidence for multiple modes. If the fluctuations are so large, then there is no reason why a particular mode should be very stable or why higher modes should not be excited. The interference of modes with different symmetry was theoretically studied by Brandenburg *et al.* (1989) employing a nonlinear dynamo model.

Since the toroidal magnetic field is much stronger than the equipartition value, it is certainly not justified to assume that the magnetic fields do not back-react on the flow. One should, therefore, ideally solve (6.1) and (6.2) simultaneously instead of proceeding with kinematic models. As this is a fairly difficult job – even by the standard of today's computers – attempts are made to include the back-reaction of the magnetic field within kinematic models instead of going to fully dynamic models. One easy way to incorporate the back-reaction in the dynamo equation (6.13) is to make the crucial quantity α decrease with the magnetic field following some prescription like

$$\alpha = \frac{\alpha_0}{1 + (\overline{B}/B_0)^2}. \tag{6.17}$$

The effect of such α-quenching on the dynamo process was first studied by several authors in the 1970s (Stix, 1972; Jepps, 1975; Ivanova and Ruzmaikin, 1977; Yoshimura, 1978). When the velocity field $\overline{\mathbf{v}}$ is specified, the dynamo equation (6.13) is a linear equation for the magnetic field $\overline{\mathbf{B}}$, provided we assume the various coefficients in the equation to be independent of $\overline{\mathbf{B}}$. If α is quenched by $\overline{\mathbf{B}}$ in accordance with (6.17), then we have a nonlinear problem. One important question is whether the irregularities of the solar cycle, as seen in Figure 6.3, can be explained with the help of nonlinear models. It seems that the simple nonlinearity introduced through (6.17) cannot cause such chaotic behaviour. Since a sudden increase in the amplitude of magnetic field would diminish the dynamo activity by reducing α and thereby pull down the amplitude again (a decrease in the amplitude would do the opposite), the α-quenching mechanism tends to lock the system in a stable mode once the system relaxes to it. In fact, Krause and Meinel (1988) argued that nonlinearities must be what make one particular mode of the dynamo so stable. Only by introducing more complicated kinds of nonlinearity (with suppression of differential rotation) in some highly truncated dynamo models, Weiss *et al.* (1984) were able to find the evidence of chaos. Jennings and Weiss (1991) presented a study of symmetry-breakings and bifurcations in a nonlinear dynamo model. Since α-quenching of the form (6.17) cannot explain the irregularities of the solar cycle, Choudhuri (1992) explored the effect of stochastic fluctuations on the mean equations and obtained some solutions resembling Figure 6.3. The subsequent papers by Hoyng (1993), Moss *et al.* (1993) and Ossendrijver *et al.* (1996) explored this possibility further. The effect of stochastic fluctuations in an advective dynamo model has been studied by Charbonneau and Dikpati (2000).

We now comment on the efforts in building fully dynamic models by solving both (6.1) and (6.2) simultaneously. This is a highly complicated nonlinear problem and can only be tackled numerically. Gilman (1983) and Glatzmaier (1985) presented very ambitious numerical calculations in which convection, differential rotation and dynamo process were all calculated together from the basic MHD equations. These calculations, however, gave results which do not agree with observational data. For example, the angular velocity was found to be constant on cylinders, whereas helioseismology found it to be constant on cones. If various diffusivities were set such that the surface rotation pattern was matched, then the dynamo waves propagated from the equator to the pole. The codes of Gilman (1983) and Glatzmaier (1985) naturally had finite grids, and the physics at the sub-grid scales was modeled by introducing various eddy diffusivities. Probably the physics at sub-grid scales is more subtle and the details of it are crucially important in determining the behaviour of the dynamo. This is generally believed to be the reason why these massive codes did not produce agreement with observations. The subsequent approach in numerical modeling has been to do dynamic calculations over cubes which correspond to a small regions of the Sun, rather than trying to build models for the whole Sun. Brandenburg *et al.* (1990) and Nordlund *et al.* (1992) followed this approach.

Finally, our increased knowledge of coronal magnetic fields in the last few years have raised the question whether we can learn something about the dynamo process by looking at the magnetic fields in the corona. It has been found that coronal loops in a particular hemisphere have a preferred sense of helicity (Pevtsov *et al.*, 1995). While Gilman and Charbonneau (1999) pointed out that a dynamo operating at the base of the convection zone would give rise to a preferred helicity, it has been shown by Longcope *et al.* (1998) that flux tubes may also receive their helical twists while rising through the convection zone. The possible connection between the solar dynamo and the structures of coronal fields needs further exploration.

6.8   Conclusion

It seems that the solar magnetic fields are generated and maintained by the dynamo process. There is only a small minority of solar physicists who would disagree with this point of view. It is, however, not easy to build a sufficiently detailed and realistic model of the dynamo process to account for all the different aspects of solar magnetism. The 1970s happened to be a period of optimism in dynamo research when various researchers were producing butterfly diagrams by choosing different forms of $\alpha(r, \theta)$ and $\Omega(r, \theta)$. It was felt that further research would narrow down the parameter space and establish a standard model of the solar dynamo. As we discussed above, this did not happen. Schmitt (1993) wrote in his review on the solar dynamo: "The original hope that detailed observational and theoretical information would yield better results of the dynamo did not prove true, on the contrary, they raised difficulties instead." Today, a few years afterwards, we can perhaps have a less pessimistic outlook. The advective models, which are closely linked to observations and which combine together some of the best ideas that came out of dynamo research in the last few decades, certainly do look promising. Although sufficiently detailed models have not yet been worked out, we hope that we are close to building kinematic models which are much more realistic and sophisticated than the kinematic models of the 1970s. Perhaps other researchers may regard this point of view as a reflection of this author's personal prejudice. Only time will tell if this prejudice is justified.

Finally we end by cautioning the reader again that this chapter should not be regarded as a comprehensive review of the whole solar dynamo problem. We have primarily concentrated on those aspects of kinematic models which have direct relevance to observations. There is no doubt that kinematic models can never fully satisfy us. The ultimate challenge is to build fully dynamic models starting from the basic equations, and then to explain both the fluid flow patterns and magnetic patterns in the interior of the Sun in a grand scheme. Although the modern computers may at last be able to handle the geodynamo problem in a dynamic fashion (Glatzmaier and Roberts, 1995), they still seem inadequate for a similar onslaught on the solar dynamo problem, where we have to deal with a highly stratified convection zone with flows and magnetic structures having widely different scales. The solar dynamo problem will certainly remain alive for years to come.

Acknowledgements

This chapter is an updated and somewhat modified version of a review article which appeared in *Current Science* (Choudhuri, 1999). I am grateful to the editor of *Current Science* for permission to reproduce material from its pages. I thank Alexander Kosovichev for valuable comments on the manuscript.

References

Alfvén, H. (1942). *Ark. f. Mat. Astr. o. Fysik*, **29B**, No. 2.

Alfvén, H. and Fälthammar, C.-G. (1963). *Cosmical Electrodynamics* (Oxford University Press).

Babcock, H. D. (1959). *Astrophys. J.*, **130**, 364.

Babcock, H. W. (1961). *Astrophys. J.*, **133**, 572.

Babcock, H. W. and Babcock, H. D. (1955). *Astrophys. J.*, **121**, 349.

Belvedere, G., Kuzanyan, K., and Sokoloff, D. (2000). *Mon. Notic. Roy. Astron. Soc.*, **315**, 778.

Belvedere, G., Lanzafame, G., and Proctor, M. R. E. (1990). *Nature*, **350**, 481.

Brandenburg, A., Krause, F., Meinel, R., Moss, D., and Tuominen, I. (1989). *Astron. Astrophys.*, **213**, 411.

Brandenburg, A., Nordlund, A., Pulkkinen, P., Stein, R. F., and Tuominen, I. (1990). *Astron. Astrophys.*, **232**, 277.

Brandenburg, A. and Tuominen, I. (1990). In *The Sun and Cool Stars: Activity, Magnetism, Dynamos*, ed. I. Tuominen, D. Moss, and G. Rüdiger, *Lecture Notes in Physics*, **380**, 223.

Braun, D. L. and Fan, Y. (1998). *Astrophys. J.*, **508**, L105.

Bumba, V. and Howard, R. (1965). *Astrophys. J.*, **141**, 1502.

Caligari, P., Moreno-Insertis, F., and Schüssler, M. (1995). *Astrophys. J.*, **441**, 886.

Carrington, R. C. (1858). *Mon. Notic. Roy. Astron. Soc.*, **19**, 1.

Chandrasekhar, S. (1952). *Phil. Mag.*, **43**, 501.

Charbonneau, P. and Dikpati, M. (2000). *Astrophys. J.*, **543**, 1027.

Charbonneau, P. and MacGregor, K. B. (1997). *Astrophys. J.*, **486**, 502.

Choudhuri, A. R. (1989). *Solar Phys.*, **123**, 217.

Choudhuri, A. R. (1990). *Astrophys. J.*, **355**, 733.

Choudhuri, A. R. (1992). *Astron. Astrophys.*, **253**, 277.

Choudhuri, A. R. (1998). *The Physics of Fluids and Plasmas: An Introduction for Astrophysicists* (Cambridge University Press).

Choudhuri, A. R. (1999). *Curr. Sci.*, **77**, 1475.

Choudhuri, A. R. and Dikpati, M. (1999). *Solar Phys.*, **184**, 61.

Choudhuri, A. R. and Gilman, P. A. (1987). *Astrophys. J.*, **316**, 788.

Choudhuri, A. R., Schüssler, M., and Dikpati, M. (1995). *Astron. Astrophys.*, **303**, L29.

Cowling, T. G. (1934). *Mon. Notic. Roy. Astron. Soc.*, **94**, 39.

Cowling, T. G. (1976). *Magnetohydrodynamics* (Adam Hilger).

DeLuca, E. E. and Gilman, P. A. (1986). *Geophys. Astrophys. Fluid Dyn.*, **37**, 85.

DeVore, C. R., Sheeley, N. R., and Boris, J. P. (1984). *Solar Phys.*, **92**, 1.

Dikpati, M. and Charbonneau, P. (1999). *Astrophys. J.*, **518**, 508.

Dikpati, M. and Choudhuri, A. R. (1994). *Astron. Astrophys.*, **291**, 975.

Dikpati, M. and Choudhuri, A. R. (1995). *Solar Phys.*, **161**, 9.

D'Silva, S. and Choudhuri, A. R. (1993). *Astron. Astrophys.*, **272**, 621.

D'Silva, S. and Howard, R. F. (1993). *Solar Phys.*, **148**, 1.

Durney, B. R. (1995). *Solar Phys.*, **160**, 213.

Durney, B. R. (1996). *Solar Phys.*, **166**, 231.

Durney, B. R. (1997). *Astrophys. J.*, **486**, 1065.

Fan, Y., Fisher, G. H., and DeLuca, E. E. (1993). *Astrophys. J.*, **405**, 390.

Ferriz-Mas, A., Schmitt, D., and Schüssler, M. (1994). *Astron. Astrophys.*, **289**, 949.

Giles, P. M., Duvall, T. L., and Scherrer, P. H. (1997). *Nature*, **390**, 52.

Gilman, P. A. (1983). *Astrophys. J. Suppl.*, **53**, 243.

Gilman, P. A. (1986), In *Physics of the Sun, Vol. I*, ed. P. A. Sturrock, p. 95 (D. Reidel).

Gilman, P. A. and Charbonneau, P. (1999). In *Magnetic Helicity in Space and Laboratory Plasmas*, ed. M. R. Brown, R. C. Canfield, and A. A. Pevtsov, *Geophysical Monograph* **111**, 75.

Gilman, P. A., Morrow, C. A., and DeLuca, E. E. (1989). *Astrophys. J.*, **338**, 528.

Glatzmaier, G. A. (1985). *Astrophys. J.*, **291**, 300.

Glatzmaier, G. A. and Roberts, P. H. (1995). *Nature*, **377**, 203.

Gokhale, M. H. and Javaraiah, J. (1990). *Mon. Notic. Roy. Astron. Soc.*, **243**, 241.

Hale, G. E. (1908). *Astrophys. J.*, **28**, 315.

Hale, G. E., Ellerman, F., Nicholson, S. B., and Joy, A. H. (1919). *Astrophys. J.*, **49**,153.

Howard, R. F. (1996). *Solar Phys.*, **169**, 293.

Howard, R. F. and LaBonte, B. J. (1981). *Solar Phys.*, **74**, 131.

Hoyng, P. (1988). *Astrophys. J.*, **332**, 857.

Hoyng, P. (1990). In *Basic Plasma Processes on the Sun, IAU-Symp. No. 142*, ed. E. R. Priest and V. Krishan, p. 48 (Kluwer).

Hoyng, P. (1993). *Astron. Astrophys.*, **272**, 321.

Ivanova, T. S. and Ruzmaikin, A. A. (1977). *Sov. Astron.*, **21**, 479.

Jennings, R. L. and Weiss, N. O. (1991). *Mon. Notic. Roy. Astron. Soc.*, **252**, 249.

Jepps, S. A. (1975). *J. Fluid Mech.*, **67**, 625.

Köhler, H. (1973). *Astron. Astrophys.*, **25**, 467.

Krause, F. and Meinel, R. (1988). *Geophys. Astrophys. Fluid Dyn.*, **43**, 95.

Krause, F., and Rädler, K.-H. (1980). *Mean-Field Magnetohydrodynamics and Dynamo Theory* (Pergamon).

Krause, F., Rädler, K.-H., and Rüdiger, G. (eds.) (1993). *The Cosmic Dynamo: IAU Symp. No. 157* (Kluwer).

Küker, M., Rüdiger, G. and Schultz, M. (2001). *Astron. Astrophys.*, **374**, 301.

Leighton, R. B. (1969). *Astrophys. J.* **156**, 1.

Longcope, D. and Choudhuri, A. R. (2002). *Solar Phys.*, **205**, 63.

Longcope, D. and Fisher, G. H. (1996). *Astrophys. J.*, **458**, 380.

Longcope, D., Fisher, G. H., and Pevtsov, A. A. (1998). *Astrophys. J.*, **507**, 417.

Markiel, J. A. and Thomas, J. H. (1999). *Astrophys. J.*, **523**, 827.

Maunder, E. W. (1904). *Mon. Notic. Roy. Astron. Soc.*, **64**, 747.

Moffatt, H. K. (1978). *Magnetic Field Generation in Electrically Conducting Fluids* (Cambridge University Press).

Moss, D., Brandenburg, A., Tavakol, R., and Tuominen, I. (1992). *Astron. Astrophys.*, **265**, 843.

Nandy, D. and Choudhuri, A. R. (2001). *Astrophys. J.*, **551**, 576.

Nordlund, A., Brandenburg, A., Jennings, R. L., Rieutord, M., Ruokolainen, J., Stein, R. F., and Tuominen, I. (1992). *Astrophys. J.*, **392**, 647.

Ossendrijver, A. J. H., Hoyng, P., and Schmitt, D. (1996). *Astron. Astrophys.*, **313**, 938.

Parker, E. N. (1955a). *Astrophys. J.*, **121**, 491.

Parker, E. N. (1955b). *Astrophys. J.*, **122**, 293.

Parker, E. N. (1975). *Astrophys. J.*, **198**, 205.

Parker, E. N. (1979). *Cosmical Magnetic Fields* (Oxford University Press).

Parker, E. N. (1987). *Astrophys. J.*, **312**, 868.

Parker, E. N. (1993). *Astrophys. J.*, **408**, 707.

Pevtsov, A. A., Canfield, R. C. and Metcalf, T. R. (1995). *Astrophys. J.*, **440**, L109.

Priest, E. R. (1982). *Solar Magnetohydrodynamics* (D. Reidel).

Roberts, P. H. (1972). *Phil. Trans. Roy. Soc. Lond.*, **272**, 663.

Rüdiger, G. and Brandenburg, A. (1995). *Astron. Astrophys.*, **296**, 557.

Ruzmaikin, A. A. (1985). *Solar Phys.*, **100**, 125.

Schmitt, D. (1987). *Astron. Astrophys.*, **174**, 281.

Schmitt, D. (1993). In *The Cosmic Dynamo, IAU-Symp. No. 157*, ed. F. Krause, K.-H. Rädler, and G. Rüdiger, p. 1 (Kluwer).

Schmitt, D. and Schüssler, M. (1989). *Astron. Astrophys.*, **223**, 343.

Schüssler, M. (1993). In *The Cosmic Dynamo, IAU-Symp. No. 157*, ed. F. Krause, K.-H. Rädler, and G. Rüdiger, p. 27 (Kluwer).

Schüssler, M., Caligari, P., Ferriz-Mas, A., and Moreno-Insertis, F. (1994). *Astron. Astrophys.*, **281**, L69.

Schwabe, S. H. (1844). *Astron. Nachr.*, **21**, 233.

Sheeley, N. R., DeVore, C. R., and Boris, J. P. (1985). *Solar Phys.*, **98**, 219.

Spiegel, E. A. and Weiss, N. O. (1980). *Nature*, **287**, 616.

Steenbeck, M. and Krause, F. (1969). *Astron. Nachr.*, **291**, 49.

Steenbeck, M., Krause, F., and Rädler, K.-H. (1966). *Z. Naturforsch.*, **21a**, 1285.

Stenflo, J. O. (1973). *Solar Phys.*, **32**, 41.

Stenflo, J. O. and Vogel, M. (1986). *Nature*, **319**, 285.

Stix, M. (1972). *Astron. Astrophys.*, **20**, 9.

Stix, M. (1976). *Astron. Astrophys.*, **47**, 243.

Thomas, J. H. and Weiss, N. O. (1992). In *Sunspots: Theory and Observations*, ed. J. H. Thomas and N. O. Weiss, p. 3 (Kluwer).

van Ballegooijen, A. A. (1982). *Astron. Astrophys.*, **113**, 99.

van Ballegooijen, A. A. and Choudhuri, A. R. (1988). *Astrophys. J.*, **333**, 965.

Wang, Y.-M., Nash, A. G., and Sheeley, N. R. (1989a). *Astrophys. J.*, **347**, 529.

Wang, Y.-M., Nash, A. G., and Sheeley, N. R. (1989b). *Science*, **245**, 712.

Weiss, N. O. (1981). *J. Fluid Mech.*, **108**, 247.

Weiss, N. O., Cattaneo, F., and Jones, C. A. (1984). *Geophys. Astrophys. Fluid Dyn.*, **30**, 305.

Yoshimura, H. (1975). *Astrophys. J. Suppl.*, **29**, 467.

Yoshimura, H. (1978). *Astrophys. J.*, **226**, 706.

Zeldovich, Ya B., Ruzmaikin, A. A., and Sokoloff, D. D. (1983). *Magnetic Fields in Astrophysics* (Gordon and Breach).

7

Spectro-polarimetry

J.O. Stenflo

Institute of Astronomy, ETH Zurich, and
Faculty of Mathematics and Science, University of Zurich, Switzerland

7.1   Remote sensing of the Sun's magnetic field: an introduction

All of observational astrophysics is remote sensing: the electromagnetic radiation that reaches us from far-away objects is our source of information about the physical nature of these objects. The information about the Sun and stars is encoded in a rich but subtle way in the multitude of spectral lines throughout their spectra. Most astrophysicists have in the past used the measured intensity distribution throughout the spectra to explore the physical conditions on the celestial objects. However, the often neglected polarization of the radiation enriches the diagnostic possibilities with new dimensions. While intensity is a scalar, polarization can be represented by a 4-vector, the Stokes vector with components I, Q, U, and V, where I represents the ordinary intensity, Q and U the different linear polarizations, and V the circular polarization (see Sect. 7.2). In comparison with spectroscopy, spectro-polarimetry thus increases the dimensionality of information space by a factor of four.

Polarization carries information about broken symmetries in the object under study. These asymmetries can have various physical origins, but in solar physics the most important effect is the broken spatial symmetry for the radiating atoms caused by the presence of magnetic fields. It leads to various magnetically induced polarized spectral signatures, which may be classified as due to the Zeeman effect, the Hanle effect, or to polarizing optical pumping. By observing and interpreting these polarized signatures remote sensing of magnetic fields becomes possible.

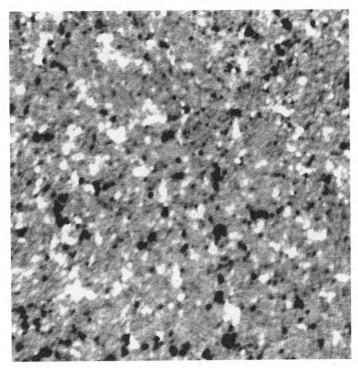

Figure 7.1. Map of the line-of-sight component of the magnetic field vector in a quiet region at the centre of the solar disc. The field of view is 280×280 arcsec^2, which corresponds to a mere 2.7% of the solar disc. The magnetic fluxes of opposite polarities are represented by the brighter and darker patches against a neutral grey background. The recording was made on July 1, 1998, by Jongchul Chae at Big Bear Solar Observatory (BBSO). Courtesy of BBSO / New Jersey Institute of Technology.

Observationally the general aim of spectro-polarimetry is to record the Stokes vector as accurately as possible with the highest spectral, spatial and temporal resolution. The interpretation problem is to invert these data to deduce the magnetohydrodynamic state of the plasma with its magnetic field. In general the relation between polarization and the physical state of the atmosphere is highly non-linear, which makes such inversion a complex problem. There is however another fundamental obstacle: A substantial fraction of the magnetohydrodynamic structures on the Sun occurs on scales that are much smaller than the resolution of current telescopes.

Figure 7.1 illustrates how the surface of the "quiet" solar disc is seething with magnetic fields of mixed polarities. Progress is being made in numerical simulations of magnetoconvection (Emonet and Cattaneo 2001), but since turbulence on the Sun has a dynamic range orders of magnitude larger than the grid resolution of the simulations, idealizations in the theoretical model are necessary to obtain results that resemble the real Sun. Only through observations the validity of the model assumptions can be tested.

We may conceptually divide the spatial scales into the spatially resolved domain, which is directly observed, and the unresolved domain, which relies on indirect diagnostic methods, similar to those used to interpret stellar observations. In the resolved domain the theories and simulations can be checked and calibrated by the direct observations. The so calibrated theory predicts what happens in the unresolved domain. Although the observations cannot reveal the morphology of the structures in the unresolved domain, indirect diagnostic techniques allow us to place observational constraints on the theoretical predictions and to determine the free parameters in the theoretical model.

The most used and established framework for diagnosing the Sun's magnetic field is based on the Zeeman effect (cf. Sect. 7.3). With the advent of new, highly sensitive imaging polarimeters a rich world of previously unknown polarization phenomena due to coherent scattering on the Sun has been discovered, which has opened new diagnostic windows to explore magnetic fields in parameter domains not accessible to the Zeeman effect. The new magnetically induced polarization signatures are due to the Hanle effect and to polarizing optical pumping (cf. Sects. 7.4 and 7.5). For recent overviews of spectro-polarimetry, see Nagendra and Stenflo (1999), Sigwarth (2001), and Trujillo Bueno *et al.* (2002).

7.2   Observational techniques and their limitations

It is most convenient to describe polarized light in terms of the Stokes vector $\vec{I}$ consisting of the four Stokes parameters I, Q, U, V, or $S_k, k = 0, 1, 2, 3$:

$$\vec{I} = \begin{pmatrix} S_0 \\ S_1 \\ S_2 \\ S_3 \end{pmatrix} \equiv \begin{pmatrix} I \\ Q \\ U \\ V \end{pmatrix}. \tag{7.1}$$

The meaning of these parameters can be visualized and defined by considering the effect on a light beam by the four idealized filters F_k in Figure 7.2. Filter F_0 represents empty space, F_1 and F_2 transmit linear polarization with the electric vector at position angles 0 and 45°, respectively, while F_3 transmits right-handed circular polarization. Let I_k be the intensity of the beam behind each filter. Then $I_k = \frac{1}{2}(I_0 + S_k)$, or

$$S_k = 2I_k - I_0, \tag{7.2}$$

which can be seen as a definition of the Stokes parameters S_k.

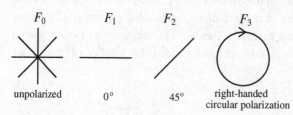

F_0 F_1 F_2 F_3

unpolarized   0°   45°   right-handed
                                          circular polarization

Figure 7.2. Symbolic properties of the four idealized filters F_k used in the operational definition of the four Stokes parameters.

If we instead insert filters F_k that transmit the orthogonal polarization state, then Q, U, and V change sign. From this it follows that Stokes I represents the intensity, Stokes Q the intensity difference between horizontal and vertical linear polarization, Stokes U the intensity difference between linear polarization at $+$ and $-45°$, Stokes V the intensity difference between right- and left-handed circular polarization.

In the Stokes formalism the polarizing effect of a medium can be described in terms of a 4×4 matrix, the Mueller matrix, by which the Stokes vector is multiplied to transform it into another Stokes vector. Such a Mueller matrix can be defined not only for optical components like filters, modulators, or beam splitters, but it can be formulated for a medium with atomic transitions that include the Zeeman effect, or for coherent scattering including the Hanle effect. In the latter cases the Mueller matrix contains line profile and geometric information describing the direction of the magnetic field vector and the direction of the incident and scattered radiation.

For an optical system, like a telescope, spectrograph, liquid crystal modulator system, etc., one obtains the total polarization matrix $\vec{M}$ of the system through simple matrix multiplication of the Mueller matrices $\vec{M}_i$ of each individual optical component:

$$\vec{M} = \vec{M}_n \vec{M}_{n-1} \dots \vec{M}_2 \vec{M}_1 . \tag{7.3}$$

The whole optical system can then be treated as a "black box" characterized by a single 4×4 matrix, which may be calibrated by inserting polarizing filters in front of the "black box". For more details, see Stenflo (1994).

It is a common misunderstanding that the Sun always provides us with enough photons for our observations, in contrast to stellar observations, which are photon starved. Instead, the surface brightness of a stellar disk is independent of distance and depends only on the effective temperature of the surface. If we would barely resolve a solar-type star with one angular resolution element, then we would receive from that distant star the same number of photons as we receive from the nearby solar disk within the same resolution element. The difference is that we in contrast to stellar disks can fit many angular resolution elements inside the solar disk.

We may characterize an observational program in terms of a 4-D parameter space, spanned by the three resolutions (angular, spectral, and temporal), and the polarimetric (or photometric) accuracy. In the design of observing programs we always make major trade-offs between these four observing parameters, depending on the scientific objectives. Even for the largest solar telescopes that will be built in the foreseeable future (the next few decades) major trade-offs will still be necessary. Polarimetry with an accuracy of 10^{-5} requires according to Poisson statistics 10^{10} collected electrons, or, taking into account the typical optical transmission and detector efficiency, on the order of 10^{12} photons per spectral resolution element. This can only be achieved by making large trade-offs with the temporal and spatial resolutions, as illustrated in Figure 7.3.

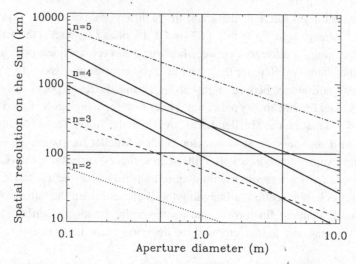

Figure 7.3. Overview of the necessary trade-offs between spatial resolution and polarimetric accuracy, for an assumed spectral resolution of 300,000. The thin slanted lines labeled by parameter n indicate a polarimetric accuracy of 10^{-n}. The thick, slanted lines give the diffraction limit for wavelengths $1.56\,\mu$m (upper line) and 5000 Å (lower line). The vertical lines represent telescope sizes 0.5, 1, and 4 m. From Stenflo (1999).

The aim in the design of a polarimeter is to eliminate all spurious effects so that the polarimetric accuracy is exclusively limited by photon statistics. The two dominating sources of spurious noise come from seeing fluctuations in the earth's atmosphere and from pixel-to-pixel sensitivity variations (gain table) in the detector. Since the seeing distortions vary with frequencies below a few hundred Hz, we can avoid spurious seeing-induced polarization effects by modulating the state of polarization in the kHz range. This demands that the detector system is equipped with fast buffers, between which the images in the different polarization states can be shifted at kHz frequencies. Such systems now exist even for large-size CCD detectors (cf. Povel 1995, 2001). They allow the elimination of not only seeing but also gain table noise in the fractional polarization images Q/I, U/I, and V/I, since the identical pixels are used for the different polarization states. In this way it has been possible to routinely obtain polarization accuracies better than 10^{-5} in the degree of polarization in combination with high spectral resolution (Stenflo and Keller 1996, 1997). At this level of accuracy entirely new physical effects on the Sun are brought out, which otherwise would be drowned by noise.

Spurious seeing-induced polarization signals can also be avoided without modulation by using a polarizing beam splitter that gives two orthogonally polarized images, which have identical seeing distortions since they are recorded simultaneously. Therefore the difference image would be zero in the absence of intrinsic polarization. However, although the fractional polarization obtained as the ratio between the difference and the sum image should then be free from seeing-induced features, it is affected by the gain table, which cannot be determined by flat-field calibration to much better than 0.5%.

The gain-table problem may be eliminated by making a second exposure for which the polarization states of the two images have been reversed by a half-wave plate, and combining the four images in a certain way (Semel *et al.* 1993; Semel 1995). For this technique to work the two beams have to have identical distortions by the aberrations in the optical system. Also one beam splitter and two separate exposures only give us Stokes I and one of the other Q, U, or V parameters, but not all four of them. For Stokes vector polarimetry it is therefore inferior to fast modulation systems.

The Stokes vector that is measured is further corrupted by the polarizing properties of the telescope optics that precedes the optical package for polarization analysis. This instrumental polarization leads to cross talk between the Stokes parameters in a way that can be described by the Mueller matrix of the telescope. Unfortunately none of the world's major solar telescopes has been designed to optimize polarimetric work, but they have obliquely reflecting elements that produce large and time-varying instrumental polarization. Determination of the instrumental cross talk to allow its removal in the data analysis cannot be done by direct calibration, since it is not feasible to cover the entire telescope aperture with large polarizers immediately before each observing sequence. Instead one is forced to rely on various indirect methods, like making use of symmetry properties of the Stokes parameters, or doing idealized modelling of the telescope.

7.3   Zeeman-effect diagnostics

With modern imaging polarimeters it is now possible to record images of the full Stokes vector as four images, one for each Stokes parameter. This can be done either with a narrow-band filter, which gives monochromatic images of the spatial morphology at a selected wavelength, or with a spectrograph, which provides us with one spectral and one spatial dimension. Examples of the spectral signatures are shown in Figure 7.4. The transverse Zeeman effect produces line profiles in Stokes Q and U that are nearly symmetric around the line centers. In contrast, the longitudinal Zeeman effect in the Stokes V images has almost anti-symmetric line profiles.

The shapes of the polarized line profiles can be studied in greater detail in spectral recordings with a Fourier Transform Spectrometer (FTS), which provides fully resolved spectra with high signal-to-noise ratio at the expense of reduced temporal and spatial resolution (cf. Figure 7.5). One striking property of the Stokes V spectra concerns the relative polarization amplitudes in different lines. Since the polarization signals seen outside active regions are generally small, one might be led to believe that the magnetic fields there are weak. If each of the σ components is Zeeman-shifted

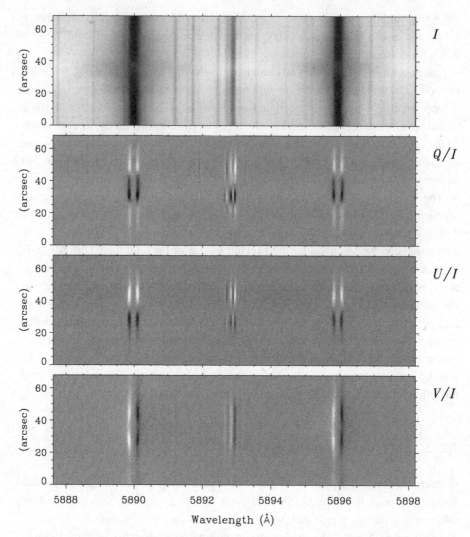

Figure 7.4. Example of imaging vector spectro-polarimetry with the characteristic signatures of the Zeeman effect. The spectrograph slit is crossing a large sunspot. The spectral region around the strong D_2 and D_1 lines of neutral sodium at 5890 and 5896 Å contains many narrow telluric lines, but it is only the solar lines that exhibit polarization signatures. From Stenflo *et al.* (2001).

by $\Delta\lambda_H$ to either side of the line center, and we denote the signals in right and left circular polarization by $I_{\pm}$, so that

$$I_{\pm} = \tfrac{1}{2}(I \pm V),\tag{7.4}$$

then for weak fields

$$I_{\pm}(\Delta\lambda) \approx I_0(\Delta\lambda \mp \Delta\lambda_H \cos\gamma),\tag{7.5}$$

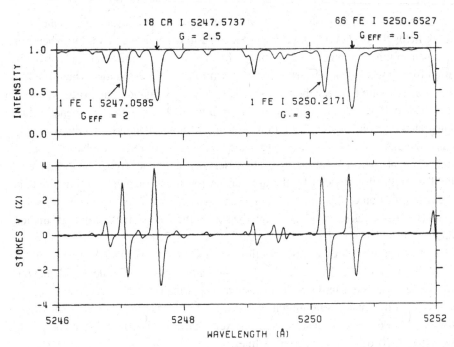

Figure 7.5. Stokes I (intensity) and V (circular polarization) spectra recorded in a strong plage near disc center with the FTS polarimeter of the National Solar Observatory. From Stenflo *et al.* (1984).

where I_0 represents the intensity profile in the absence of magnetic fields, while γ is the angle between the field vector and the line of sight. Since for weak fields the line broadening due to the Zeeman effect is insignificant, I_0 should be practically identical to the Stokes I profile. A Taylor expansion then gives

$$V \approx -\frac{\partial I}{\partial \lambda}\Delta\lambda_H \cos\gamma \,. \tag{7.6}$$

Since

$$\Delta\lambda_H = 4.67 \times 10^{-13} g_{\text{eff}} \lambda^2 B \,, \tag{7.7}$$

where the field strength B should be given in G and the wavelength in Å, we would expect that the observed Stokes V amplitudes scale with $g_{\text{eff}}\partial I/\partial\lambda$. The observations however show that this is generally not the case.

The two iron lines Fe I 5250.22 and and 5247.06 Å provide an example of such apparently "anomalous" behavior. They have the same line strength and excitation potential, and they belong to the same atomic multiplet. The only significant difference is in their Landé factors, 3.0 for the 5250 Å line, 2.0 for the 5247 Å line. One would then expect their V amplitudes to be in proportion 3:2, but a glance at the FTS spectrum shows that the observed ratio is much closer to 1:1. The reason for this discrepancy is that the magnetic fields are not spatially resolved, and much of the unresolved magnetic flux is not weak but clumped in a strong-field state. For weak fields the Stokes V signal is proportional to the field strength, but when the Zeeman splitting becomes comparable to the line width saturation sets in. The deviation from linearity comes earlier for lines with larger Landé factors. Therefore the V amplitude ratio between the 5250 and 5247 Å lines is smaller than expected under the weak-field assumption. While the V amplitudes scale with the magnetic flux integrated over the angular resolution element, the amplitude ratio in this particular case contains information on the intrinsic field strengths of the unresolved, clumped field elements. This field strength can be orders of magnitude larger than the average field strength (flux divided by the area of the resolution element). It was through the determination of this line ratio that the intermittent kG flux tube nature of solar magnetic fields was revealed, although the size of the flux elements were far smaller than the telescope resolution (Stenflo 1973).

Let us next compare the two iron lines at 5247.06 and 5250.65 Å. Although the latter line has a smaller Landé factor it has a larger Stokes V amplitude. This behavior cannot be explained in terms of Zeeman saturation, which is small for both lines, but is caused by different temperature sensitivities. The line weakening is more pronounced for the 5247.06 Å line because of its lower excitation potential (0.09 eV) as compared with that (2.20 eV) of the 5250.65 Å line.

Further differential effects can be found when comparing lines of different strengths, which are formed at different heights in the atmosphere. The height of formation also varies with wavelength within a single line profile.

All such differential effects in the spectrum are of tremendous diagnostic value, since they provide us with a large set of qualitatively different model constraints, which allow us not only to determine the physical parameters (magnetic fields, velocities, temperatures, densities, etc.) as functions of height, but also to find out how these parameters vary on small spatial scales that are beyond the resolution of the telescopes.

When we derive magnetic-field properties in the spatially unresolved domain, we need an adequate interpretational framework. A two-component concept, according to which a magnetic component with a certain filling factor is embedded in non-magnetic surroundings, has proven extremely successful and has led to self-consistent interpretations and empirical models of the magnetic elements (flux tubes). For instance the rich set of observational constraints provided by FTS spectra like the section shown in Figure 7.5 has allowed the construction of highly sophisticated flux tube models with self-consistent height variations of their thermodynamic parameters, height expansion of the magnetic field, downdrafts outside the flux tube, etc. (cf. Solanki 1993; Stenflo 1994). Such two-component models have also been consistent with all other Stokesmeter and magnetograph observations made with moderate or low spatial and temporal resolution.

With the advent of polarimetric observations in the infrared as well as Stokesmeter observations with high spatial and temporal resolution, many instances have been

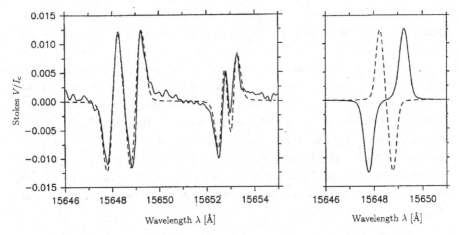

Figure 7.6. Anomalous Stokes V profiles are seen more frequently in the near infrared, where the Zeeman splitting is larger than at visible wavelengths. Their interpretation requires more than one magnetic component. In the case illustrated here two magnetic components of opposite polarities have been used, whose profile contributions are illustrated in the panel to the right. Their field strengths are 1.70 kG (solid curve) and -1.05 kG (dashed curve) at height $z = 0$ (level of continuum formation at 5000 Å). In the left panel the solid curve represents the observations, the dashed curve the model. From Rüedi *et al.* (1992).

found where the two-component model is inadequate and in need for extension by adding an extra magnetic component, which may have either the same or different polarity as compared with the main magnetic component, while its field strength, Doppler shift, and temperature structure may be different. The need for such an extension becomes obvious for "pathological" Stokes V profiles with a multiple lobe structure. Such anomalous profiles are rare in the visible but are frequently seen in infrared Stokes spectra, since the much larger Zeeman splitting in the infrared allows better separation of the components that lie on top of each other in the visible (cf. Figure 7.6).

FTS spectra (with low spatial resolution) show that the lobe in the blue wing of Stokes V profiles is larger than the red wing lobe. This Stokes V amplitude and area asymmetry varies with excitation potential, line strength and center-to-limb distance in ways that can be explained and modelled in terms of correlated spatial gradients of the velocity field and the magnetic field. More recent Stokesmeter spectra with high spatial and temporal resolution reveal large variations with frequent extreme values of these Stokes V asymmetries (Sigwarth *et al.* 1999). Thus it not only happens that one of the Stokes V wing lobes disappears entirely, but many cases are found when the two lobes have the same sign. It has been possible to show theoretically, in terms of a single flux tube model, how such profiles can arise when the line-forming region straddles both sides of a magnetopause or magnetic canopy separating magnetic and non-magnetic regions with different flow fields and temperatures (Grossmann-Doerth *et al.* 1988, 2000). In principle such anomalous Stokes *V* profiles could therefore be used as diagnostic signatures for various types of magnetopause structures (Steiner 2000).

With these amendments and extensions of the two-component model, the basic picture that Zeeman-effect observations have given us is a photosphere with only about 1% of the volume filled with strong-field kG flux tubes, but with indications of many weak-field magnetic elements of mixed polarity in between. The evolution of discrete clumps of apparently weak mixed-polarity fields can be followed in "deep" magnetograms (with long integrations) at scales down to a few arcsec (cf. Figure 7.1), with hints of many more elements of mixed polarities at smaller scales, which remain invisible due to cancellation of the opposite polarities within the spatial resolution element.

7.4   The Hanle effect

The Zeeman effect is a source of polarization regardless of whether the spectral line has been formed by coherent or incoherent processes. The Hanle effect on the other hand is exclusively a coherence phenomenon. In the absence of magnetic fields polarization can be produced by coherent radiative scattering when the incident radiation has some degree of anisotropy. The word "coherent" here means that the phase relations in the scattering process are not destroyed by depolarizing collisions. Due to the Zeeman splitting of the atomic energy levels when a magnetic field is present, the scattering polarization gets modified. All the magnetically induced modifications of the scattering polarization are covered by the term "Hanle effect". It manifests itself primarily in two ways: depolarization and rotation of the plane of linear polarization (cf. Moruzzi and Strumia 1991).

On the Sun the anisotropy of the emergent radiation comes mainly from the limb darkening, which means that the intensity is not constant over the half sphere in the outwards direction, but is more peaked in the vertical direction. For non-magnetic scattering the polarization at disk center is zero for symmetry reasons, and it increases monotonically as we approach the solar limb. The polarization is linear, and in the non-magnetic case the electric vector is usually, with few exceptions, parallel to the nearest limb (perpendicular to the radius vector from disk center).

A description of the Hanle effect in terms of a damped classical oscillator is illustrated in Figure 7.7. We assume that in the absence of a magnetic field a vertical oscillation is induced by the anisotropic excitation. If we introduce a magnetic field, the damped oscillation will precess around the field vector and form a rosette pattern. The amount of depolarization and rotation of the plane of linear polarization (which is computed by Fourier transformation of the rosette pattern) depends on the ratio between the Larmor precession frequency (which is proportional to the field strength) and the damping rate (inverse life time of the excited atomic state). The Hanle effect

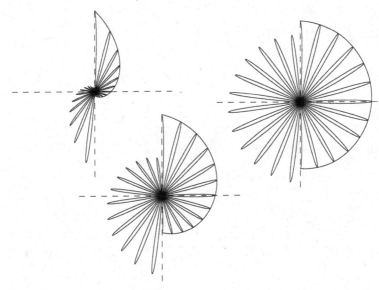

Figure 7.7. Rosette motion of a damped classical oscillator, which results in Hanle depolarization and rotation of the plane of linear polarization. The magnetic field is assumed to be directed along the line of sight. The field strength increases from left to right.

therefore has its largest sensitivity to magnetic fields when the Zeeman splitting is comparable in magnitude to the damping width. In contrast, the ordinary Zeeman-effect polarization depends on the ratio between the Zeeman splitting and the Doppler width of spectral lines. Since the Doppler width is usually orders of magnitude larger than the damping width, the Hanle effect is sensitive to much weaker fields than the ordinary Zeeman effect. The sensitivity range depends on the life time of the concerned atomic state. For excited states of allowed atomic transitions the typical range is 1–100 G.

Figure 7.8 gives an example of Hanle and Zeeman signatures side by side. The spectrograph slit has been placed parallel to and 2.5 arcsec inside the limb, such that half the slit covers a small facular region, while the other half lies outside it. In the facular region we see the characteristic signatures of the transverse Zeeman effect in practically all the atomic lines, while outside it the only significant signal in Q/I (linear polarization parallel to the limb) is scattering polarization in the Sr I 4607 Å line. As we move closer to the facular region, the amplitude of Q/I in the Sr I line decreases, which is evidence for Hanle depolarization by magnetic fields.

In the previous section we saw how Zeeman-effect observations have led to the concept of an extremely intermittent solar magnetic field with kG flux tubes as its basic building blocks. While most of the photospheric flux is carried by these intense flux tubes according to the interpretative models used, they occupy only on the average 1% of the photospheric volume. The remaining 99% does not appear to contribute

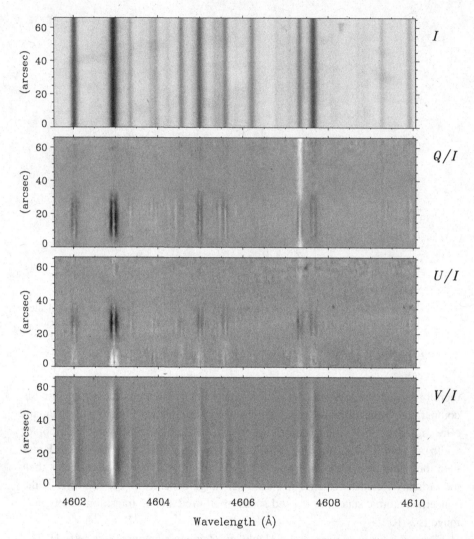

Figure 7.8. Coexistence of scattering polarization (in Q/I for the Sr I 4607 Å line) with the transverse and longitudinal Zeeman effect. The spectrograph slit was placed parallel to and 2.5 arcsec inside the SW solar limb, where there was some minor facular activity. From Stenflo (2001).

significantly to the Zeeman-effect signals. The absence of observable magnetic flux through a spatial resolution element however does not imply the absence of magnetic flux, since the Zeeman-effect is almost "blind" to a tangled field, if the magnetic polarities are mixed on a subresolution scale.

This kind of blindness does not apply to the Hanle effect, which for mixed-polarity fields reduces the amount of scattering polarization from its non-magnetic value. The depolarization does not depend on field polarity. This unique property was first used to constrain the strength of a volume-filling, turbulent photospheric magnetic field to be somewhere in the range 10–100 G (Stenflo 1982). With detailed numerical modelling of polarized radiative transfer in the Sr I 4607 Å line it has been possible to derive more precise values for the turbulent field, assuming that it is single-valued with an isotropic distribution of field vectors (although in reality we must have a continuous distribution of field strengths). Thus values in the range 10–30 G have been found (Faurobert-Scholl 1993; Faurobert-Scholl *et al.* 1995). Similar values, although slightly smaller (5–10 G) and for lines formed higher in the atmosphere (Ca I 4227 Å, Sr II 4078 Å) are found with a more direct, empirical and statistical approach that avoids extensive radiative-transfer modelling (Bianda *et al.* 1998a,b, 1999). However, it is also found that the strength of the turbulent field varies significantly with the phase of the solar cycle, being stronger when the Sun is active (Stenflo *et al.* 1998; Faurobert *et al.* 2001).

In contrast to the Zeeman effect, the Hanle effect is practically blind to the flux tube field, because this field has such a small filling factor, and the Hanle effect is insensitive to vertical magnetic fields (since then the illumination of the scattering particles is axially symmetric). Almost the entire contribution to the Hanle effect comes from the 99% of the volume to which the ordinary Zeeman effect is almost blind. The Hanle and Zeeman effects therefore ideally complement each other.

Although the Hanle effect opens new diagnostic possibilities that are not available with the Zeeman effect, it has the disadvantage that it does not lend itself to direct mapping of the magnetic field but instead constrains the field properties in more convolved ways. A fundamental reason for this is that the Hanle effect shows up in two observed parameters, Q and U, while the magnetic field vector needs three parameters to fully constrain its three vector components. The field vector is therefore not uniquely constrained by Hanle observations alone, but needs some additional constraint, either from theory or other types of observations (e.g. from the longitudinal Zeeman effect in Stokes V). The Zeeman effect on the other hand can in principle constrain the full vector, the line of sight component from Stokes V, the two transversal components from Q and U (in the case of spatially resolved fields).

With sensitive imaging polarimeters we can now begin to map the spatial variations of the Hanle effect across the solar surface. Figure 7.9 illustrates such spatial fluctuations in the core of the Na I D_2 5890 Å line. The varying amplitude of the core peak in Q/I is due to varying amount of Hanle depolarization, while the varying U/I signal is due to Hanle rotation of the plane of linear polarization. The existence of a signal in Stokes U implies that the magnetic fields must be at least partially spatially resolved, so that there is a net orientation of the field after averaging over the spatial resolution element. The cores of strong lines like the D_2 line are formed rather high in the atmosphere (in the lower chromosphere), where the field lines from the bundled flux tubes are expected to spread out almost horizontally to form so-called canopies over the more turbulent atmosphere below. The fluctuations that we see in Figure 7.9 are caused by fluctuations of this chromospheric field.

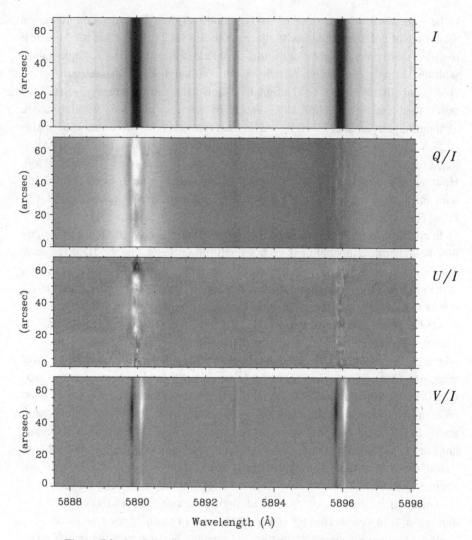

Figure 7.9. Spatially varying Hanle effect in the core of the Na I D$_2$ line. Q/I shows varying Hanle depolarization, U/I varying Hanle rotation of the plane of polarization, V/I the longitudinal Zeeman effect. The spectrograph slit was placed parallel to and 5 arcsec inside the SE solar limb in a region with moderate magnetic activity. From Stenflo *et al.* (2001).

7.5   Optical pumping

The linearly polarized solar spectrum that is due to coherent scattering phenomena is called the "second solar spectrum" (Stenflo and Keller 1997), since it is as richly structured as the ordinary intensity spectrum but has a very different appearance. It is as if the Sun has presented us with an entirely new spectrum, and we have to start over again to identify the unfamiliar structures that we see. Due to the small amplitudes of

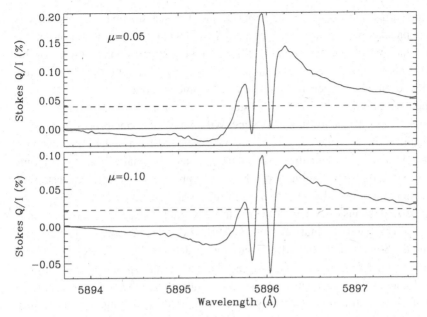

Figure 7.10. Q/I profiles of the enigmatic Na I D$_1$ line in a very quiet limb region. The spectrograph slit has been placed parallel to the limb at two limb distances, represented by their μ (cosine of the heliocentric angle) values. $\mu = 0.05$ corresponds to a limb distance of a mere 1.2 arcsec, $\mu = 0.10$ to a limb distance of 5 arcsec. From Stenflo *et al.* (2000a).

its polarization signals the "second solar spectrum" was inaccessible to observations until a few years ago, when imaging polarimeters of sufficient sensitivity opened the window to this new and rich diagnostic world of polarization physics, which was previously unknown to astrophysics. When exploring this new world we find numerous unexpected, "anomalous" polarization features that appear to contradict our understanding of quantum mechanics (Stenflo *et al.* 2000b).

One example is the observed polarization profile of the Na I D$_1$ 5896 Å line, which is illustrated in Figure 7.10. It remained an enigma for several years, since the total angular momentum quantum numbers J are $\frac{1}{2}$ for both the upper and lower level. According to quantum mechanics such transitions should be intrinsically unpolarizable, in contrast to the observed profile, which exhibits a sharp polarization peak in the line core.

A possible resolution of this enigma was suggested by Landi Degl'Innocenti (1998, 1999) in terms of a combination of hyperfine structure splitting and optical pumping. Due to coupling between the electronic angular momentum and the nuclear spin, the atomic levels get split in hyperfine structure components with different total angular momentum quantum numbers, and these sublevels are polarizable in principle. This is however not sufficient to make the scattering transition polarizable. We need in addition an optical pumping process to polarize the lower atomic state.

In standard scattering theory the initial atomic state is assumed to be unpolarized. Anisotropic exciting radiation induces a dipole moment (atomic polarization) of the excited state, which gets imprinted on the emitted radiation as polarization. For the Na I D_1 line, however, this process alone leads to zero polarization in the scattered radiation, even when hyperfine structure is taken into account.

When however the polarized excited state spontaneously decays, it transfers some of its polarization to the lower atomic state. With many scattering processes a statistical equilibrium is reached, in which the lower state has been optically pumped by spontaneous decay from the upper state to acquire atomic polarization. A scattering process that starts from an initially polarized state produces very different scattering polarization as compared with scattering from an unpolarized ground state.

Figure 7.10 is thus evidence for the existence of ground-state atomic polarization in sodium. This opens qualitatively new possibilities for the diagnostics of solar magnetic fields: Since the life time of atomic ground states is 2–3 orders of magnitude longer than the life time of excited states (this factor corresponds to the ratio between the spontaneous and stimulated emission rates), the typical Hanle sensitivity range of 1–100 G for excited states is down in the milligauss range for the ground states.

The application of these new diagnostic possibilities has to be done with care, since the Hanle effect depends not only on the strength but also on the direction of the field. Maximum depolarization (which becomes complete in the limit of strong fields) occurs for fields directed along the line of sight. In contrast, the Hanle effect vanishes when the field is vertical (because then the exciting radiation is symmetrical around the magnetic field vector).

If the magnetic fields in the solar chromosphere were predominantly horizontal, then the polarization signal in the D_1 line core would already for magnetic fields stronger than a few tens of milligauss be too much suppressed to be compatible with the polarization amplitudes that are observed for the quiet Sun. Since it is implausible that most of the lower chromosphere is occupied by such weak fields, we are led to the conclusion that the magnetic field in the lower chromosphere has an angular distribution that favors the vertical direction in the solar regions, where a significant D_1 core polarization peak is seen. This conclusion is of special significance, since it contradicts the previously prevailing view from modelling of expanding magnetic flux tubes on the quiet Sun that the fields in the lower chromosphere are largely horizontal (canopy-like).

Ground-state atomic polarization can also be unambiguously distinguished from other polarization phenomena by observing the relative polarization amplitudes in spectral lines that belong to the same multiplet. An example is given in Figure 7.11 for the three members of multiplet no. 2 of Mg I. They have a common upper level with $J = 1$, while for the lower levels J varies from 0 to 1 to 2 when we go from the left to the right diagram in Figure 7.11. With conventional theory that neglects lower-level polarization the intrinsic polarizabilities determined by the respective J quantum

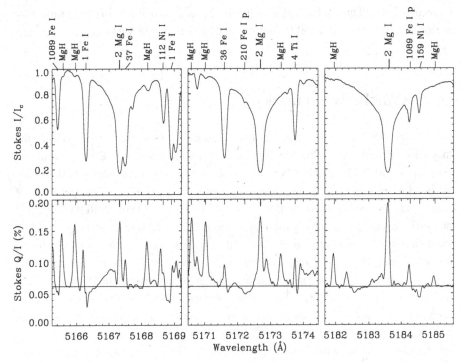

Figure 7.11. Evidence from the relative polarization amplitudes of the three strong Mg I lines for lower-level atomic polarization produced by optical pumping. In particular the pronounced peak in the 5184 Å line would not be present in the absence of optical pumping. Note the many surrounding polarization peaks due to molecular MgH lines. The recording was made 5 arcsec inside the solar north pole on April 4, 1995, near the minimum of solar activity. From Stenflo *et al.* (2000b).

numbers imply that the amplitude of the 5184 Å line should be smaller by between one and two orders of magnitude (when accounting for not only the resonant but also for the fluorescent transitions) as compared with the amplitude of the 5167 Å line. Instead the observations show that the amplitudes are similar. There is no possibility to account for this order-of-magnitude effect by playing around with free parameters in radiative transfer modelling. If one however takes into account lower level atomic polarization produced by optical pumping, then the observed amplitude ratios within the multiplet are reproduced without the need to adjust free parameters (Trujillo Bueno 2001). Since these polarization effects are sensitive to magnetic fields in the milligauss range, they extend our possibilities for magnetic field diagnostics.

Another feature noticed in Figure 7.11 is the abundance of polarization peaks from molecular MgH lines. Comparison of the behaviour of the molecular and atomic lines in different magnetic regions on the Sun has shown that while the polarization amplitudes of the atomic lines fluctuate greatly due to varying Hanle depolarization, the molecular lines appear to be unaffected by magnetic fields. An explanation for this has recently been found by Berdyugina *et al.* (2002), who could show that these molecular lines all have a Landé factor that is almost zero, while at the same time their intrinsic scattering polarizability (determined by their J quantum numbers) remains high.

7.6   Concluding remarks

Solar magnetic fields are structured on various scales, from the global scales to the magnetoturbulent scales that are far smaller than the resolution limit of available telescopes. The kG magnetic flux tubes, which are generally considered to be the basic building blocks of the photospheric magnetic flux, are only now at the verge of being spatially resolved (due to their small size of about 100 km). Still their detailed properties could be explored by indirect methods during the past three decades. While the boundary between the spatially resolved and unresolved domains is being pushed towards ever smaller scales, recently with the aid of adaptive optics, powerful imaging Stokes polarimeters with sensitivities only limited by photon statistics have come into use. With the new, previously not available polarimetric sensitivities the door has been opened to the "second solar spectrum", a world of polarization phenomena produced by scattering processes on the Sun. The diagnostic signatures of magnetic fields due to the Zeeman effect have thereby been complemented by the imprints left by magnetic fields in the "second solar spectrum" due to the Hanle effect and polarizing optical pumping. These various effects are highly complementary to each other, since they respond to magnetic fields in entirely different parameter regimes. While the new diagnostic tools have considerable future potential, their use is only in its infancy, since we are still busy identifying the underlying physics and building the required interpretative tools. The diagnostic objective is to explore the magnetized solar plasma on all scales, including its detailed properties in the spatially unresolved domain.

References

Berdyugina, S.V., Stenflo, J.O., and Gandorfer, A. (2002). *Astron. Astrophys.*, **388**, 1062.

Bianda, M., Solanki, S.K., and Stenflo, J.O. (1998a). *Astron. Astrophys.*, **331**, 760.

Bianda, M., Stenflo, J.O., and Solanki, S.K. (1998b). *Astron. Astrophys.*, **337**, 565.

Bianda, M., Stenflo, J.O., and Solanki, S.K. (1999). *Astron. Astrophys.*, **350**, 1060.

Emonet, T. and Cattaneo, F. (2001). *Astrophys. J.* **560**, L197.

Faurobert, M., Arnaud, J., Vigneau, J., and Frisch, H. (2001). *Astron. Astrophys.*, **378**, 627.

Faurobert-Scholl, M. (1993). *Astron. Astrophys.*, **268**, 765.

Faurobert-Scholl, M., Feautrier, N., Machefert, F., Petrovay, K., and Spielfiedel, A. (1995). *Astron. Astrophys.*, **298**, 289.

Grossmann-Doerth, U., Schüssler, M. and Solanki, S.K. (1988). *Astron. Astrophys.*, **206**, L37.

Grossmann-Doerth, U., Schüssler, M., Sigwarth, M., and Steiner, O. (2000). *Astron. Astrophys.*, **357**, 351.

Landi Degl'Innocenti, E. (1998). *Nature*, **392**, 256.

Landi Degl'Innocenti, E. (1999). In *Solar Polarization*, ed. K.N. Nagendra and J.O. Stenflo. *ASSL*, **243**, p. 61 (Kluwer).

Moruzzi, G. and Strumia, F. (eds.) (1991). *The Hanle Effect and Level-Crossing Spectroscopy* (Plenum).

Nagendra, K.N. and Stenflo, J.O. (eds.) (1999). *Solar Polarization. ASSL*, **243** (Kluwer).

Povel, H.P. (1995). *Opt. Eng.*, **34**, 1870.

Povel, H.P. (2001). In *Magnetic Fields across the Hertzsprung-Russel Diagram*, ed. G. Mathys, S.K. Solanki, and D.T. Wickramasinghe. *ASP Conf. Ser.*, **248**, 543.

Rüedi, I., Solanki, S.K., Livingston, W., and Stenflo, J.O. (1992). *Astron. Astrophys.*, **263**, 323.

Semel, M. (1995). In *3D Spectroscopic Methods in Astronomy*, ed. G. Comte and M. Marcelin. *ASP Conf. Ser.*, **71**, 340.

Semel, M., Donati, J.-F., and Rees, D.E. (1993). *Astron. Astrophys.*, **278**, 231.

Sigwarth, M. (ed.) (2001). *Advanced Solar Polarimetry – Theory, Observation, and Instrumentation. ASP Conf. Ser.*, **236**.

Sigwarth, M., Balasubramaniam, K.S., Knölker, M., and Schmidt, W. (1999). *Astron. Astrophys.*, **349**, 941.

Solanki, S.K. (1993). *Space Sci. Rev.*, **63**, 1.

Steiner, O. (2000). *Solar Phys.*, **196**, 245.

Stenflo, J.O. (1973). *Solar Phys.*, **32**, 41.

Stenflo, J.O. (1982). *Solar Phys.*, **80**, 209.

Stenflo, J.O. (1994). *Solar Magnetic Fields – Polarized Radiation Diagnostics* (Kluwer).

Stenflo, J.O. (1999). In *Solar Polarization*, ed. K.N. Nagendra and J.O. Stenflo. *ASSL*, **243**, p. 1 (Kluwer).

Stenflo, J.O. (2001). In *Advanced Solar Polarimetry – Theory, Observation, and Instrumentation*, ed. M. Sigwarth. *ASP Conf. Ser.*, **236**, 97.

Stenflo, J.O. and Keller, C.U. (1996). *Nature*, **382**, 588.

Stenflo, J.O. and Keller, C.U. (1997). *Astron. Astrophys.*, **321**, 927.

Stenflo, J.O., Harvey, J.W., Brault, J.W., and Solanki, S.K. (1984). *Astron. Astrophys.*, **131**, 333.

Stenflo, J.O., Keller, C.U., and Gandorfer, A. (1998). *Astron. Astrophys.*, **329**, 319.

Stenflo, J.O., Gandorfer, A., and Keller, C.U. (2000a). *Astron. Astrophys.*, **355**, 781.

Stenflo, J.O., Keller, C.U., and Gandorfer, A. (2000b). *Astron. Astrophys.*, **355**, 789.

Stenflo, J.O., Gandorfer, A., Wenzler, T., and Keller, C.U. (2001). *Astron. Astrophys.*, **367**, 1033–1048.

Trujillo Bueno, J. (2001). In *Advanced Solar Polarimetry – Theory, Observation, and Instrumentation*, ed. M. Sigwarth. *ASP Conf. Ser.*, **236**, 161.

Trujillo Bueno, J., Moreno Insertis, F., and Sánchez, F. (eds.) (2002). *Astrophysical Spectropolarimetry* (Cambridge Univ. Press).

8

Solar photosphere and convection

Å. Nordlund

Niels Bohr Institute for Astronomy, Physics, and Geophysics,
Copenhagen University, Denmark

8.1 Introduction

The surface layers of stars are of crucial importance for stellar structure and evolution; it is the properties of the surface layers that ultimately determine the relations between the luminosity, the surface acceleration of gravity, and the effective temperature of the radiation from the star. This is indeed the case also for the Sun; at, and immediately below the visible surface, a transition takes place to a convective and almost adiabatic envelope. The entropy jump across those shallow surface layers determines both the depth of the whole convection zone, and the relation between radius and luminosity of the Sun (cf. Figure 8.1, and Rosenthal *et al.*, 1999). In traditional stellar structure and evolution modeling the entropy jump is controlled by a 'mixing length' parameter, or by some other parameterization of convection, which leads to an unpleasant dependence on parameters that need to be calibrated.

The occurrence of a very complex surface layer, whose properties are hard to predict a priori, may be considered unfortunate from the point of view of the theory of stellar structure and evolution. However, from the point of view of astrophysical fluid dynamics, the presence of convection in the photosphere of the Sun presents a unique "test bench", which may be used both to benchmark numerical models and to benchmark our qualitative understanding of fluid dynamics and magnetohydrodynamics.

When constructing models of the surface layers of stars it is necessary to take into account the detailed physics of these layers, including such effects as atomic ionization, molecular dissociation, and the influence of the numerous atomic and molecular spectral lines on the radiative energy transport (Nordlund 1982; Stein and Nordlund, 1989, 1998).

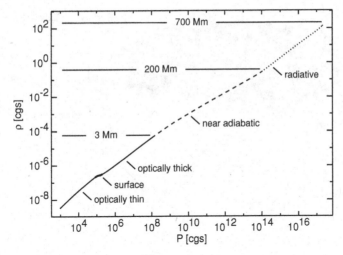

Figure 8.1. The overall stratification of the Sun. The properties of the thin transition region (photosphere) between the optically thin layers of the solar atmosphere and the convection zone are crucial for the structure of the whole envelope.

The resulting models have no free parameters, other than those specifying the effective temperature, surface acceleration of gravity, abundances of the chemical elements, and the numerical resolution (or equivalently, the amount of numerical diffusion). At sufficiently high numerical resolution (in space as well as in the numerical representation of the radiation field), synthetic observational data derived from the models have been found to agree closely with data from real observations (Asplund *et al.*, 1999a, 2000b,a,c). This occurs at numerical resolutions where a short turbulent inertial range is developing. One does indeed, from Kolmogorov's theory, expect that the numerical resolution (or numerical diffusion) should not be important when an inertial range is present.

Note that even if, for the Sun, spatially and temporally resolved observations provide plentiful information about details of the dynamics, such non-spatially resolved properties as spectral line width, shape, and convective blue-shift are also of great importance, since comparison of such quantities with observations bypasses the difficulties associated with atmospheric seeing and finite instrumental resolution.

It is, for example, very difficult to measure, with sufficient accuracy, the rms amplitude of the vertical velocity in the solar photosphere, or the rms amplitude of the continuum intensity fluctuations. Even with the best of instruments, and with excellent seeing conditions, corrections for seeing are of the order of a factor two, and are intrinsically uncertain. On the other hand, direct comparisons of the widths and shapes of synthetic and observed spectral lines are straight-forward.

Helioseismology provides additional opportunities to test the accuracy of numerical models of convection. The depth of the solar convection zone, for example, depends sensitively on the temperature structure of the surface layers, and is known to high precision from analysis of solar p-mode oscillations (Basu and Antia, 1997) .

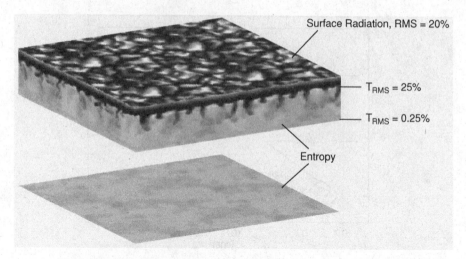

Figure 8.2. A perspective rendering of entropy fluctuations in the solar surface
layers (Stein and Nordlund, 1998). The top surface shows the radiation intensity
emergent from the top boundary of the model, while the drop-down copy of the
bottom boundary illustrates the very weak entropy fluctuations there. Root mean
square temperature fluctuations are also indicated.

Surface structure also directly influences the frequencies of solar p-modes, and comparison of observed and theoretical p-mode frequencies thus provides yet another
observational test of the models.

Once it is established that the dynamics of the photosphere is well understood,
and is even quantitatively pinned down to high precision, one has a firm ground to
stand on with respect to understanding the implications of the surface dynamics for
other layers, above and below the surface.

8.2   Dynamic and thermal properties of the solar photosphere

Figure 8.2 illustrates the structure of entropy fluctuations in the solar surface layers.
As indicated in the figure, the entropy fluctuations correspond to root mean square
temperature fluctuations of about 25% in the very surface layers (actual temperatures
vary between about 4,000 K and 10,000 K). The temperature fluctuations near the
bottom of the model, at a depth below the surface of some 2.5 Mm, are only $\sim 0.25\%$,
and are thus about two orders of magnitude smaller than at the surface. Note that the
size of the convection cells (granules) visible at the surface is an order of magnitude
larger than the pressure and density scale heights there. Patterns on the drop-down
bottom boundary are again substantially larger than those at the surface.

The top face of the cube shows the solar "granulation"; a cellular pattern of
brightness fluctuations with a root mean square amplitude of some 20%. Note that

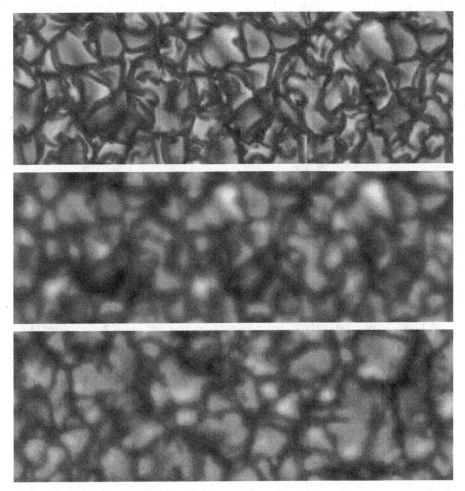

Figure 8.3. Images of surface radiation intensity (granulation) from a
$253 \times 253 \times 163$ simulation (top panel – rms amplitude $\sim 20\%$), the same data,
folded with a point spread function that reduces the rms amplitude to $\sim 10\%$ (mid
panel), comparable to the rms fluctuation amplitude of observations taken at the
Swedish Solar Telescope at La Palma (bottom panel – data courtesy of G. Scharmer).

the intensity rms is only about a fifth of what would be expected from an object with
25% temperature fluctuations emitting a black body radiation $\sim T^4$ – the strong temperature sensitivity of the continuum opacity ($\kappa \sim T^{10}$) hides the strong temperature
fluctuations.

Raw images of the emergent surface intensity of models with a horizontal resolution of 253×253 show more detail than is visible in even the best observations of
the solar granulation. After folding with a point spread function (PSF) that reduces
the rms contrast to the observed values there is a remarkable similarity in appearance
between the observations and the simulations (Figure 8.3).

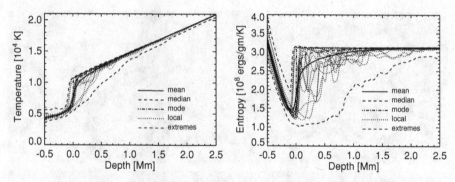

Figure 8.4. Temperature (left) and entropy (right) as a function of depth, for a subsample of the horizontal grid points in a numerical simulation.

The intensity power spectra of observations and PSF convolved model images are also consistent. However, Nordlund *et al.* (1997) have shown that a power spectrum quite similar to the observed intensity power spectrum may be produced by suitably smearing the edges of the word TURBULENCE, thus illustrating that a power spectrum has very little information content.

A characteristic property of strongly stratified convection is revealed by plotting entropy as a function of depth for a sample of horizontal points. As shown by Figure 8.4, the upper envelope of the entropy distribution is almost totally flat. Ascending fluid is nearly isentropic, while descending fluid has entropy fluctuations that decrease systematically with depth, asymptotically approaching the entropy of the ascending fluid. The reason for this is that mass conservation puts a severe constraint on the flow. Within one scale height, most of the ascending fluid must overturn, and since the surface layers of stars span a large number of scale heights, only a tiny fraction of the ascending fluid at some depth below the surface actually makes it to the top.

Conversely, the descending fluid is constantly "diluted" by overturning, nearly isentropic fluid. The contribution of low entropy fluid from the surface is thus rapidly overwhelmed by turbulent mixing with nearly isentropic fluid.

8.3   Spectral line synthesis

The Doppler broadening of photospheric spectral lines is a direct manifestation of the photospheric velocity field. In particular, spectral lines from heavy atoms such as iron are significantly influenced by Doppler broadening, because their purely thermal profiles are quite narrow. Iron lines are thus ideal diagnostics of photospheric dynamics, also because of the large number of iron lines that have accurately measured absorption coefficients and wavelengths. Many photospheric lines are blended, but because of the large number of iron lines, it is still possible to find dozens of relatively clean lines of FeI and FeII in the solar spectrum.

Non-LTE effects are potentially important, both for the degree of ionization, and for the population levels within each ionization species. FeII is the dominant ionization stage in the lower photosphere, and FeII lines are therefore less sensitive to non-LTE ionization effects than FeI lines. Non-LTE excitation effects are generally smallest for weak lines (being dependent on the ratio of radiative to collisional transition rates). The most trustworthy iron lines are therefore the weak FeII lines.

Various aspects of spectral lines directly reflect key properties of the convection motions; the spectral line width measures the vertical velocity amplitudes, the convective blue-shift measures the velocity-intensity correlation, and the spectral line asymmetry is the result of a complicated convolution of effects; correlations between velocity, temperature, and density, as reflected in the local spectral line absorption coefficients through the non-local effects of spectral line formation in a 3-D medium.

Spectral line shapes thus provide ideal "finger-prints" of convection in the surface layers of stars. As finger-prints they are very hard to forge; it would take a number of arbitrarily assigned parameters to obtain a good fit to just one of the observed spectral lines, and it is unlikely that the same set of parameters would results in a good fit for a different spectral line. In contrast, the 3-D models contain no arbitrary parameters. In addition to the traditional parameters; effective temperature, surface acceleration of gravity, and chemical abundances, only the numerical resolution, and the associated treatment of viscous effects on scales below of a few grid zones enters in the numerical parameterization.

In traditional 1-D models non – thermal broadening can only be accounted for by introducing arbitrary "micro-" and "macro-turbulence" parameters. Micro-turbulence represents the small scale end of the turbulent spectrum and is applied to the spectral line absorption coefficient profile; it thus enhanced the total absorption (equivalent width) of the spectral lines. Macro-turbulence represents the large scale end of the turbulent spectrum, and is applied to the resulting spectrum; it increases the spectra line width (e.g., the full-width-at-half-maximum) without increasing the equivalent with. Spectral line asymmetries and shifts cannot be accounted for at all in 1-D models.

To illustrate the extent to which the photospheric velocity field influences the width and shape of photospheric iron lines, Figure 8.5 a shows what a weak FeII line would look like if there were no macroscopic photospheric velocities. The line is much too deep and narrow, and it is perfectly symmetric, in contrast to the slightly asymmetric observed spectral line. The much larger width of the observed spectral line is the result of spatial and temporal averaging.

Figure 8.5b shows comparisons of the observed and synthetic spectral line profiles when the photospheric velocity field is included. The width and shape of the synthetic spectral line match those of the observed line very closely. Figure 8.6 illustrates the disparity of the contributions that go into the spatial averaging, for a single snapshot in time.

Figure 8.7 shows a corresponding comparison for a spectral line from neutral iron; here, the right hand side panel shows spectral line bisectors (Asplund et al., 2000b).

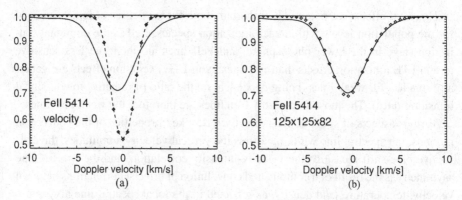

Figure 8.5. a) The weak FeII line λ5414, as observed and as it would appear if there were no photospheric velocities. b) Comparison of observed and synthetic FeII λ5414 spectral line profiles, based on a 253 × 253 × 163 simulation.

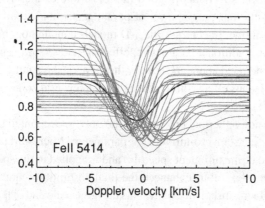

Figure 8.6. Observed line profile, and a subsample of the contributions from individual horizontal grid points in a snapshot.

Because of the correlation between vertical velocity and horizontal temperature fluctuations (surface brightness) there is a net ("convective") blue-shift of spectral lines, and the spectral line bisectors are C-shaped.

Note that the very good agreement between model and observations exemplified in Figure 8.5b and Figure 8.7 is obtained on an *absolute* wavelength scale; the absolute wavelength of solar spectral lines is known to high precision, and the absolute wavelength of the spectral line absorption profile is known to similar precision from laboratory measurements.

Stronger FeI and FeII lines again show good agreement between synthetic and observed spectral line profiles over most of the line profiles. There are small, systematic differences in the cores of strong lines, that tend to be slightly too deep, and to have slightly too weak blue shifts. These differences may be caused by non-LTE effects, or they may be caused by slightly too low temperatures in the upper layers of the photosphere, where the temperature balance depends critically on the energy balance in relatively few, strong spectral lines.

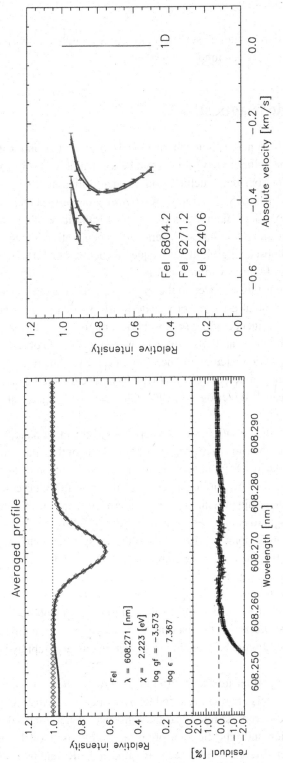

Figure 8.7. Comparison of observed and synthetic FeI λ6082 spectral line profiles (left) and spectral line bisectors for three FeI lines (right).

The confrontation of stellar surface convection simulations with stellar spectroscopy is discussed by Asplund *et al.* (1999a,b).

8.4   P-mode Diagnostics

Important constraints on 3-D photospheric models (or important applications of them – depending on the point of view) come from helioseismology, which provides high precision diagnostics of solar structure. Again, the information is 'coded' into 'finger prints', which one may choose to model with inverse or forward methods.

P-modes are excellent diagnostics of the photospheric structure because their upper reflection occurs in the surface layers, with a penetration into the photosphere that varies from almost negligible at low frequency (below 2 mHz) to nearly complete penetration at high frequency (above the acoustic cut-off frequency). The structure of the photosphere is thus encoded in the frequency dependence of p-modes.

Since even non-radial p-modes propagate nearly vertically near their upper reflection points the influence of the photospheric structure is similar over a range of horizontal wave numbers, and differences between models and observations having to do with photospheric structure are therefore primarily a function of frequency. A systematic frequency discrepancy of that type has indeed bothered helioseismology for a number of years (Monteiro *et al.*, 1996; Christensen-Dalsgaard and Thompson, 1997).

The lower reflection point of p-modes depends, on the other hand, sensitively on horizontal wave number, and such properties as the depth of the solar convection zone may therefore be derived from observations with good precision, irrespective of the discrepancies related to the surface layers. The depth of the solar convection zone has indeed been determined with great precision from helioseismic measurements (Christensen-Dalsgaard *et al.*, 1991; Basu and Antia, 1997).

From the modeling point of view both of these aspects are directly useful; the upper turning point behavior can be computed in 3-D models, and the depth of the convection zone is mainly determined by the entropy jump near the surface, which is again something that may be computed from the 3-D models.

The main part of these effects can be related to the horizontally averaged structure of the models; horizontal fluctuations become quite small in the interior of the convection zone (cf. Figure 8.2) and the vertical structure also rapidly becomes very nearly adiabatic.

Frequency changes that are due to differences between the mean vertical structure of 3-D and 1-D models are often referred to as 'model' (or 'extrinsic') differences, while remaining differences, related to the influence of horizontal fluctuations on p-mode propagation and reflection are often referred to as 'modal' (or 'intrinsic') differences. Model differences have been investigated by Rosenthal *et al.* (1999) and are briefly summarized below, while modal differences still remain to be explored.

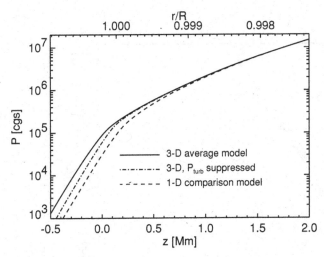

Figure 8.8. Elevation of the photosphere by turbulent pressure and 3-D effects. The full drawn curve shows pressure as a function of depth for an averaged 3-D model, the dashed curve shows the pressure for a comparison 1-D envelope model, and the dot-dashed curve shows the pressure for a 3-D model where the gradient of the turbulent pressure has been artificially removed.

Figure 8.8 compares the average stratification of a 3-D model with that of a standard envelope model (model S of Christensen-Dalsgaard *et al.*, 1996). The photosphere of the 3-D model is elevated by about one pressure scale height relative to the standard model, with two effects contributing about equally. As illustrated by Figure 8.8, the turbulent pressure support accounts for about half the effect.

The remaining part of the effect is a genuine 3-D effect, due mainly to the non-linear fluctuations of the opacity: Because of hydrogen ionization, the temperature sensitivity of the opacity is extremely high in the solar surface layers ($\kappa \sim T^{10}$). Therefore, the visibility of positive temperature fluctuations is reduced (they are masked by the increased opacity), while that of negative temperature fluctuations is increased (they become more visible because of the reduced opacity). As a result, the average flux emitted from a 3-D model is smaller than it would be from a 1-D model with a structure equal to the average 3-D structure. Conversely, a 3-D model that emits the required, nominal solar flux must necessarily be hotter on the average than a corresponding 1-D model, as illustrated by Figure 8.8.

Note that, in contrast to 1-D models computed with the Canuto and Mazzitelli (1991) recipe, the averaged 3-D models do not have a steeper and more narrow superadiabatic structure in the surface layers than obtained with the classical Böhm-Vitense (1958) mixing length recipe. The difference between the 1-D and average 3-D structure illustrated in Figure 8.8 is instead a result of modifications of the pressure balance as well as of modifications of the temperature structure.

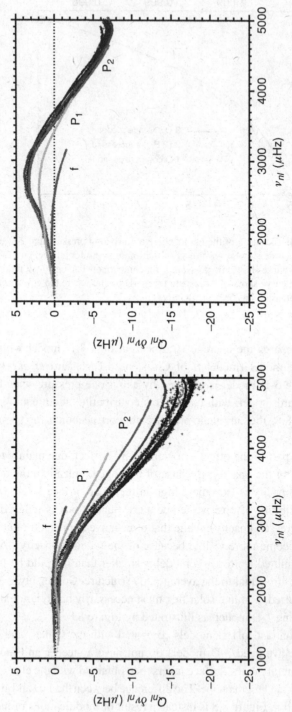

Figure 8.9. Scaled frequency differences between observations and a standard 1-D solar model (left), and between observations and a 1-D model constructed from horizontal averages of a 3-D model (right). From Rosenthal *et al.* (1999).

The elevation of the photosphere corresponds to an extension of the acoustic cavity, and hence a lowering of the frequencies of those modes that have turning points in the photosphere (Rosenthal *et al.*, 1999). The effect is illustrated in Figure 8.9, which shows scaled differences between observed frequencies and frequencies of adiabatic modes in 1-D models. The right hand side panel shows that the well known frequency discrepancy between observations and standard models is reduced significantly when the standard model is replaced by a 1-D model corresponding to a horizontally averaged 3-D model.

The remaining discrepancies are presumably related to 'modal' (intrinsic 3-D) effects; the frequencies used to construct Figure 8.9 were computed under the assumption that the modes are adiabatic, and using horizontally averaged values from the 3-D model. It would therefore be alarming rather than pleasing if there was perfect agreement; then there would be no room for non-adiabatic effects, and for differences between wave propagation in an inhomogeneous 3-D medium and the assumed 1-D, average medium.

As mentioned above, the depth of the convection zone is to a large extent controlled by near-surface properties, and is also well determined from helioseismology. The essential factor is the entropy jump in the surface layer. It is noteworthy that an accurate determination of the entropy jump requires both a detailed (3-D) modeling of convection, *and* a reasonably accurate modeling of 'spectral line blocking' of the radiative flux in the photosphere. About 12% of the continuum radiation is blocked by spectral lines in the solar spectrum; this raises the surface temperature by some 3%, or about 200 K, relative to a case with no spectral lines. If this effect was not included there would be a mismatch of about 2% in the prediction of the depth of the solar convection zone.

Rosenthal *et al.* (1999) predicted the solar convection zone depth to be $d = 0.286R$, by patching a high-resolution 3-D model of the surface layers with a 1-D envelope models of the rest of the convection zone. This is in excellent agreement with the helioseismic determinations; $d = 0.287R \pm 0.003R$ (Christensen-Dalsgaard *et al.*, 1991) and $d = 0.287R \pm 0.001R$ (Basu and Antia, 1997).

One may also use the amplitudes of solar p-modes as a diagnostic (Nordlund and Stein, 2001). The excitation of p-modes is related to non-adiabatic fluctuations of the gas pressure and the turbulent pressure, with contributions to the excitation coming mostly from a region near the optical surface (Stein and Nordlund, 2001). Estimates of the excitation power based on the numerical models are consistent with the observed excitation power.

8.5   Large scale velocity fields

There is another aspect of the solar photosphere that both reflects properties of the solar interior and is of great importance for its tenuous outer regions; the chromosphere and corona. As has been known for a long time, there are larger scale velocity fields visible on the solar surface; these are traditionally referred to as 'super-granulation', with turn-over times $\sim$ 1 day on scales $\sim$ 20 − 40 Mm (Leighton et al., 1962; Hathaway et al., 2000), and 'meso-granulation' with turn-over times $\sim$ 1 hour on scales $\sim$ 5 − 10 Mm November et al., (1981); Straus et al. (1992); Straus and Bonaccini (1997).

Conservation of mass requires that these motions, which have scales much larger than density scale heights in the photosphere, must have very small ratios of vertical to horizontal velocity amplitude there, and these motions are indeed primarily horizontal in the photosphere, with hardly measurable vertical velocities.

The temperature fluctuations associated with these scales are also very small, and as a consequence (e.g.Nordlund, 1982) the scale heights of their associated pressure fluctuations are very large. The large scale horizontal velocity patterns may thus be expected to have amplitudes that depend only weakly on depth; i.e., the surface patterns may be expected to be representative of the horizontal velocity fields below the surface.

Figure 8.10 shows a schematic view of the Sun on a global scale, with patterns typical of motions over a range of scales superimposed and warped onto cut-out surfaces. The lower two thirds of the solar convection zone has a nearly poly-tropic structure, while the upper one third is more complicated, and more strongly stratified.

Numerical simulations of stratified convection, polytropic as well as more elab-orate ones, have consistently shown motion patterns to have similar (large) aspect ratios; the solar granulation, for example, has cell sizes of the order of 1 − 3 Mm associated with vertical mass flux scale heights $\sim$ 300 km. The ratio of horizontal pattern size to vertical mass flux scale height is indeed expected to be $\sim 2\pi$ from simple considerations of mass conservation.

By a similar argument, the amplitudes of the horizontal velocities at a certain scale may be expected to scale with the vertical velocity amplitude at depths that increase in proportion to the horizontal scale. The vertical velocity amplitude, in turn, decreases in approximately inverse proportion to depth in the deep layers of 3-D numerical simulations. This is different than in mixing length models, where the scaling is more similar to the square root of depth (Spruit 1974).

The resulting prediction for the scale dependence of the horizontal velocity field in the photosphere is that the velocity amplitudes should be roughly proportional to the inverse of the scale, and that turn-over times should scale as the inverse square of the size. This is indeed consistent with the scales and turn-over times for super- and meso-granulation quoted above, and with the properties of granulation obtained in the 3-D simulations: horizontal rms amplitudes $\sim$ 3 km s^{-1} at granular scales ($\sim$ 1–3 Mm).

In terms of the power spectrum of horizontal velocities $P(k)$ (which satisfies $\int P(k)dk = \langle u_{\text{hor}}^2\rangle$), typical velocities at scales $\ell = 2\pi/k$ are $u(\ell) \sim \sqrt{k\,P(k)}$. In

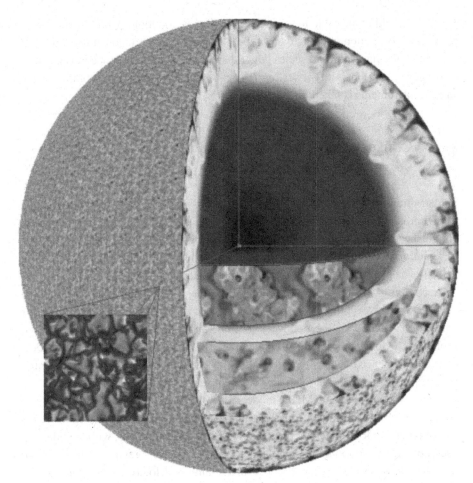

Figure 8.10. The solar photosphere in a global context, illustrating the huge range of scales of motions in the solar convection zone – these scales of motion are all to some extent visible as horizontal motions in the solar photosphere.

the case $u(\ell) \sim 1/\ell$ the corresponding power spectrum is $P(k) \sim k$ (of course only expected to hold for scales larger than granular scales).

A smooth spectrum of large scale motions, $P(k) \sim k$, is apparently at odds with the notion of distinct super- and meso-granular scales. Note, however, that such a power spectrum has the generic property that any resolution cutoff leads to an observed (but false) maximum in the power, with corresponding cellular features, at scales a few times the resolution element. Smaller scales loose visibility because of resolution, while larger scales loose visibility because their amplitudes are smaller.

The issue of the existence or not of distinct larger scales has been around since the meso-granulation concept was first introduced (November *et al.*, 1981). With accurate, long duration observations accumulating from instruments such as SOHO/MDI (Scherrer *et al.*, 1995) TRACE (Handy *et al.*, 1999), and SVST (Scharmer *et al.* 1985,

1999), it should be possible to settle the issue before too long. A preliminary compilation of correlation tracking data (in preparation) shows no indication of distinct maxima or strong features at scales larger than granules; all power spectra appear to be roughly consistent with the scaling $P(k) \sim k$, when the effective resolution of the various instruments is taken into account.

8.6 Consequences for coronal and chromospheric heating

The spectrum of photospheric horizontal motions is a crucial boundary condition for the problems of coronal and chromospheric heating. If, as is generally assumed, coronal heating and (at least the active region part of) chromospheric heating are due to some form of magnetic dissipation then a knowledge of the photospheric driving velocity is a pre-requisite for any model of the heating process.

According to the scaling formula for magnetic dissipation derived by Galsgaard and Nordlund (1996), the heating is proportional to the product of the correlation length ℓ of the boundary motions and the corresponding velocity v_ℓ. From the previous Section, this product may be expected to be roughly independent of scale, and hence one may expect a large range of scales to be contributing similar amounts to the heating (but over very different time scales!).

The first semi-realistic MHD models of solar active regions (Gudiksen and Nordlund 2002) seem to confirm the viability of maintaining coronal temperatures and of obtaining loop-like structures similar to the observed ones, with a model that uses a potential extrapolation of a magnetogram of an active region as initial condition, and evolves the initial condition with a boundary velocity field consistent with $P(k) \sim k$ for small k.

8.7 Concluding remarks

The solar photosphere is arguably the best observed example of a highly dynamic astrophysical plasma; there, we are able to observe motions over a large range of spatial and temporal scales, and we are able to pinpoint regions where the plasma betas ($\beta = P_{\mathrm{gas}}/P_{\mathrm{magnetic}}$) are both small and large. High resolution instruments are able to follow the dynamics at the scales that dominate the energy spectra, and are beginning to penetrate to even smaller scales.

As illustrated by the discussion in this Chapter, such a situation is ideal for testing our ideas and models for the interaction of radiation, dynamics, and magnetic fields in astrophysical plasmas. Numerical modeling is necessary in order to be able to make quantitative, testable predictions, but such models are also very important as sources of inspiration for improving our qualitative understanding of astrophysical plasmas – enabling, for example, the derivation of approximate scaling laws for estimating order of magnitude effects in other astrophysical contexts.

Acknowledgements

I have benefitted greatly from discussions and data browsing with Tom Berger, Dave Hathaway, Klaus Galsgaard, Boris Gudiksen, Hans-Gunter Ludwig, Michel Rieutord, Karel Schrijver, Dick Shine, Bob Stein, and Alan Title. This work was supported in part by the Danish Research Foundation, through its establishment of the Theoretical Astrophysics Center. The hospitality of ITP/UCSB, where this chapter was finalized, is gratefully acknowledged.

References

Asplund, M., Ludwig, H.-G., Nordlund, Å., and Stein, R. F. (2000a). *Astron. Astrophys.*, **359**, 669.

Asplund, M., Nordlund, Å., and Trampedach, R. (1999a). *Astron. Astrophys.*, **346**, L17.

Asplund, M., Nordlund, Å., and Trampedach, R. (1999b). In Gimenez, A., Guinan, E., and Montesinos, B., editors, *Theory and Tests of Convective Energy Transport*, volume 173 of *ASP Conference Series*, pages 221–224. American Society of Physics.

Asplund, M., Nordlund, Å., Trampedach, R., Allende Prieto, C., and Stein, R. F. (2000b). *Astron. Astrophys.* **359**, 729.

Asplund, M., Nordlund, Å., Trampedach, R., and Stein, R. F. (2000c). *Astron. Astrophys.*, **359**, 743.

Basu, S. and Antia, H. M. (1997). *Mon. Not. Roy. Astron. Soc.*, **287**, 189.

Böhm-Vitense, E. (1958). *Zs. f. Ap.*, **46**, 108.

Canuto, V. M. and Mazzitelli, I. (1991). *Astrophys. J.*, **370**, 295.

Christensen-Dalsgaard, J., Däppen, W., Ajukov, S. V., Anderson, E. R., Antia, H. M., Basu, S., Baturin, V. A., Berthomieu, G., Chaboyer, B., Chitre, S. M., Cox, A. N., Demarque, P., Donatowicz, J., Dziembowski, W. A., Gabriel, M., Gough, D. O., Guenther, D. B., Guzik, J. A., Harvey, J. W., Hill, F., Houdek, G., Iglesias, C. A., Kosovichev, A. G., Leibacher, J. W., Morel, P., Proffitt, C. R., Provost, J., Reiter, J., Jr., R., J., E., Rogers, F. J., Roxburgh, I. W., Thompson, M. J., and Ulrich, R. K., (1996). *Science*, **272**, 1286.

Christensen-Dalsgaard, J., Gough, D. O., and Thompson, M. J. (1991). *Astrophys. J.*, **378**, 413.

Christensen-Dalsgaard, J. and Thompson, M. J. (1997). *Mon. Not. Roy. Astron. Soc.*, **284**, 527.

Galsgaard, K. and Nordlund, Å. (1996). *J. Geophys. Res.*, **101** (A6), 13445.

Gudiksen, B. and Nordlund, Å. (2002). *Astrophys. J.*, submitted.

Handy, B. N., Acton, L. W., Kankelborg, C. C., Wolfson, C. J., Akin, D. J., Bruner, M. E., Caravalho, R., Catura, R. C., Chevalier, R., Duncan, D. W., Edwards, C. G., Feinstein, C. N., Freeland, S. L., Friedlaender, F. M., Hoffmann, C. H.,

Hurlburt, N. E., Jurcevich, B. K., Katz, N. L., Kelly, G. A., Lemen, J. R., Levay, M., Lindgren, R. W., Mathur, D. P., Meyer, S. B., Morrison, S. J., Morrison, M. D., Nightingale, R. W., Pope, T. P., Rehse, R. A., Schrijver, C. J., Shine, R. A., Shing, L., Strong, K. T., Tarbell, T. D., Title, A. M., Torgerson, D. D., Golub, L., Bookbinder, J. A., Caldwell, D., Cheimets, P. N., Davis, W. N., Deluca, E. E., McMullen, R. A., Warren, H. P., Amato, D., Fisher, R., Maldonado, H., and Parkinson, C. (1999). *Solar Phys.*, **187**, 229.

Hathaway, D. H., Beck, J. G., Bogart, R. S., Bachmann, K. T., Khatri, G., Petitto, J. M., Han, S., and Raymond, J. (2000). *Solar Phys.*, **193**, 299.

Leighton, R. B., Noyes, R. W., and Simon, G. (1962). *Astrophys. J.*, **135**, 474.

Monteiro, M. J. P. F. G., Christensen-Dalsgaard, J., and Thompson, M. J. (1996). *Astron. Astrophys.*, **307**, 624.

Nordlund, Å. (1982). *Astron. Astrophys.*, **107**, 1.

Nordlund, Å., Spruit, H. C., Ludwig, H.-G., and Trampedach, R. (1997). *Astron. Astrophys.*, **328**, 229.

Nordlund, Å. and Stein, R. F. (2001). *Astrophys. J.*, **576**.

November, L. J., Toomre, J., Gebbie, K. B., and Simon, G. W. (1981). *Astrophys. J.*, **245**, L123.

Rosenthal, C. S., Christensen-Dalsgaard, J., Nordlund, Å., Stein, R. F., and Trampedach, R. (1999). *Astron. Astrophys.*, **351**, 689.

Scharmer, G., Owner-Petersen, M., Korhonen, T., and Title, A. (1999). in *ASP Conf. Ser. 183: High Resolution Solar Physics: Theory, Observations, and Techniques*, p. 157.

Scharmer, G. B., Pettersson, L., Brown, D. S., and Rehn, J. (1985). *Appl. Opt.*, **24**, 2558.

Scherrer, P. H., Bogart, R. S., Bush, R. I., Hoeksema, J. T., Kosovichev, A. G., Schou, J., Rosenberg, W., Springer, L., Tarbell, T. D., Title, A., Wolfson, C. J., Zayer, I., and MDI Engineering Team (1995). *Solar Phys.*, **162**, 129.

Spruit, H. C. (1974). *Solar Phys.*, **34**, 277.

Stein, R. F. and Nordlund, Å. (1989). *Astrophys. J.*, **342**, L95.

Stein, R. F. and Nordlund, Å. (1998). *Astrophys. J.*, **499**, 914.

Stein, R. F. and Nordlund, Å. (2001). *Astrophys. J.*, **585**.

Straus, T. and Bonaccini, D. (1997). *Astron. Astrophys.*, **324**, 704.

Straus, T., Deubner, F. L., and Fleck, B. (1992). *Astron. Astrophys.*, **256**, 652.

The dynamics of the quiet solar chromosphere

W. Kalkofen

Harvard-Smithsonian Center for Astrophysics, Cambridge, MA 02138, USA

S.S. Hasan

Indian Institute of Astrophysics, Bangalore-560034, India

P. Ulmschneider

University of Heidelberg, Germany

9.1 Introduction

The quiet chromosphere is bifurcated into magnetic and nonmagnetic regions. Although magnetic fields are found everywhere on the Sun they are dynamically unimportant in the interior of supergranulation cells (CI). In the magnetic network on the cell boundary (CB), fields are of decisive importance for the nature of the waves and their propagation characteristics.

In the magnetic network, the field occurs in concentrations of intense magnetic flux that are idealized as tubes in which the contribution of the field to the pressure may exceed that of the gas. A typical photospheric value of the plasma$-\beta$ (the ratio of gas to magnetic pressure, $\beta = 8\pi p/B^2$) is $\beta = 1/3$, for which the gas density in a thin tube is lower than in the ambient medium by a factor of 4.

In the thin-tube approximation, vertical tubes in pressure equilibrium with the outside medium expand upward to conserve magnetic flux. From a low filling factor of 1% in the photosphere the tubes spread to 15% in the layers of formation of the emission features in the H and K lines of ionized calcium (at a height of 1 Mm) and to 100% in the magnetic canopy, often defined as the region above the layer where $\beta = 1$. At some height the idealization of a thin flux tube ceases to be useful.

The nature of the waves that can exist in these two media depends on the magnetic field. In the field-free CI, the restoring forces in the wave equation are pressure and gravity. In the upward direction, only acoustic waves can propagate. The chromospheric oscillations in the CI are therefore due to acoustic waves. In oblique directions, gravity-modified acoustic waves can also contribute (Skartlien, Stein and Nordlund 2000). In flux tubes in the magnetic network, all three restoring forces, namely, pressure, gravity and the magnetic field, can act. But the only important wave types appear to be transverse and longitudinal flux tube waves; the former are mainly magnetic, and the latter, mainly acoustic. Internal gravity waves appear to play no role in any of the chromospheric oscillations. The main effect of gravity is to provide the stratification of the atmosphere which, in turn, is responsible for the dispersion of all waves traveling in the vertical direction, and hence for the existence of cutoffs and limits on the frequency ranges in which the waves can either propagate or are evanescent.

In the cell interior, acoustic waves have a period of three minutes (corresponding to a frequency of $\nu = 5$ mHz), and kink waves in the magnetic network have typical periods of seven minutes ($\nu = 2.5$ mHz), depending on the strength of the magnetic field; longitudinal flux tube waves also have three-minute period and only a weak dependence on field strength. In all these cases, the waves are propagating if their frequency is higher than the respective cutoff frequency, and evanescent if it is lower.

Chromospheric oscillations are seen most prominently in intensity enhancements of the emission peaks in the cores of the K and H lines of Ca II. In the nonmagnetic cell interior, the bright phase of the oscillations gives rise to K_{2v} and H_{2v} bright points, so called because of the pronounced asymmetry of the line profile which favors the blue emission peaks in the line cores; these bright regions are also referred to as calcium grains. In the magnetic network, the corresponding features are called network bright points; their line profiles are much less asymmetric.

Oscillations reveal the properties of the medium through which they propagate. But whereas p–mode oscillations of the solar interior have low damping and long life time, for some modes corresponding to thousands of wave periods, and therefore reveal the internal solar structure in great detail, chromospheric oscillations are highly damped by shock dissipation and therefore live only for intervals of the order of a few times the wave period before they must be excited again. Chromospheric oscillations tell us mainly about the state of the atmosphere in the layers where the waves arise. An important difference between internal solar and chromospheric oscillations is that the former are resonance oscillations whereas the characteristic frequencies of chromospheric oscillations owe their values to the existence of cutoffs of dispersive waves propagating upward (or downward) in a stratified medium, in which waves with wavelengths much longer than the density scale height propagate with reduced group velocity and, eventually, with the group velocity approaching zero and the phase velocity approaching infinity; in that limit, a wave is evanescent and transports no energy (in the linear limit), and the atmosphere swings in unison, either up and down (acoustic waves) or sideways (transverse flux tube waves).

9.2   Oscillations in the nonmagnetic chromosphere

The most important lines for the three-minute oscillations in the nonmagnetic chromosphere have been the H and K lines, with residual central intensity of 4% the strongest lines in the visible solar spectrum. Seminal papers on ground-based observations of oscillations are: Liu (1974); Grossmann-Doerth, v.Uexküll and Kneer (1974); Cram and Dámé (1983); v.Uexküll and Kneer (1995); and on space observations by Carlsson, Judge and Wilhelm (1997). The only empirical model of chromospheric oscillations is by Carlsson and Stein (1994, 1997). Important theoretical papers are Lamb (1908), Fleck and Schmitz (1991) and Rossi *et al.* (1991) for the analytic solution of the wave equation for impulsive excitation of oscillations in a one-dimensional (1D), stratified, isothermal atmosphere; Kalkofen *et al.* (1994) for the numerical solution in an isothermal atmosphere and Sutmann and Ulmschneider (1995) in an empirical chromospheric model; and Kato (1966) and Bodo *et al.* (2000) for the analytic solution of the wave equation in three-dimensional geometry.

The wave period was proposed to be due to standing waves in the cavity formed by the chromospheric temperature structure (Leibacher and Stein 1981). That explanation became untenable with the simulation of bright-point oscillations in the dynamical model of Carlsson and Stein (1994). The excitation of oscillations was proposed to be due to internal p−modes (e.g. Rutten and Uitenbroek 1991, v.Uexküll and Kneer 1995, Carlsson *et al.* 1997), which were presumed to generate upward-propagating waves by superposition. Such a model calls for a source region with a linear size of the order of the wavelength of p−modes, approximately 2,000 km. But observations of the horizontal size of the propagation channel in which the waves responsible for the oscillations travel upward suggest a size varying between the width of an intergranular lane in the photosphere (Sivaraman, Bagare and November 1990), about 100 km, and about 6,000 km in high layers of the chromosphere (Carlsson *et al.* 1997). The actual area of the excitation region would thus be smaller by two orders of magnitude than that required for p−mode excitation. This mechanism also fails to explain the wave period.

The excitation of bright-point oscillations is still an unsolved problem. The most plausible explanation for the wave period is that it is the acoustic cutoff period in the upper photosphere (discussed below).

Although both calcium bright points and chromospheric heating are caused by acoustic waves, the paucity and location of bright points suggest that they arise in a process that is different from that of the general heating of the chromosphere. For a ratio of $I_{K_{2v}}/\bar{I} = 1.5$ of the intensity maximum at the K_{2v} emission peak and the average intensity in the cell interior (in a band of 0.3 Å centered on K_{2v}), v.Uexküll and Kneer found that bright points occur only in 5% to 10% of the CI. Nevertheless, the process of heating the chromosphere is likely to cause some motions of low amplitude with a period of three minutes since the waves generated in the convection zone by the Lighthill mechanism carry a substantial signal at the acoustic cutoff period (Theurer, Ulmschneider and Kalkofen 1997). Thus, a plausible scenario of three-minute oscillations consists of calcium bright points that are caused by discrete excitation processes in intergranular lanes, and of a low-level background of waves

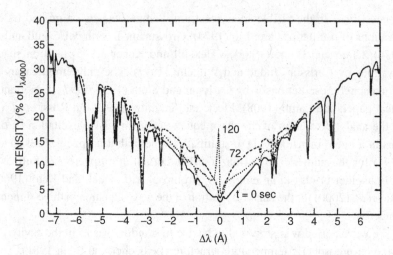

Figure 9.1. Profile of the K line at three phases in the evolution of an oscillation – from Liu (1974).

with the cutoff period as a signature that are caused by the general heating of the chromosphere.

The oscillations seen in calcium bright points reveal the dynamics and the underlying structure of the chromosphere. An instructive example is the K line observed by Liu (1974) and shown in Figure 9.1 at three phases during the evolution of a wave. At time $t = 0$, the wave is deep in the photosphere and has low amplitude. The intensity profile is symmetric and shows the temperature structure of the undisturbed atmosphere, mapped from depth to wavelength. Following the line intensity inwards from the wings, the temperature drops through the photosphere to symmetrically placed minima in the K line profile, which are formed at the temperature minimum between photosphere and chromosphere. The intensity then rises inward, corresponding to the outward rise of the kinetic temperature, until photons from the line can escape from the atmosphere. At that point the line profile forms the K_2 maxima, where the excitation temperature of the two combining states in the line transition separates from the kinetic temperature and the source function drops below the Planck function; the intensity continues to drop to the line center, forming the K_3 minimum.

At the time of $t = 120$ s, the wave has reached the layer of formation of the K_2 maxima and led to a very large, asymmetrical increase of the intensity of the blue emission peak, K_{2v}. At the same time, the K_3 minimum is shifted towards the red side of the spectrum. The simultaneous intensity enhancement at K_{2v} and the redshift of K_3 is the signature of K_{2v} bright-point oscillations that must be explained by a model of chromospheric dynamics.

The empirical dynamical model by Carlsson and Stein (1994) reproduced these features. Starting from cospatial observations by Lites, Rutten and Kalkofen (1993) of the Doppler velocity of a photospheric Fe I line and of the simultaneous evolution of the H line, Carlsson and Stein drove the lower boundary in their simulation with

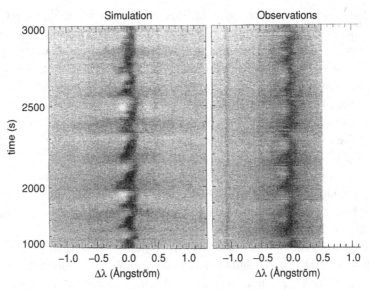

Figure 9.2. The H line intensity as a function of wavelength and time: left panel, simulated; right panel, observed – from Carlsson (1994).

the observed Doppler velocity. The comparison of the simulated and observed H line intensities as functions of wavelength and time (Figure 9.2: profiles, respectively, on the left and the right) shows broad agreement in the signature feature for the simulated bright point, namely, the simultaneous occurrence of enhancement of the H_{2v} intensity peak and redshift of the line center H_3. These features are formed when the upward-propagating shock wave, which heats the layer where the H_2 emission peaks are formed, meets downward-streaming gas above this layer, which had been lifted upwards by the preceding wave. Although this feature has been explained before, from a study of waves in an isothermal atmosphere (Kalkofen *et al.* 1994), the significance of the Carlsson and Stein simulations is that the intimate relation between waves in the photosphere and subsequent dynamics in the chromosphere was shown in an empirical model. This model established that calcium bright points are caused by propagating acoustic waves and the intermittent formation of shocks. The value of this demonstration is not diminished by the flaws of the model, which are apparent in Figure 9.2, which is taken from the frontispiece of the Oslo proceedings booklet.

There are three differences between the simulated and the observed intensities. They are the late arrival of the shock in the layer of formation of the blue emission peak, the H_3 intensity formed above the shock that may be low by an order of magnitude, and the H_{2v} intensity at maximal brightness that may be high by an order of magnitude.

The reason for the late arrival of the shock is not understood. The low intensity of the line center was traced to the intermittent heating in the dynamics and the absence of general heating (Kalkofen, Ulmschneider and Avrett 1999, Kalkofen 2001). And the excess intensity at H_{2v} is due to the topology of wave propagation in bright-point oscillations, which the model assumes to be in the form of plane waves, where the energy of the waves is trapped in vertical cylinders, whereas the Sun shows them to be in the form of spherical waves, where the energy spreads horizontally in upward propagation. This spreading is apparent in the size of the area disturbed by the shock and varies from the 100 km in the photosphere, noted above, to 0.5 to 1 Mm at the base of the chromosphere (Foing and Bonnet 1984), to 2 to 3 Mm in the layer of formation of the H_{2v} emission peak (Cram and Damé 1983).

The main properties of the oscillations can be understood from the linearized hydrodynamic equations. The (1D) equations for plane waves illuminate the role played by the acoustic cutoff frequency, and the 3D solutions show the horizontal spreading of the wave energy.

For the discussion of the equations it is convenient to separate the exponential variation of the physical quantities and to write the equations in terms of their corresponding "reduced" quantities. Thus, instead of the physical velocity $v(z, t)$ in terms of height z and time t we write the hydrodynamic equations for a reduced velocity $u(z, t)$, defined by

$$v(z, t) = u(z, t)\exp(z/2\mathcal{H}). \qquad (9.1)$$

The equation expresses the property that the physical velocity grows in the vertical direction with an e-folding distance of twice the density scale height. This vertical growth compensates for the exponential decrease of the density and insures that energy flux is conserved.

The hydrodynamic equations for dimensionless variables z' and t' can be cast into the form of a wave equation, called Klein-Gordon equation,

$$\frac{\partial^2}{\partial z'^2}u - \frac{\partial^2}{\partial t'^2}u - u = 0. \qquad (9.2)$$

With the ansatz $u(z', t') \sim \exp(ikz' - i\omega t')$, we obtain the dispersion relation for disturbances propagating in this stratified, isothermal medium,

$$k^2 = \omega^2 - 1, \qquad (9.3)$$

and the phase and group velocities

$$v_{ph} = \frac{\omega}{k} = \frac{\omega}{\sqrt{\omega^2 - 1}}, \quad v_g = \frac{\partial \omega}{\partial k} = 1/v_{ph}, \quad v_{ph}v_g = 1. \qquad (9.4)$$

At the cutoff frequency, $\omega = 1$, the phase velocity becomes infinite and the group velocity becomes zero. The product of phase and group velocities equals the square of the sound speed, which is unity here.

The analytic solution of the equations for a velocity pulse (see Kalkofen et al. 1994), which is due to Lamb (1909), shows the pulse propagating with sound speed

in the vertical direction, followed by a wake, which in the limit of late times results
in a decaying oscillation at the cutoff frequency,

$$u(z', t') \propto \frac{\cos(t')}{\sqrt{t'}}.$$ (9.5)

It is interesting to note that transverse and longitudinal waves in a thin flux tube
embedded in an isothermal medium obey the same wave equation (9.2); only the
transformation (9.1) is modified to allow for the upward expansion of the flux tube to
conserve magnetic flux, leading to an e-folding distance for the growth of the velocity
amplitude of tube waves of $4\mathcal{H}$.

The three-dimensional hydrodynamic equations describe internal gravity waves in
addition to acoustic waves (see Kato 1966). But internal gravity waves are excluded
from the vertical direction, and carry little energy in the horizontal direction. Acoustic
waves in the horizontal direction suffer only geometrical dilution and are therefore
also unimportant for chromospheric oscillations. In the vertical direction, the growth
of the "wave amplitude" (for pressure p or v^2) is still proportional to the factor
$\exp(z/\mathcal{H})$.

There are two major differences between solutions for acoustic waves in 1D and
3D. In the spherical case, the wave amplitude decreases with the square of the distance
from the origin suggesting flux conservation through a spherical surface, and varies
along the perimeter of the circle about the source. This is seen in the upper panel
of Figure 9.3, where the top curve displays the pulse at the time when its apex has
reached the reference height of 15 $\mathcal{H}$ that represents the layer of formation of the
H_{2v} and K_{2v} emission peaks, showing the wave amplitude increasing from the pole
to the equator (and decreasing again to the south pole). For the oscillations in the

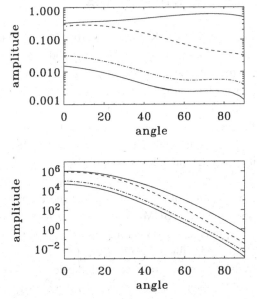

Figure 9.3. Wave
amplitudes in arbitrary units
as functions of polar angle
in a vertical cut, at the times
when the apex of the pulse
(top curve) and the apices of
the waves in its wake
(ordered below the pulse)
have reached the target
height. Top panel: wave
amplitudes without the
exponential-growth factor;
bottom panel: with the
factor – from Bodo *et al.*
(2000).

wake of the pulse, the energy is reduced – in the third wave, for example, by an order of magnitude – and the amplitude decreases with increasing zenith angle. Thus, the profile of the upward-propagating wave narrows and more energy is concentrated towards the axis of the channel. The lower panel shows the same curves, but with the exponential factor $\exp(z/\mathcal{H})$. The steeper decrease of the amplitudes with angle reflects the lower height reached by parts of the wave at larger angle.

The results from the linear, analytic solution cannot be carried over to the non-linear regime immediately. In the 1D case, for example, the nonlinear solution no longer has most of the energy in the initial pulse, but in the second or third oscilla-tion in its wake. Similarly, in the numerical simulation of bright-point dynamics by Carlsson and Stein (1994), a calculation for the atmosphere initially at rest shows little correspondence with the observations until a time interval of the order of the wave period.

Among the important results of the 3D calculation is the increase of the phase velocity for the oscillation behind the pulse (only in the vertical direction), the finite width of the propagation channel, and the narrowing of the channel for the later oscillations, with some modification expected for the nonlinear behavior of the waves on the Sun. A prediction from the 3D solution is that a bright point would begin at the center and grow outward. It would fade the same way and leave behind a bright ring. Another unique feature of calcium bright points in the field-free medium is that the axis of the propagation channel is controlled only by the direction of gravity, and should therefore be vertical, without being inclined as can be the case for waves propagating in magnetic flux tubes.

9.3   Oscillations in the magnetic network

Ground-based observations of the Ca II H and K lines, which are formed in the low chromosphere, show similar emission from network and internetwork regions. While instantaneous bright points from the internetwork may outshine network bright points (see Figure 1 of Lites *et al.* 1993), the long-time average intensity shows total calcium emission from the network to be more important (see Figure 1 of von Uexküll and Kneer 1995). In addition to the higher intensity of the network bright points, their period is longer, ~ 7 minutes (Lites *et al.* 1993, Curdt and Heinzel 1998), and the time variation of their intensity profile is much less peaked.

Space-based observations of UV spectral lines and continua provide important constraints on the structure and dynamics of the chromosphere and chromosphere-corona transition region. Observations with SUMER indicate that network regions are brighter than internetwork regions and show strong oscillatory power only at lower frequencies (Judge, Carlsson and Wilhelm 1997). Transition region lines from the network show persistent redshifts and the line widths indicate the presence of subsonic, unresolved nonthermal Doppler motions of several kilometers per second

(Dere and Mason 1993; Peter 2000, 2001). Furthermore, there is a strong correlation between high intensity and redshift (Hansteen, Betta and Carlsson 2000). Curdt and Heinzel (1998) found evidence for upward propagating waves within the network (also see Heinzel and Curdt 1999). However, the wave modes responsible for these oscillations have not yet been identified.

In this section, we shall focus on dynamical processes occuring in the network and their role in heating the magnetic chromosphere. A complete model must explain the nature and period of the oscillations observed in the network as well as its heating. Furthermore, the model must be compatible with observations.

The magnetic field in the network can be idealized in terms of isolated vertical flux tubes in the photosphere which fan out with height. It is well known that flux tubes support a variety of wave modes. The detailed behavior of these modes for thin flux tubes has been extensively studied (for a recent review see Roberts and Ulmschneider 1998). The modes that we shall be concerned with are the sausage or longitudinal mode (Defouw 1976; Roberts and Webb 1978) and the kink or transverse mode (Ryutov and Ryutova 1976; Parker 1979; Spruit 1982).

The earliest studies on MHD wave excitation were based on extensions of the Lighthill (1952) mechanism (Osterbrock 1961; Musielak and Rosner 1987; Collins 1989, 1992). More recently, Musielak et al. (1989, 1995), Huang, Musielak and Ulmschneider (1995) and Ulmschneider and Musielak (1998) examined the generation of longitudinal and transverse waves in a flux tube through turbulent motions in the convection zone. An alternative scenario, based on observations of granule motions and G-band bright points in the network by Muller and Roudier (1992) and Muller et al. (1994), suggests that transverse waves can be generated through the impulse imparted by granules to magnetic flux tubes (Choudhuri, Auffret and Priest 1993; Choudhuri, Dikpati and Banerjee 1993; Steiner et al. 1998). These investigations suggested that there is sufficient energy flux in MHD waves to account for chromospheric heating.

In this work, we consider in some detail consequences of MHD wave excitation in magnetic flux tubes through the buffeting action of convective motions (granulation) in the surrounding medium. Such waves are likely to play an important role in heating the magnetic chromosphere and also possibly the corona.

Consider a vertical magnetic flux tube extending through the photosphere, which we assume to be "thin" and isothermal. It is convenient again to use the "reduced" displacement, $Q(z, t)$, which for a thin flux tube is related to the physical Lagrangian displacement, $\xi(z, t)$, by $Q(z, t) = \xi(z, t)e^{-z/4H}$.

It can be shown that Q_α ($\alpha = \kappa$ for transverse waves and $\alpha = \lambda$ for longitudinal waves) satisfies a Klein-Gordon equation (Hasan and Kalkofen 1999, henceforth HK), similar to the case of the non-magnetic medium.

$$\frac{\partial^2 Q_\alpha}{\partial z^2} - \frac{1}{c_\alpha^2}\frac{\partial^2 Q_\alpha}{\partial t^2} - k_\alpha^2 Q_\alpha = F_\alpha, \qquad (9.6)$$

where $k_\alpha = \omega_\alpha/c_\alpha$, ω_α is the cutoff frequency for the wave and c_α is the wave propagation speed in the medium and F_α is a forcing function (see HK for details). The speeds for the transverse and longitudinal waves are, respectively,

$$c_\kappa^2 = \frac{2}{\gamma}\,\frac{c_s^2}{1+2\beta}\,,$$

$$c_\lambda^2 = \frac{c_s^2}{1+\gamma\beta/2}\,,$$

where c_s is the sound speed, γ is the ratio of specific heats ($\gamma = 5/3$), $\beta = 8\pi p/B^2$, p is the gas pressure inside the tube and B is the magnitude of the vertical component of the magnetic field on the tube axis.

The cutoff frequencies for transverse and longitudinal waves are, respectively,

$$\omega_\kappa^2 = \frac{g}{8\mathcal{H}}\,\frac{1}{1+2\beta}\,, \tag{9.7}$$

$$\omega_\lambda^2 = \omega_{BV}^2 + \frac{c_\lambda^2}{\mathcal{H}^2}\left(\frac{3}{4}-\frac{1}{\gamma}\right)^2, \tag{9.8}$$

where $\omega_{BV}^2 = g^2\,(\gamma-1)/c_s^2$ is the Brunt-Väisälä frequency.

The solutions of Eq. (9.6) can easily be developed using Green's functions (for details see HK). The generic behavior for the impulsive excitation of transverse and longitudinal waves by granular motions in the magnetic network is the same: the buffeting action due to a single impact excites a pulse that propagates along the flux tube with the kink or longitudinal tube speed. For strong magnetic fields ($\beta < 1$), most of the energy goes into transverse waves, and only a much smaller fraction into longitudinal waves. After the passage of the pulse, the atmosphere gradually relaxes to a state in which it oscillates at the cutoff period of the mode. These results show that the first pulse carries most of the energy and after this pulse has passed the atmosphere oscillates in phase without energy transport. The period observed in the magnetic network is interpreted as the cutoff period of transverse waves, which leads naturally to an oscillation at this period (typically in the 7-minute range) as proposed by Kalkofen (1997).

For weaker magnetic fields the energy fluxes in the two modes are comparable. From the absence of a strong peak at low frequencies in the power spectrum of the cell interior (CI) we conclude that both transverse and longitudinal waves must make a negligible contribution to K_{2v} bright point oscillations. The absence of the magnetic modes then implies that the waves in the CI are probably acoustic waves, and the observed 3 minute period is therefore the acoustic cutoff period – and not the cutoff period of longitudinal flux tube waves. This implies that the magnetic field structure in the CI is likely to be different from that of flux tubes in the magnetic network.

The above analysis has considered the buffeting of flux tubes as a single impact. In reality, we expect the excitation of waves in a tube to occur not as a single impact

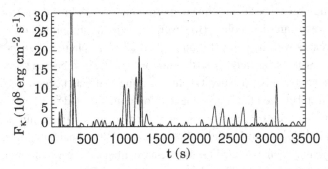

Figure 9.4. Time variation of the vertical energy flux in transverse waves in a single flux tube at $z = 750$ km due to footpoint motions, taken from observations, excited in an isothermal flux tube with $T = 6650$ K, $\beta = 0.3$.

but continually due to the highly turbulent and stochastic motion of granules. It is interesting to examine the consequences of this interaction for chromospheric heating. Such an investigation was carried out by Hasan, Kalkofen and van Ballegooijen (2000, hereafter HKB), who modeled the excitation of waves in the magnetic network due to the observed motions of G-band bright points, which were taken as a proxy for footpoint motions of flux tubes. Using high resolution observations of G band bright points in the magnetic network, the energy flux in transverse waves was calculated in a large number of magnetic elements.

Figure (9.4) shows the vertical energy flux in transverse waves versus time at a height $z = 750$ km for a typical magnetic element in the network. We find that the injection of energy into the chromosphere takes place in brief and intermittent bursts, lasting typically 30 s, separated by longer periods (longer than the time scale for radiative losses in the chromosphere) with lower energy flux. The peak energy flux into the chromosphere is as high as 10^9 erg cm^{-2} s^{-1} in a single flux tube, although the time-averaged flux is $\sim 10^8$ erg cm^{-2} s^{-1}. However, from an observational point of view, such a scenario for heating the magnetic network would yield a high variability with time in Ca II emission, which appears incompatible with observations. A possible remedy to this difficulty would be to postulate the existence of other high-frequency motions (periods 5-50 s) which cannot be detected as proper motions of G-band bright points (HKB). Adding such high-frequency motions to the simulations HKB obtained much better agreement with the persistent emission observed from the magnetic network. For a filling factor of 10% at $z = 750$ km, the predicted flux is $\sim 10^7$ erg cm^{-2} s^{-1}, which is sufficient to balance the observed radiative loss of the chromospheric network (see Model F' of Avrett 1985). Therefore, for transverse waves to provide a viable mechanism for *sustained* chromospheric heating, the main contribution to the heating must come from high-frequency motions, with typical periods 5-50 s. HKB speculated that the high-frequency motions could be due to turbulence in intergranular lanes, but whether the level of turbulence is sufficiently high remains to be investigated.

The above studies were based on a linear approximation, in which the longitudinal and transverse waves are decoupled. However, the velocity amplitude $v(z)$ for the two modes increases with height z (for an isothermal atmosphere $v \propto \exp(z/4\mathcal{H})$, where $\mathcal{H}$ is the pressure scale height), so the motions are likely to become supersonic higher up in the atmosphere. At such heights, nonlinear effects become important, leading to coupling between the transverse and longitudinal modes. Some progress on this question has been made using the nonlinear equations for a thin flux tube (Ulmschneider, Zähringer and Musielak 1991; Huang, Ulmschneider and Musielak 1995). This work has been extended to include a treatment of kink and longitudinal shocks (Zhugzhda, Bromm and Ulmschneider 1995).

Recently, Hasan, Kalkofen and Ulmschneider (2001) carried out preliminary adiabatic calculations of nonlinear kink waves in a thin, isothermal flux tube. The footpoints are impulsively shaken with a transverse velocity of the form $v_x(0, t) = v_0 \exp[-(t - t_0)^2/\tau^2]$, where v_0 is the specified velocity amplitude, t_0 is the time of maximum velocity, and τ is the duration of the impulse (the longitudinal velocity at the base is assumed to be zero). This impulse generates a transverse wave that propagates upwards with a phase speed $c_\kappa \approx 7.9$ km s^{-1}. Figure (9.5a) shows the transverse and longitudinal velocity components as functions of height z at various times. We find that in the photospheric layers, where the transverse velocity amplitude is small compared to the kink wave speed, the longitudinal component of the velocity is negligible. However, as the pulse propagates upward the transverse velocity increases and longitudinal motions are generated due to nonlinear effects when $v_x \sim c_\kappa$. The longitudinal motions, being compressive, steepen into shocks. These results are similar to those found by Hollweg, Jackson and Galloway (1982), who studied the nonlinear coupling of torsional Alfvén waves and longitudinal waves. However, in the present calculations we find wakes following the passage of the initial pulse. Figure (9.5b) shows the velocity as a function of time at a fixed height. Note that at late times the transverse and longitudinal components oscillate with different periods that closely match the cutoff periods (310 s for the kink wave, 230 s for the longitudinal wave). At this stage the velocity amplitudes are small and the two modes are nearly decoupled. This suggests that a power spectrum of network oscillations should detect peaks corresponding to these periods. There is a hint that such peaks may be present in the observations of Lites et al. (1993).

To summarize the main conclusions emerging from the nonlinear calculations: When the transverse velocities are significantly less than the kink wave speed (i.e. the linear regime), there is essentially no excitation of longitudinal waves. However, at heights where $V_x \approx c_\kappa$, longitudinal wave generation becomes efficient, leading to the modes having comparable amplitudes; a large-amplitude transverse pulse generates a longitudinal pulse, which eventually generates wakes that have low amplitudes and represent decoupled longitudinal and transverse waves, oscillating at their respective cutoff periods. We have examined the coupling between the two modes and find that V_z increases quadratically with V_x at low Mach number M

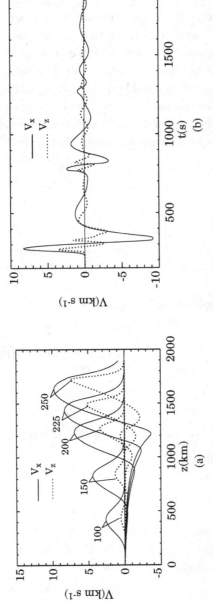

Figure 9.5. Nonlinear coupling of transverse and longitudinal waves in a flux tube: (**a**) Transverse velocity v_x (*solid curves*) and longitudinal velocity v_z (*dashed curves*) as functions of height z at various times, for $t_0 = 50$ s, $\tau = 20$ s, $v_0 = 1.0$ km s^{-1}. The numbers besides the curves denote time in seconds. (**b**) Velocities as function of time t at a fixed height $z = 1800$ km for $v_0 = 0.75$ km s^{-1}.

(with respect to c_k), and linearly with V_x for $M \to 1$. Transverse waves lose energy due to mode coupling. The fractional wave energy in longitudinal motions increases rapidly at first with the forcing transverse velocity V_0, before eventually saturating at a value of about 0.4, which is close to equiparition of energy between the two modes. For a forcing amplitude of $V_0 = 1.5$ km s^{-1}, when there is almost equipartition of energy, the transverse energy flux entering the transition region is approximately 10^7 erg cm^{-2} s^{-1}. This estimate is clearly an upper bound since we need to consider two effects: first, footpoint motions with this velocity occur on average with a frequency of 0.1, and second, there is an attenuation of the flux as it propagates through the transition region, which could lead to a further reduction by a factor of about 10. Hence, we estimate that the net energy flux entering the corona is about 10^5 erg cm^{-2} s^{-1}. Large-amplitude longitudinal waves generated in the upper photosphere steepen and form shocks in the chromosphere. They are likely to be important for chromospheric heating.

References

Avrett E.H. (1985). in *Chromospheric Diagnostics and Modelling*, B. Lites Ed., Sunspot NM.

Bodo, G., Kalkofen,W., Massaglia, S., and Rossi, P. (2000). *Astron. Astrophys.*, **354**, 296.

Bodo, G., Kalkofen,W., Massaglia, S., and Rossi, P. (1991). in *Chromospheric and Coronal Heating Mechanisms*, ed. P. Ulmschneider, E. Priest and R. Rosner, p. 353.

Carlsson, M. (1994). in *Chromospheric Dynamics*, Proc. Mini-Workshop, Inst. Theor. Astroph., Oslo, frontispiece.

Carlsson, M., Judge, Ph., and Wilhelm, K. (1997). *Astrophys. J.*, **486**, L63.

Carlsson, M. and Stein, R. F. (1994). in *Chromospheric Dynamics*, Proc. Mini-Workshop, Inst. Theor. Astroph., Oslo, 47.

Carlsson, M. and Stein, R. F. (1997). *Astrophys. J.*, **481**, 500.

Choudhuri, A. R., Auffret, H., and Priest, E. R. (1993). *Solar Phys.*, **143**, 49.

Choudhuri, A. R., Dikpati, M., and Banerjee, D. (1993). *Astrophys. J.*, **413**, 811.

Collins, W. (1989). *Astrophys. J.*, **337**, 548.

Collins, W. (1992). *Astrophys. J.*, **343**, 499.

Cram, L. and Damé, L. (1983). *Astrophys. J.*, **272**, 355.

Curdt, W., and Heinzel, P. (1998). *Astrophys. J.*, **503**, L95.

Defouw, R. J. (1976). *Astrophys. J.*, **209**, 266.

Dere, K. P. and Mason, H. E. (1993). *Solar Phys.*, **144**, 217.

Fleck, B. and Schmitz, F. (1991). *Astron. Astrophys.*, **250**, 235.

Foing, B. and Bonnet, R. M. (1984). *Astrophys. J.*, **279**, 848.

Grossmann-Doerth, U., Kneer, F., and v. Uexküll, M. (1974). *Solar Phys.*, **37**, 85.

Hansteen, V. H., Betta, R., and Carlsson, M. (2000). *Astron. Astrophys.*, **360**, 742.

Hasan, S. S. and Kalkofen, W. (1999). *Astrophys. J.*, **519**, 899.

Hasan, S. S., Kalkofen, W., and van Ballegooijen, A. A. (2000). *Astrophys. J.*, **535**, L67.

Hasan, S. S., Kalkofen, W., and Ulmschneider, P. (2001). *AGU Spring Meeting*, abstract SH41B-01.

Heinzel, P. and Curdt, W. (1999). in *Third Advances in Solar Physics Euroconference: Magnetic Fields and Oscillations*, eds. B. Schmieder, A. Hofmann and J. Staude, 201.

Hollweg, J. V., Jackson, S., and Galloway, D. (1982). *Solar Phys.*, **75**, 35.

Huang, P., Musielak, Z. E., and Ulmschneider, P. (1995). *Astron. Astrophys.*, **297**, 579.

Judge, P., Carlsson, M., and Wilhelm, K. (1997). *Astrophys. J.*, **490**, L195.

Kalkofen, W. (1997). *Astrophys. J.*, **486**, L148.

Kalkofen, W. (2001). *Astrophys. J.*, **557**, 376.

Kalkofen, W., Rossi, P., Bodo, G., and Massaglia, S. (1994). *Astron. Astrophys.*, **284**, 976.

Kalkofen, W., Ulmschneider, P., and Avrett, E. H. (1999). *Astrophys. J.*, **521**, L141.

Kato, S. (1966). *Astrophys. J.*, **144**, 326.

Lamb, H. (1909). *Proc. London Math. Soc.*, **7**, 122.

Leibacher, J. W. and Stein, R. F. (1981). in: *The Sun as a Star*, NASA SP-450, 263.

Lighthill, M. J. (1952). *Proc. Roy. Soc. Lond.*, A**211**, 564.

Lites, B. W., Rutten, R. J., and Kalkofen, W. (1993). *Astrophys. J.*, **414**, 345.

Liu, S.-Y. (1974). *Astrophys. J.*, **189**, 359.

Muller, R. and Roudier, Th. (1992). *Solar Phys.*, **141**, 27.

Muller, R., Roudier, Th., Vigneau, J., and Auffret, H. (1994). *Astron. Astrophys.*, **283**, 232.

Musielak, Z. E. and Rosner, R. (1987). *Astrophys. J.*, **315**, 371.

Musielak, Z. E., Rosner, R., and Ulmschneider, P. (1989). *Astrophys. J.*, **337**, 470.

Musielak, Z. E., Rosner, R., Gail, H. P., and Ulmschneider, P. (1995). *Astrophys. J.*, **448**, 865.

Osin, A., Volin, S., and Ulmschneider, P. (1999). *Astron. Astrophys.*, **351**, 359.

Osterbrock, D. E. (1961). *Astrophys. J.*, **134**, 347.

Parker, E. N. (1979). *Cosmic Magnetic Fields*, Clarendon Press, Oxford.

Peter, H. (2000). *Astron. Astrophys.*, **360**, 761.

Peter, H. (2001). *Astron. Astrophys.*, **374**, 1108.

Roberts, B. and Webb, A. R. (1978). *Solar Phys.*, **56**, 5.

Roberts, B. and Ulmschneider, P. (1998). in *Lecture Notes in Physics*, Springer Verlag, Heidelberg, Vol. 489, p. 75.

Rutten, R. and Uitenbroek, H. (1991). *Solar Phys.*, **134**, 15.

Ryutov, D. D. and Ryutova, M. P. (1976). *Sov. Phys. JETP*, **43**, 491.

Sivaraman, K.R., Bagare, S.P., and November, L.J. (1990). in *Basis Plasma Processes on the Sun*, Dordrecht: Kluwer, 102.

Skartlien, R., Stein, R. F., and Nordlund, Å (2000). *Astrophys. J.*, **541**, 468.

Spruit, H. C. (1982). *Solar Phys.*, **75**, 3.

Steiner, O., Grossmann-Doerth, U., Knölker, M., and Schüssler, M. (1998). *Astrophys. J.*, **495**, 468.

Sutmann, G. and Ulmschneider, P. (1995). *Astron. Astrophys.*, **294**, 241.

Theurer, J., Ulmschneider, P., and Kalkofen, W. (1997). *Astron. Astrophys.*, **324**, 717.

Ulmschneider, P. and Musielak, Z. E. (1998). *Astron. Astrophys.*, **338**, 311.

Ulmschneider, P., Zähringer, K., and Musielak, Z. E. (1991). *Astron. Astrophys.*, **241**, 625.

von Uexküll, M. and Kneer, F. (1995). *Astron. Astrophys.*, **294**, 252.

Zhugzhda, Y. D., Bromm, V., and Ulmschneider, P. (1995). *Astron. Astrophys.*, **300**, 302.

Heating of the solar chromosphere

P. Ulmschneider

University of Heidelberg, Germany

W. Kalkofen

Harvard-Smithsonian Center for Astrophysics, Cambridge, MA 02138, USA

10.1 Introduction

At the start of a total solar eclipse when the photosphere has just vanished behind the rim of the Moon, the Fraunhofer spectrum changes abruptly from absorption to emission; it is referred to as a *flash spectrum*. After the fading of the weak metal lines, the spectrum is dominated by Balmer lines of hydrogen, the H and K lines of Ca II, and He lines, all emanating from a quickly vanishing layer, called *chromosphere*, which extends over only a few thousand km above the solar limb; the chromosphere is named after the intense red color of the Hα line at 6563 Å. A few seconds later the chromosphere disappears again and the *corona*, which extends to many solar radii, becomes visible.

Grotrian (1939) and Edlén (1941) were surprised to discover that the corona is extremely hot when they identified the red and green coronal emission lines as lines of Fe X and XIV, which are emitted by gas with temperatures of $1 - 2 \cdot 10^6$ K. The existence of intermediate temperatures was shown by the OSO satellites in the 1960's, which observed the full sequence of ions, e.g., from O II to O VI in the UV, indicating regions with temperatures from 10^4 to 10^5 K.

Based on observations with the IUE and Einstein satellites launched in the late 1970's it was found that essentially all late-type stars, which have surface convection zones, have such hot chromospheric layers, where the temperature increases in the outward direction from low photospheric values to about 10^4 K. From the top of the chromosphere (in most stars) the temperature then rises rapidly through a thin region, the so-called *transition layer*, to coronal values. For the Sun, this entire temperature range of the chromosphere and corona is now well observed in the UV and XUV parts of the spectrum with the SOHO satellite, launched in 1995 (Fleck, Domingo and Poland 1995).

The quiet solar chromosphere, i.e., outside active regions, is bifurcated into magnetic network and internetwork chromospheres. The network is characterized by strong magnetic fields, which are organized in tubes of intense magnetic flux, where the field strength reaches 1500 Gauss in the photosphere (Stenflo 1994). The field makes a contribution to the pressure that can exceed that of the gas. These magnetic flux tubes expand exponentially in the upward direction exponentially, with the radius of a cylindrical flux tube growing with an e-folding distance of four times the pressure scale height. The filling factor of the network increases from 1% in the photosphere to 15% in the layers of formation of the emission features in the H and K lines (Foukal, SPD meeting 2000) and to 100% in the so-called *magnetic canopy*. In the internetwork region the magnetic field is dynamically unimportant.

The network is brighter than the internetwork chromosphere. The intensity ratio depends on the criterion and the height where it is measured. A typical value for the average brightness ratio of network to internetwork is 1.27 for Ca II K line emission (Skumanich, Smythe and Frazier 1975). In the region of formation of the Lyman continuum in the upper chromosphere, the Skylab observations indicate a range of variation of the intensity by a factor of nearly 10, corresponding to a temperature variation of about 10% of the ambient value (e.g., Vernazza, Avrett and Loeser 1981).

The temperature structure of the chromosphere is described by empirical models of the brightness seen in chromospheric emission (see Figure 10.1, discussed in the next section). These models do not explicitly separate magnetic from nonmagnetic regions, but one may suppose that the coolest model refers mainly to the nonmagnetic chromosphere in the interior of supergranulation cells and the hottest gives greatest weight to the magnetic network. Although modeling on the basis of brightness distributions does not guarantee that the various models describe physically connected regions, the deduction that the temperature profile implies a heating mechanism provides support for these models as physical models (see the section on Wave Heating).

A deduction from the similarity of the temperature structures (see Figure 10.3 for models A - F of Fontenla, Avrett and Loeser 1993) is that the heating mechanisms in the magnetic and nonmagnetic media is essentially the same, i.e., the dissipation has properties that appear to be independent of the magnetic field, although this leaves open the question of the mode of energy transport to the point of dissipation.

10.2   Empirical chromosphere models

The structure of the quiet chromosphere can be described by empirical models that match predictions against observations of radiation from the Sun (Vernazza, Avrett

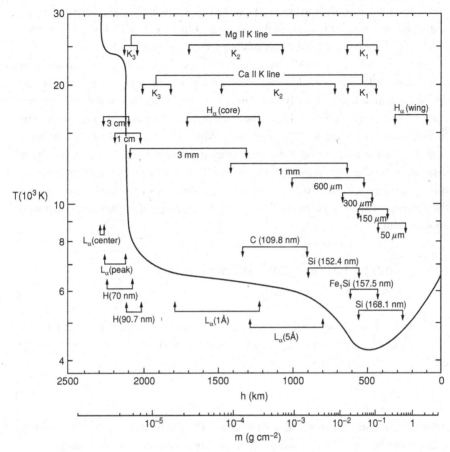

Figure 10.1. The average model of the quiet solar chromosphere, VAL81-C, from Vernazza *et al.* (1981).

and Loeser 1981, Avrett 1985; Fontenla, Avrett and Loeser 1993). For the construction, the temperature is adjusted until a satisfactory fit is achieved between observed and predicted radiation fields. The pressure distribution is then determined from the hydrostatic equilibrium equation, which also yields density and height, resulting in a model that gives temperature and pressure as functions of height. Because of the combination of empirical adjustments and the use of an equation such models are referred to as semiempirical. Figure 10.1 shows the average model of the quiet chromosphere, labeled VAL81-C, by Vernazza *et al.* (1981). From the determined temperature and pressure distribution using a multilevel atom radiative transfer code the radiation losses due to the lines and continua have been computed resulting in a total loss of $4.6 \cdot 10^6$ erg cm^{-2} s^{-1} (mainly by Ca II H + K + IRT and Mg II h + k lines), which needs to be balanced by mechanical heating.

The variation of the chromospheric emission over the solar surface was described in six brightness models by Vernazza *et al.* (1981). The deviation from the coolest to the hottest model from the average one, model C, is about 7% at the height of 1 Mm; and, at heights where the fractional ionization of hydrogen is 10%, the temperature departs by maximally 4% from that of the average model.

Andersen and Athay (1989) analyzed the temperature structure of model C and found that Fe II lines are additional strong emitters which raise the total chromospheric radiation loss to $1.4 \cdot 10^7$ erg cm^{-2} s^{-1}. In addition the cooling rate per unit mass increases from a low value in the photosphere to practically a constant value in the temperature plateau region between 1 Mm and 2 Mm. Since heat conduction is negligible the energy that is radiated away must be delivered by a transport and dissipation mechanism to the point of emission. Consequently, the heating mechanism must have a dissipation rate per gram that increases from negligible values in the photosphere to a constant and high value in the region of the temperature plateau above 1 Mm.

10.3   Energy balance and the necessity of mechanical heating

To understand the chromospheric energy balance consider a gas element. An amount of heat dQ (erg cm^{-3}) flowing into the element across its boundaries raises the entropy S (erg g^{-1}K^{-1}) by

$$dS = \frac{dQ}{\rho T} ,$$ (10.1)

where T is the temperature and ρ the density. The thermodynamic variable entropy is used here because it has a bookkeeping quality for monitoring the energy flows in and out of the gas element. In adiabatic cases no energy flows occur, while rising S indicates that energy enters into the element. For a gas element moving with wind velocity v through the chromosphere one can write an entropy conservation equation (in the Lagrange frame)

$$\frac{dS}{dt} = \frac{\partial S}{\partial t} + v\frac{\partial S}{\partial z} = \frac{dS}{dt}\bigg|_R + \frac{dS}{dt}\bigg|_J + \frac{dS}{dt}\bigg|_C + \frac{dS}{dt}\bigg|_V + \frac{dS}{dt}\bigg|_M .$$ (10.2)

Here t is time and z is height in a plane-parallel atmosphere. The five terms on the right hand side are called radiative, Joule, thermal conductive, viscous and mechanical heating. Mechanical heating denotes all processes which convert nonradiative, nonconductive, hydrodynamic or magnetic energy propagating through the gas element into heat (microscopic random thermal motion).

Let us estimate the rough sizes of the individual heating terms in Eq. (10.2). Consider a typical acoustic or magnetohydrodynamic (MHD) disturbance in the solar chromosphere with characteristic perturbations of size $L = 200$ km, temperature $\Delta T = 1000$ K, velocity $\Delta v = 3$ km/s and magnetic field $\Delta B = 10$ G. Using appropriate values for the thermal conductivity $\kappa_{th} = 10^5\,erg\,cm^{-1}s^{-1}K^{-1}$, viscosity

$\eta_{vis} = 5 \cdot 10^{-4} \, dyn \, s \, cm^{-2}$ and electrical conductivity $\lambda_{el} = 2 \cdot 10^{10} \, s^{-1}$ we find for the *thermal conductive heating rate* $(erg \, cm^{-3}s^{-1})$

$$\Phi_C = \rho T \left.\frac{dS}{dt}\right|_C = \frac{d}{dz}\kappa_{th}\frac{dT}{dz} \approx \frac{\kappa_{th}\Delta T}{L^2} \approx 3 \cdot 10^{-7} \quad , \tag{10.3}$$

the *viscous heating rate*

$$\Phi_V = \rho T \left.\frac{dS}{dt}\right|_V = \eta_{vis}\left(\frac{dv}{dz}\right)^2 \approx \frac{\eta_{vis}\Delta v^2}{L^2} \approx 1 \cdot 10^{-7} \quad , \tag{10.4}$$

and the *Joule heating rate*

$$\Phi_J = \rho T \left.\frac{dS}{dt}\right|_J = \frac{j^2}{\lambda_{el}} = \frac{c_L^2}{16\pi^2\lambda_{el}}(\nabla \times B)^2 \approx \frac{c_L^2\Delta B^2}{16\pi^2\lambda_{el}L^2} \approx 7 \cdot 10^{-5} \quad . \tag{10.5}$$

Here j is the current density and c_L the light velocity. For the *radiative heating rate* we have in the gray case

$$\Phi_R = \rho T \left.\frac{dS}{dt}\right|_R = 4\pi\bar{\kappa}(J - B) \tag{10.6}$$

where J is the frequency-integrated mean intensity, B the frequency-integrated Planck function and $\bar{\kappa}$ the gray Rosseland opacity, mainly due to the H^- ion, which can be approximated (with p the gas pressure) by

$$\frac{\bar{\kappa}}{\rho} = 1.38 \cdot 10^{-23} p^{0.738} T^5 \quad cm^2/g \quad . \tag{10.7}$$

The heating rates (10.3) to (10.5) show that normally these processes are insufficient to balance the empirical chromospheric cooling rate of $-\Phi_R = 10^{-1} erg \, cm^{-3} \, s^{-1}$, determined from VAL81-C.

As the chromosphere exists on the Sun for billions of years, there should be steady state, and the left hand side of Eq. (10.2) must be zero. This can be seen from computing the entropy gradient $\partial S/\partial z \approx g/T$, where g is the solar gravity and v

Table 10.1. *Temperature T, density ρ, sound speed c_S, wind velocity v and molecular mean free path l in the VAL81-C model as a function of height z*

z (km)	T (K)	ρ (g/cm³)	c_S (cm/s)	v (cm/s)	l (cm)
0	6400	$2.7 \cdot 10^{-7}$	$8.3 \cdot 10^5$	$4.0 \cdot 10^{-5}$	$2.9 \cdot 10^{-2}$
500	4200	$6.0 \cdot 10^{-9}$	$6.7 \cdot 10^5$	$1.8 \cdot 10^{-3}$	1.1
1000	5900	$7.5 \cdot 10^{-11}$	$7.9 \cdot 10^5$	0.28	$1.1 \cdot 10^2$
1500	6400	$2.5 \cdot 10^{-12}$	$8.3 \cdot 10^5$	4.4	$3.3 \cdot 10^3$
2100	9200	$1.2 \cdot 10^{-13}$	$1.5 \cdot 10^6$	92	$1.0 \cdot 10^3$
2543	447000	$2.3 \cdot 10^{-15}$	$1.0 \cdot 10^7$	4800	$1.9 \cdot 10^6$

the solar wind flow speed, which from the solar mass loss rate of $\dot{M} = 10^{-14} M_\odot/y$ can be derived using $v = \dot{M}/(4\pi\rho R_\odot^2) \approx 1.1 \cdot 10^{-11}/\rho$ of which typical values are shown in Tab. 10.1. In this table also the mean free path $l \approx 1/(3 \cdot 10^{-16} n_H)$ is given, valid for non-ionized gas, with n_H the hydrogen number density.

In the chromosphere we thus find

$$\frac{dS}{dt}\bigg|_R + \frac{dS}{dt}\bigg|_M = 0 \ , \tag{10.8}$$

which can be written

$$\frac{4\pi\overline{\kappa}}{\rho T}(J - B) + \frac{dS}{dt}\bigg|_M = 0 \ . \tag{10.9}$$

Conclusion: *In stellar chromospheres the main energy balance is between radiation and mechanical heating.*

In the special case of *radiative equilibrium* where there is no mechanical heating and the energy generated within the Sun is transported purely by a steady flow of radiation, one has $dS/dt|_M = 0$ and obtains $J = B$ (the absorbed radiative energy $\overline{\kappa}J$ is equal to the emitted radiative energy $\overline{\kappa}B$). With $J = \sigma T_{eff}^4/2\pi$, valid for the surface of a star, roughly represented by a black body of effective temperature T_{eff} (the factor 1/2 comes from the fact that at the solar surface there is only radiation going away from the star), and $B = \sigma T^4/\pi$, where σ is the Stefan-Boltzmann constant, one gets from Eq. (10.9) for the outer stellar layers the boundary temperature $T_b = 2^{-1/4} T_{eff} \approx 0.8\, T_{eff}$ For the Sun one has $T_{eff} = 5770\,$K and $T_b = 4600\,$K. This shows that in absence of mechanical heating we would expect to have temperatures of the order of the boundary temperature in the regions above the stellar surface.

However, as a chromosphere is a layer where the temperature is observed to rise in outward direction to values $T \gg T_{eff}$, it is clear that one must have $B \gg J$ and from Eq. (10.9) therefore $dS/dt|_M \gg 0$. This shows that *for chromospheres, mechanical heating is essential.* As for the transition layer and corona conductive and wind losses become important in addition to the radiative losses, and because these combined losses cannot be balanced by thermal conduction from a reservoir at infinity, but must ultimately be supplied from the stellar interior, there is an even stronger conclusion: that *for the existence of chromospheres and coronae, mechanical heating is essential.*

Moreover, chromospheres and coronae can only be maintained if mechanical heating is supplied *without interruption.* The time scale, in which the excess chromospheric temperature will cool down to the boundary temperature if the mechanical heating were suddenly disrupted, is given by the *radiative relaxation time* for which we have

$$t_{Rad} = \frac{\Delta E}{-\Phi_R} = \frac{\rho c_v \Delta T}{16\overline{\kappa}\sigma T^3 \Delta T} = \frac{\rho c_v}{16\overline{\kappa}\sigma T^3} \approx 1.1 \cdot 10^3\, s \ . \tag{10.10}$$

Here ΔE is the heat content of the gas element, c_v the specific heat, and typical chromospheric values have been used. It is seen that in timescales of a fraction of an hour the chromosphere would cool down to the boundary temperature if mechanical heating would suddenly stop.

How can mechanical heating processes be visualized? From Eqs. (10.3) to (10.5) it is seen that these terms would only become significant compared to the observed radiative cooling rate $-\Phi_R$, if the length scale L is considerably decreased. For acoustic waves as well as longitudinal MHD tube waves, this is accomplished by *shock formation*, where the temperature jump occurs over the very short molecular mean free path l (see Tab. 10.1). For magnetic cases, by the formation of *current sheets*, where the magnetic field jumps occur over very small distances. In summary one can conclude that shock waves and current sheets are efficient mechanical heating mechanisms. For the corona (and possibly in the magnetic chromosphere at high altitude) additional heating mechanisms, like e.g. the dissipation of high-frequency gyrokinetic plasma waves operate (Marsch, Vocks and Tu 2001).

10.4   Overview of the heating mechanisms

Table 10.2 gives an overview of the mechanisms which are thought to provide a steady supply of mechanical energy to balance the chromospheric and coronal losses (for an extensive review of heating mechanisms see e.g. Narain and Ulmschneider 1996). The term *heating mechanism* comprises three physical aspects, the *generation* of a carrier of mechanical energy, the *transport* of mechanical energy into the chromosphere and corona and the *dissipation* of this energy.

Table 10.2 shows the various proposed energy carriers which can be classified into two main categories as *hydrodynamic* and *magnetic* heating mechanisms. Both the hydrodynamic and the magnetic mechanisms can be subdivided further according to frequency. Acoustic waves are high frequency hydrodynamic fluctuations with periods $P < P_A$ and pulsational waves have periods $P > P_A$. Here P_A is the acoustic cut-off period

$$P_A = \frac{4\pi c_S}{\gamma g} \quad , \tag{10.11}$$

where c_S is the sound speed, g the gravity and $\gamma = 5/3$ the ratio of specific heats. A typical value for the acoustic cut-off period for the Sun ($g = 2.74 \cdot 10^4 \ cm/s^2$, $c_S = 7$ km/s) is $P_A \approx 190 \ s$.

The magnetic mechanisms are subdivided into high frequency wave- or *AC (alternating current)-mechanisms* and current sheet- or *DC (direct current)-mechanisms* where one has time variations of low frequency. Also in Tab. 10.2 the mode of dissipation of these mechanical energy carriers is indicated.

Ultimately the mechanical energy carriers derive their energy from the nuclear processes in the stellar core from where the energy is transported by radiation and convection to the stellar surface. In late-type stars the mechanical energy generation

Table 10.2. *Proposed chromospheric heating mechanisms*

energy carrier	dissipation mechanism
hydrodynamic heating mechanisms	
acoustic waves, $P<P_A$	shock dissipation
pulsational waves, $P>P_A$	shock dissipation
magnetic heating mechanisms	
1. alternating current (AC) or wave mechanisms	
slow mode mhd waves, longitudinal mhd tube waves	shock dissipation
fast mode mhd waves	Landau damping
Alfvén waves (transverse, torsional)	mode-coupling resonance heating compressional viscous heating turbulent heating Landau damping
magnetoacoustic surface waves	mode-coupling phase-mixing resonant absorption
2. direct current (DC) mechanisms	
current sheets	reconnection (turbulent heating, wave heating)

arises from the gas motions of the surface convection zones. These gas motions are largest in the regions of smallest density near the top boundary of the convection zone. Due to this, the mechanical energy carriers, particularly the acoustic and MHD waves, are generated in a narrow surface layer.

10.5   Search for the important heating mechanisms

Although the heating mechanisms listed in Tab. 10.2 are known to work in terrestrial applications, the question is, which of these will operate in the solar or stellar situations and particularly, what are the major mechanisms which generate chromospheres. Because the Sun is so close to Earth compared to other stars one might think that this question can be decided by solar observations. Actually we are extremely lucky when we resolve spatial structures of 0.1 arc sec on the Sun, that is, linear scales of 70 km which is much bigger than the shock structure which is of the size of a molecular mean free path l (see Tab. 10.1) or the current sheet thickness which have sizes in the m range.

For many years it was therefore not possible to decide whether the main heating mechanism of the solar chromosphere is acoustic waves or rather an unresolved magnetic mechanism. It is very surprising that only by observing distant stars, which are mere point sources on the celestial sphere, could this question be definitively decided. The reason for this is, that stars differ in four basic parameters from the Sun: metallicity Z_M, effective temperature T_{eff}, gravity g and rotation period P_{Rot}, and that these differences have a large influence on the acoustic and magnetic wave generation.

a. Acoustic heating

All late-type stars have surface convection zones, and every turbulent flow field generates acoustic waves. Propagating down the steep density gradient (see Tab. 10.1) of the stellar surface layers the acoustic waves suffer rapid amplitude growth due to wave energy flux conservation. With the velocity amplitude v the wave energy flux is given by $F_A = \rho v^2 c_{Ph}$, where c_{Ph} is the propagation speed which is close to the sound speed, c_S. Despite of strong radiation damping an amplitude growth $v \propto e^{z/2H}$ occurs due to the exponential density decrease. As the wave crests of large amplitude waves move faster than the wave troughs, shocks form which develop a sawtooth shape and typically have velocity amplitudes $v < c_S$. The shocks convert the wave energy into heat. This is called *acoustic heating theory*. The heating comes from the entropy jump ΔS at the shock front,

$$\Phi_M = \rho T \left. \frac{dS}{dt} \right|_M = \frac{\rho T}{P} \Delta S , \qquad (10.12)$$

where P is the period of the sawtooth wave into which the acoustic wave develops.

Figure 10.2 shows a radiatively damped monochromatic acoustic wave with a period $P = 45\ s$ and an initial acoustic flux $F_A = 2.0 \cdot 10^8\ erg\ cm^{-2}s^{-1}$ in the solar

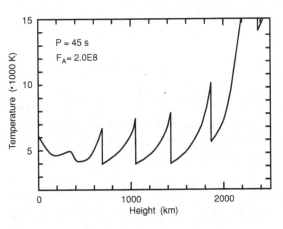

Figure 10.2. Temperature distribution for a monochromatic acoustic wave with period $P = 45\ s$ and initial acoustic flux $F_A = 2.0 \cdot 10^8\ erg\ cm^{-2}s^{-1}$ in the solar atmosphere.

atmosphere. It is seen that the wave amplitude $\Delta T \sim \Delta v$ grows rapidly with height. But the actual growth is not so rapid because of radiation damping, which removes wave energy (and thus decreases the amplitude) by radiation losses. The amplitude growth at about 500 km causes shock formation and the wave subsequently grows into a sawtooth wave. Here the shock-jump dissipates the wave energy according to Eq. (10.12). It is seen that the shock-jumps become similar in size, independent of height. This is the so-called '*limiting shock strength*' behaviour which is common for monochromatic waves in gravitational atmospheres. The magnitude of this temperature jump depends only on the wave period P and it can be shown that the dissipation of such waves can balance the observed chromospheric radiation losses (Ulmschneider 1990, 1991).

When in a time-dependent situation a shock moves through a gas element, the entropy jump ΔS introduced to the gas by the wave gets subsequently radiated away until the next shock comes along. If the next shock arrives before the gained thermal energy has been completely radiated away, then the gas element gets hotter after each shock. This rise of the temperature in the gas element continues until a stage is reached, where the entropy jump ΔS is completely radiated away at the moment where the next shock arrives. This state is called *dynamical equilibrium* and in this stage the mean chromospheric temperature distribution becomes time-independent. Together with the attainment of a high mean chromospheric temperature, the distribution of mass (the density) adjusts to the new scale height and to the support by the wave pressure imparted by the shocks (Gail, Cuntz and Ulmschneider 1990).

Note that monochromatic acoustic waves were chosen for the purpose of demonstrating the limiting strength behavior of shocks. In reality, the waves heating the chromosphere comprise a wide spectrum that stretches from the acoustic cutoff frequency (5 mHz) to large values (>60 mHz); the upper limit is set by radiation damping in the layers in the convection zone where the waves are generated.

Lighthill (1952) and Proudman (1952) have shown in terrestrial applications, that for free turbulence (away from any solid boundaries) one has quadrupole sound generation, in which the generated sound depends on the eighth power of the turbulent velocity. These authors have derived the expression

$$F_A = \int 38 \frac{\rho u^8}{c_S^5 H} dz \ , \tag{10.13}$$

called *Lighthill* or *Lighthill-Proudmann formula*, where $H = \Re T / \mu g$ is the scale height. Here u is the convective velocity (obtained for stars using a convection zone code), $\Re$ is the gas constant and μ is the the mean molecular weight, with $\mu = 1.3$ for the unionized solar gas. This Lighthill-formula has been well tested and confirmed in terrestrial applications.

Based on convection zone models and using the so-called Lighthill-Stein theory (see Musielak *et al.* 1994), Ulmschneider, Theurer and Musielak (1996) have

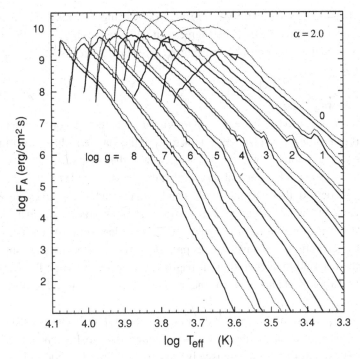

Figure 10.3. Acoustic fluxes F_A (solid) for stars versus T_{eff} for given $\log g$ and mixing length parameter $\alpha = 2.0$. Also shown (dotted) are acoustic fluxes derived from the Lighthill-formula (10.13).

computed acoustic energy fluxes F_A ($erg\ cm^{-2}s^{-1}$) for a wide range of late-type stars as shown in Figure 10.3. This figure shows that because of the u^8-dependence, F_A varies greatly with T_{eff} and g and that Eq. (10.13) despite of its simplicity gives quite reasonable results. That F_A also varies strongly with metallicity Z_M, particularly for cool stars where the opacity is not dominated by atomic hydrogen, has been shown by Ulmschneider *et al.* (1999). Yet F_A is independent of the rotation period P_{Rot} because the convection zones do not depend on rotation.

These acoustic fluxes and the wave periods derived from the acoustic spectra provided by the Lighthill-Stein theory can now be used to compute the propagation of the waves into the outer stellar atmosphere. This way theoretical chromosphere models (Figure 10.2) of non-magnetic stars or stellar regions are created. On basis of these models it is possible to compute the emission cores of the main chromospheric lines, the Ca II H and K as well as Mg II h and k lines, and evaluate the total theoretical Ca II and Mg II emission fluxes in these lines. Buchholz *et al.* (1998) were the first to show that *these theoretical line fluxes agreed quite well with the lower boundary of the observed emission fluxes in these lines, called basal flux line*. This basal flux line is indicated in Figure 10.4 by the dashed curve and fits the data of the stars with minimum emission flux.

That calculations, *starting from first principles* by using only the three parameters Z_M, T_{eff} and g, produced such a nice agreement with the purely observational basal flux line showed that *the acoustic wave heating must be a basic mechanism of stellar chromospheres.*

b. Magnetic wave heating

Let us now turn to the magnetic regions of the chromosphere. Solar observations show that appart from sunspots and plage regions most of the magnetic flux occurs in small flux tubes with a diameter of about a scale height at the solar surface (e.g., Stenflo 1978, Solanki 1993). It has been observed, that the turbulent flows of the convection zone strongly perturb these magnetic tubes and generate magnetic waves. Squeezing the tubes generates *longitudinal mhd tube waves*, shaking produces *transverse tube waves* and twisting *torsional tube waves*. If the turbulent flows of the convection zone by the generation of acoustic waves provide a basic heating mechanism for the chromosphere, could the generation of magnetic waves by the same flows not give us a basic heating mechanism for magnetic regions? Here again stellar observation give the strongest evidence.

Vaughan and Preston (1980), Noyes *et al.* (1984) and others discovered that the greater the observed emission flux deviates from the basal flux line, the more rapidly the star rotates, and the more extensively it is covered by magnetic fields. Unfortunately a solar and stellar *dynamo theory* which would allow to predict the magnetic field coverage upon specification of the four basic parameters is presently not available. But by replacing the fourth parameter P_{Rot} by an assumed *filling factor of the magnetic field at the stellar surface* f, the heating of magnetic chromospheres can be studied independently of the dynamo theory. The filling factor is the ratio of the surface area covered by magnetic fields to the total surface area.

We assume that stars are fully covered with a forest of magnetic flux tubes, similar to those on the Sun, which at the stellar surface have a diameter equal to the scale height and a magnetic field strength $B_0 = 0.85 B_{eq}$ where $B_{eq} = \sqrt{8\pi p_0}$ and p_0 is the gas pressure outside the tube. The cross-section of these tubes increases with height and at the canopy height fill out the entire available space. Ulmschneider, Musielak and Fawzy (2001) as well as Musielak and Ulmschneider (2001, 2002) have computed longitudinal and transverse wave energy fluxes for a large number of late-type stars on basis of such tubes.

For different filling factors f, stars with different magnetic flux tube forests can be constructed. The magnetic wave fluxes and wave periods can be used to compute the propagation of longitudinal waves along these flux tubes. Here the transverse wave flux is assumed to augment the longitudinal wave flux by mode-coupling while the atmosphere outside the flux tubes is heated by acoustic waves. Again the core emission fluxes of the Ca II and Mg II lines emerging from the stars covered by a flux tube forest can be calculated and compared with observations. For this we evaluate the radiative transfer along many ray paths through the tube forest taking into account contributions from magnetically and acoustically heated regions. Figures 10.4 show results by Fawzy *et al.* (2002a, b). It is seen that by varying the filling factor from the case f = 0 for pure acoustic heating to f = 0.4 (mainly magnetic wave heating) appears to account for the observed variation of the chromospheric emission in Ca II, and most of the variation in Mg II. Yet for the Mg II emission, which occurs at greater height in the chromosphere than Ca II, there seems to be a gap between our results for f = 0.4 and the largest observed emission. This indicates that although magnetic wave heating appears to be the main heating mechanism for most of the magnetic chromosphere, at the highest chromosphere an additional non-wave heating mechanism, presumably reconnective microflare heating appears to operate (Fawzy *et al.* 2002b).

10.6   Summary and outlook

The basic physical process which generates the solar chromosphere and that of other late-type stars is mechanical heating of which the main heating processes can now be identified. The base of the chromosphere is heated by pure acoustic shock waves, the lower and middle chromosphere with an increasing importance of the magnetic flux tubes are heated by longitudinal magnetohydrodynamic shock waves propagating in the vertically oriented flux tubes which spread with height and eventually fill out all the available space. In the high chromosphere one very likely has an additional non-wave magnetic heating mechanism, possibly (reconnective) microflare heating.

Continuous heating by a wide spectrum of acoustic waves as well as intermittent heating by isolated shocks are permitted, but the observed cooling requires sustained continuous heating. This excludes the model proposed by Carlsson and Stein (1994), where a shock heats the chromosphere once every three minutes provided there is a wave, and does not heat for the longer periods implied by the filling factor of 50% (Carlsson, Judge and Wilhelm 1997) of the oscillating part of the chromosphere (Kalkofen, Ulmschneider and Avrett 1999, Kalkofen 2001).

There remains the question of *what limits the chromospheres at the bottom and at the top*. The lower boundary is the temperature minimum. In cool stars, shock formation occurs at these heights and initiates the chromospheric temperature rise. For hot stars the growth of the generated shocks is delayed by strong radiation damping and the chromospheric temperature rise commences only after the zone of strong damping has been passed. The top boundary of the chromosphere occurs, when the main radiative cooling mechanisms (Lyα, Ly continuum and Mg II h + k cooling) at the high temperature end of the chromosphere cease to work because of complete ionization. In this case the heating becomes unbalanced and the temperature quickly rises up to coronal values.

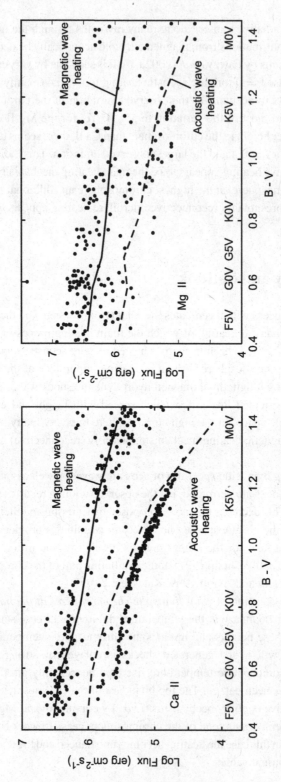

Figure 10.4. Total observed Ca II H + K (left)and Mg II h + k line core emission fluxes of minimum flux stars and active dwarfs are shown as dots, compared with theoretical fluxes for pure acoustic wave heating and for magnetic wave heating in flux tubes, with an area filling factor of f = 0.4, after Fawzy *et al.* (2002b).

References

Anderson, L. S. and Athay, R. G. (1989). *Astrophys. J.* **336**, 1089.

Avrett, A. H. (1985). in *Chromospheric Diagnostics and Modelling*, ed. B. W. Lites, Sacramento Peak, 67.

Buchholz, B., Ulmschneider, P., and Cuntz, M. (1998). *Astrophys. J.*, **494**, 700.

Carlsson, M., Judge, P. G., and Wilhelm, K. (1997). *Astrophys. J.*, **486**, L63.

Carlsson, M. and Stein, R. F. (1994). in *Chromospheric Dynamics*, Proc. Mini-Workshop, Inst. Theor. Astroph., Oslo, 47.

Edlén B. (1941). *Ark. mat., astr. och fys.*, **28 B**, No.1.

Fawzy D., Rammacher W., Ulmschneider P., Musielak Z.E., and Stępień K. (2002a). *Astron. Astrophys.*, **386**, 971.

Fawzy D., Ulmschneider P., Stępień K., Musielak Z.E., and Rammacher, W. (2002b). *Astron. Astrophys.*, **386**, 983.

Fleck, B., Domingo, V., and Poland, A. I. (1995). *The SOHO Mission*, Kluwer Academic Publ., Boston.

Fontenla, J. M., Avrett, E. H., and Loeser, R. (1993). *Astrophys. J.*, **406**, 319.

Gail H.-P., Cuntz M., and Ulmschneider P. (1990). *Astron. Astrophys.*, **234**, 359.

Grotrian W. (1939). *Naturwiss.*, **27**, 214.

Kalkofen, W. (2001). *Astrophys. J.*, **557**, 376.

Kalkofen, W., Ulmschneider, P., and Avrett, E. H. (1999). *Astrophys. J.*, **521**, L141.

Lighthill, M. J. (1952). *Proc. Roy. Soc. London*, **A211**, 564.

Marsch E., Vocks C., and Tu C.-Y. (2001). *Nonlin. Proc. in Geophys.*, 10, 1.

Musielak Z.E., Rosner R., Stein, R.F., and Ulmschneider P. (1994). *Astrophys. J.*, **423**, 474.

Musielak Z.E. and Ulmschneider P. (2001). *Astron. Astrophys.*, **370**, 541.

Musielak Z.E., and Ulmschneider P. (2002). *Astron. Astrophys.*, **386**, 606.

Narain U. and Ulmschneider P. (1996). *Space Sci. Rev.*, **75**, 453.

Noyes R.W., Hartmann L.W., Baliunas S.L., Duncan D.K., and Vaughan A.H. (1984). *Astrophys. J.*, **279**, 763.

Proudman I. (1952). *Proc. Roy. Soc. Lond.*, **A214**, 119.

Skumanich, A., Smythe, C., and Frazier, E.N. (1975). *Astrophys. J.*, **200**, 747.

Solanki S.K. (1993). *Space Sci. Rev.*, **63**, 1.

Stenflo, J.O. (1978). *Rep. Prog. Phys.*, **41**, 865.

Stenflo, J.O. (1994). *Solar Magnetic Fields*, Kluwer, Dordrecht.

Ulmschneider P. (1990). in *Cool Stars, Stellar Systems and the Sun*, Astr. Soc. Pacific Conf. Ser. 9, ed. G. Wallerstein, p. 3.

Ulmschneider, P. (1991). in *Mechanisms of Chromospheric and Coronal Heating*, ed. P. Ulmschneider, E. Priest, and R. Rosner, Springer Verlag, Berlin, 328.

Ulmschneider P., Musielak Z.E., and Fawzy D.E. (2001). *Astron. Astrophys.*, **374**, 662.

Ulmschneider P., Theurer J., and Musielak Z.E. (1996). *Astron. Astrophys.*, **315**, 212.

Ulmschneider P., Theurer J., Musielak Z.E., and Kurucz R. (1999). *Astron. Astrophys.*, **347**, 243.

Vaughan A.H. and Preston G.W. (1980). *PASP*, **92**, 385.

Vernazza J.E., Avrett E.H., and Loeser R. (1981). *Astrophys. J.*, Suppl. **45**, 635.

11

The solar transition region

O. Kjeldseth-Moe

Institute of Theoretical Astrophysics
University of Oslo, N-0315, Oslo, Norway

11.1   Introduction

The transition region may be regarded as the thermal interval between the dense and cold chromosphere and the hot and tenuous corona. New observational evidence reveals a likely new concept for the transition region as a finely structured and dynamic plasma confined inside thin strands of a filamentary magnetic field. The plasma is moving at high velocities and we find rapid time variations of all gas parameters. This differs from the old concept of a stratified and static region of very limited vertical extent located between temperatures of 3×10^4 K and slightly less than 1 MK.

The mass of the solar plasma in the transition region is small and is therefore easily disturbed. When it is pushed it moves, so to speak. The region is also well defined and spatially localized. This means that spectroscopic signatures of the energy and momentum transfer processes in the solar atmosphere are easy to detect in the transition region, where they may be "amplified" compared to the response in the more complex and extended chromosphere or corona. Thus, the transition region carry easily observable information on phenomena occurring at both higher and lower temperatures.

We shall start by describing the observations and analysis that led to the original concept of the very thin transition layer between the chromosphere and corona. We then review some of the early evidence leading to a concept of an inhomogeneous and highly fine structured transition region. Particularly we deal with results from observations with the High Resolution Telescope and Spectrograph, HRTS.

Rapid time changes in the transition region emission were noted already in the 1970's and 80's from data obtained with the spectrometers on Skylab. HRTS and the Ultraviolet Spectrometer and Polarimeter, UVSP, on OSO 8 confirmed these changes and gave evidence for the presence of high plasma velocities as well. The observations with Skylab, HRTS and UVSP were, however, isolated examples. At that early stage it was not obvious that the general state of the transition region was as dynamic as has proven to be the case.

The full picture of the transition region as a highly dynamic and time variable plasma came with the observations from SOHO, the Solar and Heliospheric Observatory. Particularly important are the observations from the two spectrometers, the Coronal Diagnostic Spectrometer, CDS, and the Solar Ultraviolet Measurements of Emitted Radiation, SUMER, and with the EUV Imaging Telescope, EIT. Later, the Transition Region and Coronal Explorer, TRACE, has contributed an even richer harvest of images in the UV that demonstrates a strongly time variable and fine structured transition region and corona.

11.2   Emission from the transition region plasma

11.2.1   The emitted intensity

The original concept of the transition region sprang from analyses of observed intensities of EUV emission lines employing emission measures. The method was developed by Pottasch (1964) and first applied by him and others, e.g. Athay (1966), Dupree and Goldberg (1967) and Dupree (1972), to find coronal abundances of elements. While the approach may be considered outdated, it introduced concepts, properties and problem approaches that are still in use and still with us.

The starting point is the expression

$$\epsilon(\lambda_{ij}) = (hc/\lambda_{ij})A_{ji}N_j \qquad (11.1)$$

for the emitted power, ϵ, per unit volume in a spectral line with wavelength λ_{ij}. Indices i and j designate the lower and upper atomic levels, A_{ji} is the Einstein transition rate and N_j the population density of the upper level.

The specific intensity, I, of a spectral line from an area on the Sun is then obtained by integrating equation (11.1) through the solar atmosphere along the line of sight, obtaining the specific intensity by dividing with $4\pi \cdot \delta S$, where δS is the surface element of the emitting volume. The result may be written as

$$I = 1.14\,10^{16}\,A_X \overline{g} f \int_T R(T)e^{-\Delta E_{ij}/kT}\,\frac{A_f}{\cos\gamma}\,P_e\,P_H\,T^{-\frac{5}{2}}|\nabla T|^{-1}\,dT. \quad (11.2)$$

Here A_X is the abundance of element X relative to hydrogen, while $\overline{g}$ and f are, respectively, the effective Gaunt factor and the oscillator strength for the line transition. We treat these as constants not varying with temperature or location in the transition region. $R(T)$ is the ionization function, $N(X^{+m})/N(X)$, for ion m of element X and is assumed to be a function of temperature only. ΔE_{ij} is the energy difference between the upper and lower levels of the transition in the atom emitting the line and is related to the line wavelength through the usual expression, $\Delta E_{ij} = hc/\lambda_{ij}$. T is the electron temperature while P_e and P_H are the partial pressures of electrons and hydrogen nuclei, respectively. These pressures derive from the corresponding densities using the ideal gas equation. Electron and proton densities enter in expressions for the collisional excitation rate and the total particle density. The quantity γ is the angle between the temperature gradient, ∇T, and the line of sight, and A_f is the filling factor for the emission. This factor takes into account that the entire volume under observation may not be filled with gas emitting the particular line. The numerical factor in front is a product of various physical constants.

Equation (11.2) will not be derived. Instead we shall comment on the most important approximations and concepts underlying its derivation, with a particular view towards the various interpretations that have been made of line intensities and profiles in terms of models and processes in the transition region. For derivations we refer to the original papers listed above and to standard textbooks, e.g. Mariska (1992). The papers of Gabriel and Jordan (1972) and Elwert and Raju (1972) are furthermore useful, respectively describing the use of collision strengths, and various ways to express the final results that facilitate the derivation of temperature gradients and conductive fluxes from the observed intensities.

11.2.2 Underlying approximations and concepts

To go from equation (11.1) to equation (11.2) implies looking at the mechanism of emission in a two level approximation at low gas density. This means that the population density in the upper level, N_j, is created exclusively by collisions from the lower level, which is the ground state of the ion. Each collision leads to the emission of a photon at the wavelength of the line. The more complicated situation when the transitions involve multiplets may be handled by employing branching ratios, and by summing intensities over all the lines in the multiplet. Downward collisions or radiative excitations can be safely disregarded for most physical conditions to be realistically imagined in the solar transition region. Optical density effects may appear, particularly for lines emitted at the lowest temperatures in the transition region. However, they will affect mainly the line profiles and the intensity ratios within multiplets. Total intensities will not be easily affected, i.e. all photons created within the gas will finally escape in the line or multiplet.

The expression furthermore relies on the formulation of collision rates by van Regemorter (1962). This do not lead to large errors if his formula is used as a means to approximate more accurate calculations, using the Gaunt factor as an adjustable parameter. Constant collision strengths, also employed in the derivation of equation (11.2), are easy to adjust for, but will lead to a more complex expression.

The presence of metastable energy levels in the ions allows populations in these levels to build up to much higher values than in similar levels with an allowed decay. This is likely to affect other level populations. Specifically, the total population of ions of stage m, can no longer be regarded as identical to the ground state population for all practical purposes.

Metastable levels will also affect the ionization equilibrium, allowing direct ionizations from and recombinations to the metastable ground states. Since collisional de-excitation may be significant for metastable levels, their populations depend on the electron density, with critical values in many cases at densities occurring in the transition region. The presence of these levels will thus lead to a density dependent term in the ionization function.

The combined density dependent effects on the ionization and on the population distribution of the energy levels within the ion are noticeable, but usually not very pronounced, and were ignored in early analyses. A model dependent analysis, where N_e varied with temperature to give a constant electron pressure through the transition region, was, however, included by Nicolas et al. (1982) and by Kjeldseth-Moe et al. (1984).

Going from equation (11.1) to equation (11.2) we have changed from a volume integration to an integration over temperature, T, leading to the inclusion of the temperature gradient in equation (11.2). This may be seen to result from the substitutions

$$dV = A_f \, d\ell = A_f \, [dh/d\ell]^{-1} \cdot [dT/dh]^{-1} \, dT. \tag{11.3}$$

Here A_f is the fractional area element on the Sun's surface that radiates in the line considered, i.e. the atmosphere (or emission) filling factor. The length, $d\ell$, is directed along the line of sight, while dh is a distance element along the temperature gradient. Thus, $dh/d\ell = \cos \gamma$ and $dT/dh = |\nabla T|$.

In this treatment there is a tacit assumption of a stratified transition region with perhaps, but not necessarily, a vertical temperature gradient. More complicated geometries can, however, be handled, albeit not with an expression as simple as this. In the analyses of Pottasch (1964), Athay (1966), Dupree and Goldberg (1967) and Dupree (1972) any effects of geometry were disregarded, i.e. $\cos \gamma$ and A_f were both set equal to 1.0. In the following discussions we shall ignore any angles between the temperature gradients and the line of sight, but spend time discussing the much more fundamental possibility of having a filling factor $A_f \ll 1.0$.

The analysis finally assumes that physical conditions are static or at least stationary. This holds if the plasma do not experience rapid changes of temperature. The ionization function $R(T)$ may then be determined from an equilibrium ionization condition. However, in a dynamic transition region with processes that we wish to detect and map, this will not be sufficient. Temperature variations caused by flows or waves may be sufficiently rapid to compare with the characteristic ionization and recombination times of several ions used in analyses of the transition region, see e.g.

Joselyn *et al.* (1979). Thus, model calculations of a dynamic transition region should for consistency include effects of non-equilibrium ionization.

11.3   Constant conductive flux and the thin transition region

Athay (1966) may have been the first to note observationally that the derived temperature gradient varied consistently with a constant thermal conductive flux, i.e. $|\nabla T| \propto T^{-\frac{5}{2}}$. This result was confirmed by the other early investigations mentioned above. Values of the order 5×10^5 erg cm^{-2} s^{-1} to 10^6 erg cm^{-2} s^{-1} were estimated by Dupree and Goldberg (1967) and by Dupree (1972). From a base level at 1×10^5 K the latter flux value gives a transition region thickness of just 24 km up to temperature 2.5×10^5 K, 190 km to 4.5×10^5 K, and 695 km at 6.5×10^5 K, confirming the thin transition region, i.e. much thinner than the pressure scale heights at the corresponding temperatures.

In a thin transition region, the pressure will be virtually constant through its entire temperature range. Anticipating the constancy of the thermal conductive flux given by

$$F_c = \kappa_0 \, T^{\frac{5}{2}} \nabla T, \tag{11.4}$$

where the numerical value of $\kappa_0 = 1.1 \times 10^{-6}$, see e.g. Ulmschneider (1970), one may write

$$F_c = \frac{1.25 \; 10^{10} \; A_X \overline{g} f \; A_f \; P_e \; P_H \int\limits_T R(T) \, e^{-\Delta E_{ij}/kT} \, dT}{I}. \tag{11.5}$$

We have simplified this expression by setting $\gamma = 0$, i.e. the line of sight falls along the temperature gradient, and assumed that the filling factor, A_f, do not depend on temperature. In the original analyses the filling factor was furthermore set equal to 1.

An excellent illustration of the constant conductive flux resulting from the analysis is found in Elwert and Raju (1972). Figure (11.1) convincingly illustrates that the flux is constant in the temperature range 1.3×10^5 K to almost 1 MK. It should be noted that an actual *value* of the flux cannot be set unless one also knows the plasma pressure and element abundances in the transition region. For their estimate, quoted above, Dupree and Goldberg (1967) used $P_e \approx 0.08$ dyn cm^{-2}, a reasonable choice for the quiet Sun transition region. Electron pressures or densities may in principle be found from density sensitive line ratios, a subject we shall not describe here.

New observational result have invalidated the static assumption on which these old models were based. Nevertheless, they are powerful reminders that *on the average the structure of the transition region mimics an atmosphere with a constant conductive*

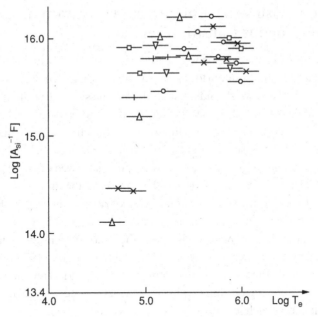

Figure 11.1. The constant thermal conductive flux in the transition region. The figure gives the flux divided by the silicon abundance relative to hydrogen as a function of temperature, using coronal element abundances. From Elwert and Raju (1972).

flux. This is a property that models featuring entirely different concepts, will have to incorporate.

11.4   The extended transition region

11.4.1   Excess emission at temperatures below 10^5 K

The approximation of a plane parallel atmosphere supporting a constant conductive flux breaks down at the low temperature end of the transition region for $T < 10^5$ K, as is seen in Figure (11.1). The emission from this temperature range is much higher than from an equilibrium atmosphere with constant conductive flux. This was a first indication that the transition zone, at least at some temperatures, deviated from the thin static layer in conductive energy equilibrium.

What is causing the extra emission? Antiochos and Noci (1986) have proposed that the lower transition region plasma resides in low loops with maximum temperatures $T_{max} < 10^5$ K that do not interface with the corona. The model explains the high emission from the low temperature layers. However, as we shall see in Section 11.6, similar models are needed for loops with maximum temperatures between 10^5 K and 5×10^5 K to explain the rest of the transition region. It is not obvious how such loops can be constructed.

11.4.2   The EUV flash spectrum: direct observation of an inhomogeneous transition region

The models of the thin transition region were based on observations with instruments that had insufficient angular resolution to give any direct information of its true extent. The thin models implied that we should not receive any emission in transition region lines at altitudes higher than 1-2 arc seconds above the chromospheric solar limb. That this is not the case became clear when the solar flash spectrum in EUV lines was recorded during the solar eclipse on 7 March 1971 (see e.g. Brueckner and Nicolas, 1973). The flash spectrum in lines emitted at temperatures between 5×10^4 K and 1.5×10^5 K was clearly visible at altitudes 3-4000 km above the limb.

The NRL UV slit spectrometer SO82B on Skylab confirmed this result, demonstrating that the emission from plasma temperatures between 3.5×10^4 K and 2.5×10^5 K peaked at 2 to 6 arc seconds above the limb, both in the quiet Sun and in coronal holes, see Doschek et al. (1976b), Feldman et al. (1976). This is obviously in strong contradiction to the thin "onion skin" transition region outlined above, where the transition region emission in the entire temperature range must arise in a layer with thickness much less than ≈ 1000 km or 1 arc second.

11.4.3   A transition region structured by the magnetic field

The model by Gabriel (1976) represents an interesting early attempt at taking into account the magnetic field in the transition region and at the same time account for the energy balance. Gabriel extended the transition region in height using a model where the filling factor was less than 1 and varied with height (or temperature) in the atmosphere. A magnetic field configuration was constructed where the magnetic flux at photospheric levels was concentrated at the boundaries of super granule cells. Higher up the field fanned out and filled the corona completely. Energy balance within the expanding magnetic field constituted an additional constraint used to derive this static model. In the Gabriel model, the atmosphere at the cell boundaries extended to altitudes high above the chromosphere, in agreement with observations.

Gabriel's result is obtained only if the magnetic field in the photosphere is unipolar over an area corresponding to the cell. This is not likely to be the case. Dowdy et al. (1986) found that many magnetic field lines close on scales smaller than 10^4 km, only a few making it into the corona. Studies with TRACE of dynamical EUV emission structures on scales of 1-3000 km, and their relation to the underlying magnetic field, also suggest that many small bipolar structures are present within an area the size of a super granule cell, see e.g. Berger et al. (1995). This is not in agreement with the picture of the field drawn by Gabriel.

11.4.4   Spicules and the transition region

Brueckner and Nicolas (1973) proposed that the explanation for their 1971 solar eclipse observation was a thin transition region wrapped on the outside of tall, spicule-like vertical structures penetrating into the hot corona. The same idea was used by Withbroe and Mariska (1976), who based an extensive analysis on data from the Harvard instrument SO55 on Skylab. They found that the intensity variations above the solar limb could be explained if 20% of the quiet Sun emission came from spicules that were distributed in area and height similarly to Hα-emitting spicules. The remaining transition region emission came from the usual thin model. Other investigations employing the same concept gave different results for the fraction of spicular transition region. While the details of Withbroe and Mariska's model later were abandoned after comparing it with higher resolution data, it might be argued that the observed height distribution of the emission outside the limb could be explained by spicule like emitting structures (Mariska *et al.* 1978).

11.4.5 An extremely fine structured transition region?

A thorough observational argument for a transition region with an extreme fine structure was presented by Dere *et al.* (1987). The possibilities of small filling factors, A_f, had been pointed out even earlier by Kjeldseth-Moe and Nicolas (1977), based on an energy conservation argument. Estimated values of $\approx 1\%$ for the filling factor were given by Feldman *et al.* (1979) and by Nicolas *et al.* (1979). However, the presentation by Dere *et al.* (1987) was more rigorous and direct and was based on better observations. Thus, they did not rely on spectra from the limb, but analyzed spatially resolved HRTS spectra from the solar disk.

The effect of the proposed fine structures is to extend the emission in altitude by putting the transition region gas into a few thin, tall structures. The structures, filling only a small fraction of the total volume, would have to be magnetically insulated from the surrounding hot corona in at least two directions to avoid being heated up by conduction.

The effect of very low filling factors, $A_f \ll 1$, is obvious from equation (11.5). With all other quantities unchanged, a smaller filling factor, A_f, leads to a lower conductive flux. This implies a smaller temperature gradient according to equation (11.4), and the extent of the transition region, $H_T \sim \delta T \cdot |\nabla T|^{-1}$, becomes correspondingly larger.

Initially Dere *et al.* (1987) note that the emission in the transition region lines extends well outside the solar limb, thus confirming the earlier SO82B observations. On the disk the emission in the C^{+3} resonance lines at 155 nm comes from features of typical size 2500 km. This is considerably larger than the instrument resolution of 1-1.5 arc second or ≈ 1000 km. It is furthermore comparable to the extent of the emission above the chromospheric limb as observed in the same HRTS spectrograms.

One may therefore regard the emission as coming from "macro" features of this size, perhaps corresponding to the extent of individual small magnetic loops in the transition region. The emitted power, EM, from such a feature may be written as

$$EM \propto A_f \int_V N_e^2 \, dV. \tag{11.6}$$

Details of how this expression is derived from equation (11.1) are again given in standard textbooks. Comparing with equation (11.2) we recognize the factor N_e^2 as deriving from the same origin as the product $P_H \, P_e$, while integration in this equation is over the volume, as is usual in definitions of the emission measure. The filling factor, A_f, is included, which is not customary in most definitions of the emission measure, but is necessary for the present purpose. A proportionality factor contains all other quantities in equation (11.2) including what is in effect an integration over the ionization and excitation terms.

Dere *et al.* (1987) use only the C^{+3} lines for their analysis and assume that they are emitted at 10^5 K. They derive the filling factor for each macro structure, assuming the electron density to be homogeneous through the entire volume emitting the C^{+3} lines. Thus, the integral in equation (11.6) is replaced by

$$EM_{C^{+3}} \propto A_f \, N_e^2 V, \tag{11.7}$$

where V is the volume of the macro feature from its observed extent. The electron density, N_e, is derived from the density sensitive O^{+3} lines near 140 nm. Applying equation (11.7) to the observed quantities result in filling factors, $A_f \approx 0.001$ - 0.01. Dere *et al.* (1987) envision this result in terms of thin plasma threads, fibrils, confined and thermally insulated inside wider magnetic macro structures, i.e. loops. From their distribution of macro feature intensities Dere *et al.* (1987) derive fibril radii in the range 4 to 40 km assuming 10 fibrils inside each macro structure.

It seems clear that Dere *et al.* (1987) imagine that the temperature in each loop reaches a maximum value in the transition region, but that each loop also contains all lower temperatures down to the temperature minimum at 4500 K. The transition region thus connects to the chromosphere in this picture, but not to the corona.

11.4.6   Unresolved fine structures

Feldman (1983, 1987) has argued that the transition region consists of unresolved fine structures. These features are *not* a part of continuous structures that encompass the entire temperature range 10^4 K to 1 MK. Instead, the unresolved fine structures are magnetically isolated from the chromosphere and the corona. They are envisioned as small shells of size 1 arc second that are somewhat opaque and pulsating, in order to account for the net redshift observed in transition region lines, see section 11.5. Most of the plasma in the temperature range from 3×10^4 K to 5×10^5 K is contained in these structures, and only a small part makes up the "true" transition region connecting the chromosphere and the corona.

Feldman has offered a number of arguments in support of the unresolved structures. Two of the more striking ones include the lack of any observed continuous structures where temperatures run from the chromosphere to the corona, and the above mentioned distribution of observed emission outside the solar limb.

11.4.7 Unresolved dynamic evolution?

The concept of unresolved fine structures is severely criticized by Wikstøl et al. (1998). These authors introduce an alternative mechanism, where a thin transition region flutters up and down in altitude over a large range as shock waves from nanoflare disturbances at the top of magnetic loops are reflected when they hit the chromosphere (Wikstøl et al. 1997). The transition region is continually in motion as it responds to a series of these nanoflare-generated pulses. The observed limb emission results from a number of such processes simultaneously going on within the resolution element of the observing instrument. There is also an average effect over time since the individual disturbances move into and out of the observed area in less than the typical integration time for a spectral exposure. This is a fully dynamic concept of the transition region. It remains to be demonstrated whether it models the detailed observed behavior of the transition region emission. A strong feature, however, is that it explains other basic dynamical properties of the transition region such as the persistent redshift or apparent downflow of the plasma discussed in Section 11.5 below.

11.5 The redshifted transition region

11.5.1 Line shifts in the transition region

A persistent redshift of the spectral lines emitted from the solar transition region plasma was first reported by Doschek et al. (1976a), based on observations with SO82B on Skylab. The lines from Si^{+3} and C^{+3} at approximately 140 nm and 155 nm, respectively, were found to be shifted corresponding to downflow velocities of 9 km s^{-1} in the quiet Sun.

Over the years several new investigations have confirmed and expanded on this rather perplexing result occurring in the Sun, but also in other stars. The redshift thus appears to be an inherent property of stellar transition regions.

Recent and representative solar measurements of the redshift, covering the temperature range of the transition region, are described by Achour et al. (1995) from HRTS observations and by Brekke et al. (1997a) and Peter and Judge (1999) from observations with SUMER.

In measurements of the transition region redshift the most reliable wavelength scale is established from the measured averaged positions of chromospheric lines, i.e. lines from neutral and singly ionized atoms. This may not seem entirely satisfactory. However, the absolute velocity of the chromospheric gas relative to the Sun appears to be small, 1 km s^{-1} according to Samain (1991). At wavelengths below 130 nm comparing with the chromosphere becomes increasingly difficult since there are few strong chromospheric reference lines, and below the Lyman continuum edge the chromospheric line spectrum disappears entirely.

In all investigations a maximum redshift, corresponding to a downflow velocity of approximately 10 km s^{-1}, is reached at a temperature around 1.3×10^5 K, while the redshift falls off rapidly at $T > 2.5 \times 10^5$ K. In active regions the behavior of the redshift with temperature is similar, but the shifts are larger by approximately a factor 2, see Achour *et al.* (1995).

Whether there are redshifts in lines emitted from the corona has been a matter of some debate. The results of Brekke *et al.* (1997a) give average net redshifts of 5-6 km s^{-1} between 3×10^5 K and 1.5 MK, while Peter and Judge (1999) obtain blueshifts, or upward velocities of 2 to 6 km s^{-1}, in lines emitted at temperatures 6×10^5 K to 1.5 MK. This difference is caused entirely by different wavelength calibrations and do not reflect any other observational difference.

If the flows causing the redshifts are predominantly vertical on the Sun a center to limb variation of the observed shifts will be expected. Various investigations in the past have given different answers to the question of center to limb effects. Peter and Judge (1999) state quite decisively that there is a $\cos \theta$ variation if the shifts are averaged over spatial scales larger than 50 arc seconds. The variations on small scales are, however, pronounced. Thus, the flow field should have a vertical predominance, but with a strongly varying horizontal component.

The redshifts are clearly associated with the magnetic field. If we order the measured shifts into groups according to the magnitude of the magnetic flux density, $|B|$, and calculate conditional probabilities, we find that profiles with large redshifts occupy an increasing fraction of the area when $|B|$ increases. This result is contrasted by blueshifted line profiles, which indicate a decrease in area fraction with increasing magnetic flux density (Brynildsen *et al.* 1996).

The exact connection between the redshifts and the magnetic field is not easy to derive. Brynildsen *et al.* (1996) point out that continuous gas flows in magnetic loops, where the mass flux must be equal in the upflowing and downflowing parts of the tube, appear to be incompatible with the observations unless there are asymmetries in the filling factor between the two flow directions. Instead they propose that Alfvèn wave pulses, traveling from the corona towards the chromosphere, may provide an explanation. This idea has not been worked out. It relates, however, to similar explanations for the redshifts as being caused by downward running waves or disturbances, see Section 11.5.3.

11.5.2   Red- or blueshifts from siphon flows and spicules?

What causes the pronounced redshifts of the transition region lines? A naive explanation of a net mass flow is clearly not tenable. The mass fluxes implied by the observed shifts are sufficient to empty the above lying corona of all its mass in less than an hour.

Difficulties with interpreting redshifts as a result of siphon flows in loops have been briefly mentioned above. Calculations by Mariska (1988) and by Spadaro *et al.* (1990) give both red- and blueshifted lines, which are not in agreement with the observations.

An early idea for the redshift was emission from downflowing, cooling plasma that had been ejected into the corona by spicules, see Pneuman and Kopp (1978) and Athay (1984). However, calculations by Mariska (1987), who produced flows by heating and cooling the footpoint regions of loops, could not reproduce the emission and velocities.

The rebound shock model, another mechanism possibly involving spicules, has been invoked to explain the redshifts. Here, a shock train is generated by an impulsive energy release in the photosphere. It propagates upward and lifts and heats the chromosphere and transition region. Cheng (1992) found that the rebound shocks give rise to an average downward material velocity. This is correct, but calculations by Hansteen and Wikstøl (1994) demonstrate that the resulting line emission is nevertheless blueshifted. The discrepancy between *average* line shifts and material velocities is due to the correlation between velocity and density in the direction of the wave propagation. Hansteen and Wikstøl conclude that upwardly moving shock fronts cannot explain average redshifts unless the average velocity in the descending portion of the wave is at least a factor of four larger than the average velocity in the ascending phase.

11.5.3 Redshifts as signatures of downward propagating waves

Instead of upward propagating waves Hansteen (1993) interprets the observed redshift as caused by downward propagating acoustic waves. Nanoflares releasing relatively small amounts of energy near the top of magnetic loops give rise to disturbances running downward along the loop legs. As in the case of the rebound shock calculations the correlation between velocity and density in the propagation direction is causing the net lineshift towards longer wavelengths.

Hansteen computed intensities and line profiles for the resonance lines from several ions and included time dependent ionization in his procedures. He found that the time averaged profiles of transition region lines emitted between 10^5 K (C^{+3}) and 2.7×10^5 K (O^{+5}) were redshifted, although less than the observed amounts. Hotter lines from Ne^{+7}, emitted at 6.5×10^5 K, displayed little average shift. In a later study Hansteen *et al.* (1997) included the effect of downward propagating MHD waves being reflected on the chromosphere. This increased the transition region redshifts, although coronal blueshifts became too large. Nevertheless, downward propagating waves seem like a promising explanation that is sufficiently general to explain why redshifts are observed also in other stars.

An investigation by Judge *et al.* (1998) lends strong observational support to the picture of downward running waves or disturbances. They used a connection between Doppler shifts and asymmetries in the ratio between the profiles of two density sensitive lines from the same multiplet. The observed ratios of the line profiles from the O^{+3} lines at 139.9774 nm and 140.1156 nm, showed an asymmetric tilt that implied an association between redshifted gas and higher mean electron densities. This is exactly what one would expect in a downward propagating disturbance.

11.6   The dynamic and time dependent transition region

Observations with the instruments on the Solar and Heliospheric Observatory, SOHO, have shown that the solar transition region is thoroughly dynamic. This is easiest to observe in active areas, where high plasma velocities, several tens of kilometer per second, are the rule and the emission varies with time on scales from minutes to a few hours. However, from the pervading presence of magnetic fields in the solar atmosphere one may surmise that strong and presently unresolved dynamical conditions are likely in quiet regions as well.

In the following we shall describe the new dynamic transition region revealed with SOHO, particularly in the active region loops. We shall leave out descriptions of several interesting, well observed and probably important phenomena. Some of these may represent basic processes relating to the heating and structure of the transition region and corona. Examples are explosive events, waves propagating in the transition region, sunspot plumes and flows and 3 minute oscillations in sunspot transition regions. Explosive events and waves may be important indicators of processes heating the solar atmosphere. Other new phenomena left out are "blinkers" and rotating spicule structures. Descriptions of these two new topics are given by Harrison (1997), Harrison *et al.* (1999) and Pike and Mason (1998).

11.6.1   Morphology of transition region loops

Skylab observations demonstrated that the solar transition region and corona so to speak consisted of loops, see e.g. Tousey *et al.* (1973). Referring mainly to the new observations of transition region loops from SOHO, particularly those obtained with the Coronal Diagnostic Spectrometer, CDS, we shall note three characteristic morphological properties.

Firstly, loops emitting over a range in temperatures from 10^4 K to 5×10^5 K are generally co-located within the CDS angular resolution, 3-4 arc sec. Co-location may extend up to 1 MK (Mg^{+8} emission). This do not imply that *all* loops emit over the full temperature range.

Secondly, the distribution of emitted intensity along a loop may vary with temperature. And thirdly, the loops often appear "complete", i.e. at transition region temperatures they emit along their full length or along a significant fraction of their length in any particular transition region line. This gives them an isothermal appearance, a property already noted by Foukal (1976) for sunspot loops. The combined first and third property means that the macrostructures seen with CDS must contain plasma in several fine structure elements at different temperatures.

11.6.2 Velocities in transition region loops

Using CDS on SOHO Brekke *et al.* (1997b) reported large Doppler shifts in the lines emitted from loops at transition region temperatures above active regions. Their result was the first indication that velocities comparable to the speed of sound occurred in the transition region outside of special phenomena such as explosive events. They found two sections in two separate loops in the same loop system where Doppler shifts corresponding to velocities of 50-60 km s^{-1} went in opposite directions. In a third section the plasma appeared to be at rest.

Investigations of loop velocities have continued. Presently 70-80 loop systems have been investigated. Results from the first 20 systems that were studied, were published by Kjeldseth-Moe and Brekke (1998) and further descriptions are found in Kjeldseth-Moe *et al.* (1998), Kjeldseth-Moe and Brekke (1999), Kjeldseth-Moe (2000) and Fredvik and Kjeldseth-Moe (2002). These and other recent results may be summarized as follows.

Large Doppler shifts are a general feature always observed in lines emitted from active region loops at transition region temperatures. If interpreted as flow velocities the shifts correspond to flows along the loops with typical velocities $v \approx$ 20-100 km s^{-1} for $T < 5 \times 10^5$ K. In the corona, $T \approx$ 1 - 2.7 MK, systematic Doppler shifts are generally not found, i.e. $v < 5$ km s^{-1}, see Figure 11.2. Velocity variations occur on a time scale of 10 min or less. This temporal variability of the shifts matches the variability of the emission, see section 11.6.3 below.

Flow patterns may be of many types. Thus, red- and blueshifts may occur in opposite legs of loops, but examples of the same shift along the full length of the loop are also seen. Red- and blueshifts can co-exist within the same part of a loop, as if the structure is rotating or twisted. The highest velocities or Doppler shift values often occur at the outer edges of loops or where the emission is weak, but in general we do not find any pronounced connection between intensity and Doppler shifts. Occasionally, regular changes between strong redshifts and strong blueshifts occur within limited regions, as if the plasma is "sloshing" back and forth.

What possible mechanisms may cause these high velocities at transition region temperatures? Calculations by Mariska (1987) of asymmetric heating and cooling in the loop legs give velocities that are generally smaller than 20 km s^{-1}. This is significantly less than the observed velocity values. In Mariska's calculations the heating was abruptly reduced to 1% of the value required to keep the loop in its equilibrium state. Calculations by Schrijver (2001) show that catastrophic cooling, where the heating at the top of loops is reduced by a factor 30 and the plasma cools

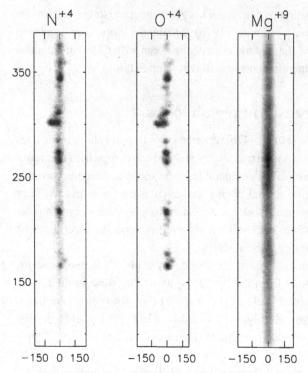

Figure 11.2. Doppler shifts in an active region loop in two transition region lines from N^{+4} and O^{+4} emitted at 1.7×10^5 K and 2.4×10^5 K, respectively, compared with the simultaneous spectrum of a coronal line from Mg^{+9} emitted at 1.1 MK. The abscissa gives the velocity scale in km s^{-1} while the ordinate gives position along the SUMER slit in arc seconds measured from the bottom end of the slit. In this observation the SUMER slit was placed outside the solar limb through the loop system. One note the complete lack of velocities or wavelength shifts in the coronal line.

to transition region temperatures, will cause the plasma to fall down the legs of the loops at speeds of up to 100 km s^{-1}. These calculations are supported by observations from TRACE.

Much larger, supersonic velocities are obtained by Orlando *et al.* (1995), Peres (1997) and Reale *et al.* (2000) from their heating and cooling calculations. Up to several hundred km s^{-1} may occur, especially in a coronal loop. It is not clear what energy input mechanism provides the acceleration to such high velocities.

Alfvén disturbances may be another possible mechanism for producing high line shifts. The downward propagating disturbance in a flux tube generated by a nanoflare at the top may give lineshifts comparable to those observed, owing to the compressible, but particularly to the Alfvénic part, of the disturbance, see Wikstøl *et al.* (1998). In their present form none of these calculations reproduce the observations well, and it is not easy to decide if any of the mechanisms can provide realistic explanations.

11.6.3 Rapid time changes in the emission

An unexpected result of the survey to find whether high velocities were a common feature of the loops, was the discovery of rapid time changes in the emission. The emission in the large loop structures of 100 Mm or more turn out to be extremely transient, with life times ranging from 10 min to a few hours. The emission also appear and disappear rapidly, i.e. in 10 min or less.

The surprise should perhaps have been less. The variability of the emission in active region loops was noted already in the observations from the NRL and Harvard spectrometers on Skylab. Among early observations, one may mention Sheeley (1980) who found a typical loop life time of 30 min in the emission from Ne^{+6} at 5×10^5 K, and Habbal *et al.* (1985) who discovered that transition region and coronal loops changed on different characteristic time scales, i.e. 10 min and 30 min, for plasma temperatures less than and greater than 1 MK, respectively.

A first description of the time variability of the active region loops from the survey with CDS is found in Kjeldseth-Moe and Brekke (1998). Additional results have been given by Kjeldseth-Moe *et al.* (1998), Kjeldseth-Moe and Brekke (1999), Kjeldseth-Moe (2000) and Fredvik and Kjeldseth-Moe (2002).

These and more recent observations show that the emission from the loops at transition region temperatures, i.e. $T \leq 5 \times 10^5$ K, is strongly time variable. The intensity distribution along the loops may change in minutes in a way consistent with the flow of emission "fronts" or "blobs" along loops at coronal temperatures (1-1.5 MK), see e.g. Schrijver (2001).

Life times of loops emitting at transition region temperatures typically range from 10 min to 3 hours. But whether they last for minutes or hours the loop emission at transition region temperatures form and disappear suddenly, typically in less than 11 minutes, the raster cadence time used for most of the loop studies with CDS. The changing appearance of a loop system is illustrated in Figure 11.3, which shows the emission from a system of loops in the 62.9 nm line from O^{+4} emitted at 2.4×10^5 K. The interval between images in this sequence is 17 min and the entire sequence range over 7 hours.

Simultaneous observations of the loops are made with CDS in several lines emitted from the plasma over the full range of transition region and coronal temperatures. There are no signs of the loops running through a time sequence of temperatures, as if a heating or cooling process is going on for the entire structure. This could be due to the observing cadence being too slow to catch rapid heating or cooling, see Mariska (1987). However, recent observations run at a cadence of 5 min still give the same result.

The variability at transition region temperatures is persistent, i.e. it may last for hours and days. This is not a trivial point since there are cases where the entire set of loop structures emitting at transition region temperatures disappear from an active region and are absent for 2-3 hours or even longer.

Time variations exist in loops at coronal temperatures as well, particularly at 1 MK in Mg^{+8} and occasionally also at 2.7 MK in Fe^{+15}. Variations at these temperatures are less frequent and pronounced than those at transition region temperatures. Perhaps coronal loops change more slowly. However, comparing difference images of EIT loops in the 19.5 nm band, emitted at 1.5 MK, with rapidly changing loops in the CDS rasters, leaves the impression that the same rapid changes occur at coronal temperatures. However, the coronal loops that change are those that occupy the same

Figure 11.3. Variation with time of an active region loop system emitting in the O^{+4} line at 62.9 nm observed with CDS on 14 September 1997. The field of view is 240" × 240" and the interval between each image is 17 minutes. The series start in the upper left corner and proceeds row by row downward. From Kjeldseth-Moe and Brekke (1998).

general spatial location as the active transition region loops. Most coronal structures appear stable and not associated with loops at transition region temperatures. The overall impression of quiet coronal conditions applies to the large number of overlapping, quiescent coronal loops.

11.7   Conclusion – a new concept for the transition region

What is the solar transition region like? The view of a static, thin transition region has long been left behind, but a new model is not generally agreed upon. The observational

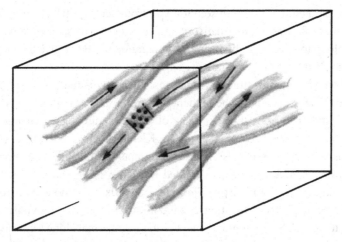

Figure 11.4. Fine strands in thick magnetic flux ropes may make up the solar transition region and corona. The box has a size comparable to best obtainable resolution, i.e. 1 arc second or 725 km on the Sun. Arrows designate different flows inside the individual strands. From Kjeldseth-Moe and Brekke (1998).

facts and theoretical considerations, however, consistently point towards a strongly dynamic solar plasma.

The high emission at low temperatures, $T \leq 10^5$ K, requires that a large amount of plasma is present at these temperatures. Cool loops, see Antiochos and Noci (1986), may be a likely possibility, but the picture should allow for the velocities observed in the transition region also in this low temperature range.

High velocities are directly observed in active regions, but are likely to be present and unresolved in quiet regions as well. This is indicated by the large non-thermal line widths in lines radiated at temperatures between 3×10^4 K and 3×10^5 K, see e.g. Kjeldseth-Moe and Nicolas (1977). Waves running through these cool loops may give the required line widths or shifts.

How can the extension in altitude of the transition region be explained? We have seen two different ways: 1) by the static fine structures of Dere *et al.* (1987) extending high above the solar surface, or 2) having a much thinner transition region carried aloft, either by upstreaming material in spicules or by waves and shocks propagating downward and then reflected from the chromosphere.

The wave models seem very promising. Downward running disturbances obviously characterizes the transition region (Judge *et al.* 1998). Waves may also explain the net redshift, another truly basic property of the transition region.

Will a wave mechanism *by itself* reproduce the distribution of observed intensities at the solar limb? Considering the present wave models, a small observing aperture pointed outside the solar limb, would receive brief and intermittent emission in a particular line as plasma with the right temperature passes that altitude. However, it is not demonstrated that a high resolution spectrometer, like HRTS or SUMER, will observe these flashes, or, alternatively, that a sufficient number of waves pass the aperture along the line of sight during an integration period to give the observed distribution in altitude of the emission.

The two concepts of fine structures and unresolved dynamic processes do not necessarily exclude each other. The evidence from the active region loops seems to point to a combination of the two ideas of fine structures and dynamic wave processes or flows, as has been pointed out by Kjeldseth-Moe and Brekke (1998). Thus, a magnetic loop may consist of a large number of thin fine structures, like the strands inside a thick rope, as is depicted in Figure 11.4. The strands will be thermally insulated from their surroundings since heat is not conducted across the magnetic field. Along field lines heating, conduction and transport by moving material distribute the temperature. Each strand may be quasi-isothermal along a considerable distance. A large number of different strands will cause the extended isothermal appearance seen for many loops as well as account for the co-location of loops spanning a wide range in temperature. The observed dynamic effects result from flows, waves and shocks inside individual strands, that twist around and reconnect with each other.

References

Achour, H., Brekke, P., Kjeldseth-Moe, O., and Maltby, P. (1995). *Astrophys. J.*, **453**, 945.

Antiochos, S. K. and Noci, G. (1986). *Astrophys. J.*, **301**, 440.

Athay, R. G. (1966). *Astrophys. J.*, **145**, 784.

Athay, R. G. (1984). *Astrophys. J.*, **287**, 412.

Berger, T. E., Schrijver, C. J., Shine, R. A., Tarbell, T. D., Title, A. M., and Scharmer, G. (1995). *Astrophys. J.*, **454**, 531.

Brekke, P., Hassler, D. M., and Wilhelm, K. (1997a). *Solar Phys.*, **175**, 349.

Brekke, P., Kjeldseth-Moe, O., and Harrison, R. A. (1997b). *Solar Phys.*, **175**, 511.

Brynildsen, N., Kjeldseth-Moe, O., and Maltby, P. (1996). *Astrophys. J.*, **462**, 534.

Brueckner, G. E. and Nicolas K. R. (1973). *Solar Phys.*, **29**, 301.

Cheng, Q. Q. (1992). *Astron. Astrophys.*, **262**, 581.

Dere, K. P., Bartoe, J.-D. F., Brueckner, G. E., Cook, J. W., and Socker, D. G. (1987). *Solar Phys.*, **114**, 223.

Dowdy, J. F. Jr., Rabin, D., and Moore, R. L. (1986). *Solar Phys.*, **105**, 35.

Doschek, G. A., Bohlin, J. D., and Feldman, U. (1976a). *Astrophys. J. Lett.*, **205**, L177.

Doschek, G. A., Vanhoosier, M. E., Bartoe, J.-D. F., and Feldman, U. (1976b). *Astrophys. J. Suppl. Ser.*, **31**, 417.

Dupree, A. K. (1972). *Astrophys. J.*, **178**, 527.

Dupree, A. K. and Goldberg, L. (1967). *Solar Phys.*, **1**, 229.

Elwert, G. and Raju, P. K. (1972). *Solar Phys.*, **25**, 319.

Feldman, U. (1983). *Astrophys. J.*, **275**, 367.

Feldman, U. (1987). *Astrophys. J.*, **320**, 426.

Feldman, U., Doschek, G. A., Vanhoosier, M. E., and Purcell, J. D. (1976). *Astrophys. J. Suppl. Ser.*, **31**, 445.

Feldman, U., Doschek, G. A., and Mariska, J. T. (1979). *Astrophys. J.*, **229**, 369.

Foukal, P. V. (1976). *Astrophys. J.*, **210** 575.

Fredvik, T. and Kjeldseth-Moe, O. (2002). *Adv. Space Res.* (forthcoming).

Gabriel, A. H. (1976). *Roy. Soc. (Lond.), Phil. Trans., Ser. A*, **281**, 339.

Gabriel, A. H. and Jordan, C. (1972). Interpretation of spectral intensities from laboratory and astrophysical plasmas, in *Case Studies in Atomic Collision Physics*, **vol. 2.**, ed. McDaniel and McDowell (Amsterdam: North-Holland), 209.

Habbal, S. R., Ronan, R., and Withbroe, G. L. (1985). *Solar Phys.*, **98**, 323.

Hansteen, V.H. (1993). *Astrophys. J.*, **402**, 741.

Hansteen, V.H. and Wikstøl, Ø. (1994). *Astron. Astrophys.*, **290**, 995.

Hansteen, V. H., Maltby, P. and Malagoli, A. (1997). *ASP Conf. Ser.*, **111**, Magnetic reconnection in the Solar Atmosphere, ed. R. D. Bentley and J. T. Mariska (San Francisco: ASP), 116.

Harrison, R. A. (1997). *Solar Phys.*, **175**, 467.

Harrison, R. A., Lang, J., Brooks, D. H., and Innes, D. E. (1999). *Astron. Astrophys.*, **351**, 1115.

Joselyn, J. A., Munro, R. H., and Holzer, T. E. (1979). *Astrophys. J. Suppl. Ser.*, **40**, 793.

Judge, P. G., Hansteen, V., Wikstøl, Ø., Wilhelm, K. Schühle, U., and Moran, T. (1998). *Astrophys. J.*, **502**, 981.

Kjeldseth-Moe, O. and Nicolas, K. R. (1977). *Astrophys. J.*, **211**, 579.

Kjeldseth-Moe, O., Andreassen, Ø., Maltby, P., Bartoe, J.-D. F., Brueckner, G. E., and Nicolas, K. R. (1984). *Adv. Space Res.*, **4**, 63.

Kjeldseth-Moe, O. and Brekke, P. (1998). *Solar Phys.*, **182**, 73.

Kjeldseth-Moe, O., Brekke, P. and Haugan, S. V. H. (1998). *A Crossroads for European Solar and Heliospheric Physics*, Conference on Recent Achievements and Future Mission Possibilities, Tenerife March 23-27, 1998, ed. E.R Priest, F. Moreno-Insertis, R.A. Harris, ESA SP-417, 153.

Kjeldseth-Moe, O. and Brekke, P. (1999). *Third Advances in Solar Physics Euroconference - ASPE98: Magnetic Fields and Oscillations*, Potsdam 22-25 September 1998, ed. B. Schmieder, A. Hofmann, and J. Staude. *ASP Conf. Ser.*, **184**, 286.

Kjeldseth-Moe, O. (2000). *Adv. Space Res.*, **25(9)**, 1713.

Mariska, J. T., Feldman, U., and Doschek, G. A. (1978). *Astrophys. J.*, **226**, 698.

Mariska, J. T. (1987). *Astrophys. J.*, **319**, 465.

Mariska, J. T. (1988). *Astrophys. J.*, **334**, 489.

Mariska, J. T. (1992). *The Solar Transition Region* (Cambridge University Press, Cambridge).

Nicolas, K. R., Bartoe, J.-D. F., Brueckner, G. E., and vanHoosier, M. E. (1979). *Astrophys. J.*, **233**, 741.

Nicolas, K. R., Kjeldseth-Moe, O., Bartoe, J.-D. F., and Brueckner, G. E. (1982). *Solar Phys.*, **81**, 253.

Orlando, S., Peres, G., and Serio, S. (1995). *Astron. Astrophys.*, **300**, 549.

Peres, G. (1997). *Fifth SOHO Workshop: The Corona and Solar Wind Near Minimum Activity* Held at the Institute of Theoretical Astrophysics. University of Oslo, Norway, 17-20 June, 1997, ed. by A. Wilson, ESA SP 404, p. 55.

Peter H. and Judge P. G. (1999). *Astrophys. J.*, **522**, 1148.

Pike, C. D. and Mason, H. E. (1998). *Solar Phys.*, **182**, 333.

Pneuman, G. W. and Kopp, R. A. (1978). *Solar Phys.*, **57**, 49.

Pottasch, S.R. (1964). *Space Sci. Rev.*, **3**, 816.

Reale, F., Peres, G., Serio, S., Betta, R. M., DeLuca, E. E., and Golub, L. (2000). *Astrophys. J.*, **535**, 423.

Samain, D. (1991). *Astron. Astrophys.*, **244**, 217.

Schrijver, C. J. (2001). *Solar Phys.*, **198**, 325.

Sheeley, N. R., Jr. (1980). *Solar Phys.*, **66**, 79.

Spadaro, D., Noci, G., Zappala, R. A., and Antiochos, S. K. (1990). *Astrophys. J.*, **355**, 342.

Tousey, R., Bartoe, J. D. F., Bohlin, J. D., Brueckner, G. E., Purcell, J. D., Scherrer, V. E., Sheeley, N. R., Jr., Schumacher, R. J., and Vanhoosier, M. E. (1973). *Solar Phys.*, **33**, 265.

Ulmschneider, P. (1970). *Astron. Astrophys.*, **4**, 144.

van Regemorter, H. (1962). *Astrophys. J.*, **136**, 906.

Wikstøl, Ø., Judge, P. G., and Hansteen, V. (1997). *Astrophys. J.*, **483**, 972.

Wikstøl, Ø., Judge, P. G., and Hansteen, V. (1998). *Astrophys. J.*, **501**, 895.

Withbroe, G. L. and Mariska, J. T. (1976). *Solar Phys.*, **48**, 21.

12

Solar Magnetohydrodynamics

E.R. Priest

School of Mathematics and Statistics,
University of St. Andrews,
Fife KY16 9SS, Scotland, UK

12.1   Introduction

The classical view of the Sun was of a well understood object with a spherically symmetric atmosphere and a magnetic field that is negligible except in sunspots; the atmosphere was heated by sound waves and an excess pressure drove a spherically symmetric expansion, the solar wind. Many features of this old view have been completely transformed because of high-resolution observations from the ground and space and because of advances in theoretical modelling. We now realise that the plasma atmosphere of the Sun is highly structured and dynamic and that most of what we see in the corona is caused by the magnetic field.

There is a similar change of thinking in astrophysics, where now the magnetic field is realised to be crucial in star formation, stellar activity, magnetospheres of compact objects, jets and accretion discs. However, the Sun continues to be a Rosetta stone for astronomy because here we can study many of the basic physical phenomena in depth. Furthermore, many key topics do not yet have a definitive explanation such as the generation of the Sun's magnetic field, coronal heating, the origin of the solar wind, and the causes of eruptions and of solar flares.

Much of the structure and interesting behaviour on the Sun is produced by the magnetic field, so we need a theory for the interaction between the solar atmosphere and its magnetic field - magnetohydrodynamics (or MHD for short). The solar atmosphere is not a normal gas but a *plasma*. We are all familiar on Earth with the three states (solid, liquid and gas) of matter. You change from one state to another (such as ice to water) by heating, and if you raise the temperature of a gas sufficiently it changes to the fourth state, plasma. In the plasma state the atoms have split into positive ions and negative electrons, which can flow around freely, so the gas becomes electrically conducting and a current can flow.

Plasma is very important since most of the universe is in this state. Indeed, on Earth we are in an extremely unusual part of the cosmos, a tiny island of solid, liquid and gas. But, as soon as we go up to the ionosphere, the plasma universe begins, including the whole of the region between Earth and Sun and the whole of the Sun itself.

MHD is then the study of the interaction between a magnetic field and a plasma, treated as a continuous medium (e.g., Priest, 1982) so that we are not concerned with individual particles. It builds partly on electromagnetism and partly on fluid mechanics. The assumption of a continuous medium is strictly valid for length-scales much larger than the mean-free path

$$\lambda_{mpf} \approx 300 \left(\frac{T}{10^6 K} \right)^2 \left(\frac{n}{10^{17} m^{-3}} \right)^{-1} m,$$

which is typically 3 cm in the chromosphere and 30 km in the corona, but it can also be applied when such length-scales are instead larger than the ion gyro-radius.

The magnetic field has several physical effects:

(i)   it exerts a force, which may accelerate plasma or create structure;
(ii) it stores energy, which may later be released as an eruption or a solar flare;
(iii) it acts as a thermal blanket, which, when wrapped around a cool prominence, say, may protect it from the surrounding corona;
(iv) it channels fast particles and plasma;
(v)   it drives instabilities and supports waves.

Each of these effects is crucial for magnetic phenomena on the Sun such as coronal heating, prominences and solar flares. In this chapter our aim is to describe the basic MHD equations and their basic physical significance (Section 12.2) and then to show how to model equilibria (Section 12.3) and waves (Section 12.4). Lack of space does not permit us to describe also the basics of MHD instabilities and reconnection, which can be found elsewhere (e.g., Bateman, 1978; Priest, 1982; Benz, 1993; Choudhuri, 1998; Priest and Forbes, 2000). The ways in which MHD can be used to build models is covered in several of the other chapters, including the dynamo (Chapter 6), atmospheric heating (Chapters 10 and 17), solar activity (Chapter 13) and coronal oscillations (Chapter 16).

12.2   Magnetohydrodynamic equations

12.2.1   Flux tubes

A *magnetic field line* is such that the tangent at any point is in the direction of the field. Its equation is given in two dimensions by

$$\frac{dy}{dx} = \frac{B_y}{B_x}, \tag{12.1}$$

or in three dimensions by $dx/B_x = dy/B_y = dz/B_z$.

A *magnetic flux tube* is the surface generated by the set of field lines which intersect a simple closed curve. Flux tubes are the building blocks of a magnetic configuration, but they must not be thought of as independent isolated structures, since the corona, for instance, is filled with a continuous spectrum of flux tubes communicating very efficiently with one another.

The *strength (F)* of a flux tube is the amount of magnetic flux crossing a section S, i.e.,

$$F = \int_S \mathbf{B} \cdot \mathbf{dS}. \tag{12.2}$$

It is constant along a tube. If the cross-sectional area (A) of a flux tube is small, then $F \approx BA$. Thus, as the magnetic field lines become closer together, so A becomes smaller and, since F is constant, B increases in value, and vice versa.

12.2.2   Basic equations

Historically, the MHD equations were constructed as a unification of the equations of slow electromagnetism and fluid mechanics. In MKS units Maxwell's equations are

$$\nabla \times \mathbf{H} = \mathbf{j} + \frac{\partial \mathbf{D}}{\partial t}, \quad \nabla \cdot \mathbf{B} = 0, \tag{12.3a,b}$$

$$\nabla \times \mathbf{E} = -\frac{\partial \mathbf{B}}{\partial t}, \quad \nabla \cdot \mathbf{D} = \rho_c, \tag{12.4a,b}$$

where for a vacuum or a low-density plasma $\mathbf{B} = \mu\mathbf{H}$ and $\mathbf{D} = \epsilon\mathbf{E}$.

Here $\mathbf{H}$ is the magnetic field, $\mathbf{B}$ the magnetic induction (although it is loosely referred to as the magnetic field), μ the magnetic permeability of free space, $\mathbf{E}$ the electric field, $\mathbf{D}$ the electric displacement, ϵ the electrical permittivity of free space, ρ_c the charge density, $\mathbf{j}$ the electric current density. These are supplemented by Ohm's Law

$$\mathbf{E} = \mathbf{j}/\sigma, \tag{12.5}$$

where σ is the electrical conductivity.

In contrast, the equations of fluid mechanics are

$$\rho\frac{d\mathbf{v}}{dt} = -\nabla p, \quad \frac{d\rho}{dt} + \rho\nabla \cdot \mathbf{v} = 0, \quad p = \mathcal{R}\rho T \tag{12.6a,b,c}$$

and an energy equation, where ρ is the plasma density (the mass per unit volume), **v** the plasma velocity, p the plasma pressure, T the temperature, $\mathcal{R}$ the gas constant. The operator $d/dt = \partial/\partial t + \mathbf{v} \cdot \nabla$ is the total (or material) derivative. It represents the time rate of change moving with an element of plasma and may be compared with $\partial/\partial t$, which represents the time rate of change at a fixed point of space.

Equation (6a) is the equation of motion, which represents the principle that the mass times acceleration of a moving element of plasma equals the sum of the forces acting on the element (here just the pressure gradient, $-\nabla p$). Equation (6b) is the equation of mass continuity and is simply an expression of the fact that no plasma is created or destroyed. Equation (6c) is the perfect gas law, stating that the pressure of a plasma is proportional to its density and temperature.

In a gas the electromagnetic (3–5) and fluid dynamic (6) equations are decoupled and so the electromagnetic and fluid properties are independent. However, in MHD, we modify the above equations in three ways:

(i)   A plasma feels an extra force, the so-called Lorentz force ($\mathbf{j} \times \mathbf{B}$), which is added on to the right-hand side of (6a). Its presence follows by analogy with the behaviour of an element **dl** of wire carrying a current J in a magnetic field which feels a force $J\mathbf{dl} \times \mathbf{B}$ perpendicular to the wire and to the field. The analogy implies then that a plasma element of volume dV carrying a current of density **j** per unit volume feels a force $\mathbf{j}dV \times \mathbf{B}$.

(ii) Ohm's Law states that the electric field in a frame moving with the plasma is proportional to the current, but the total electric field on moving plasma is $\mathbf{E} + \mathbf{v} \times \mathbf{B}$, where **E** is the field acting on plasma at rest, so (5) is modified by adding $\mathbf{v} \times \mathbf{B}$ to the left-hand side.

(iii) We consider processes with plasma speeds much slower than the speed of light ($v \ll c$), so that the displacement current ($\partial \mathbf{D}/\partial t$) in (3a) is negligible.

12.2.3   Induction equation

With the above assumptions, Equations (3), (4) and (5) become

$$\mathbf{j} = \nabla \times \mathbf{B}/\mu, \quad \frac{\partial \mathbf{B}}{\partial t} = -\nabla \times \mathbf{E}, \qquad\qquad (12.7\text{a,b})$$

$$\mathbf{E} = -\mathbf{v} \times \mathbf{B} + \mathbf{j}/\sigma, \quad \text{where } \nabla \cdot \mathbf{B} = 0. \qquad\qquad (12.8\text{a,b})$$

We may therefore eliminate **j** and **E** by substituting for **j** from (7a) in (8a) and for **E** from (8a) in (7b), with the result that $\partial \mathbf{B}/\partial t = \nabla \times (\mathbf{v} \times \mathbf{B}) - \eta \nabla \times (\nabla \times \mathbf{B})$, where $\eta = 1/(\mu\sigma)$ is the *magnetic diffusivity* (here assumed uniform). By expanding out the triple vector product in the last term and using (8b), we obtain finally

$$\frac{\partial \mathbf{B}}{\partial t} = \nabla \times (\mathbf{v} \times \mathbf{B}) + \eta \nabla^2 \mathbf{B}, \qquad\qquad (12.9)$$

which is known as the *induction equation*.

This is the basic equation for magnetic behaviour in MHD: it determines **B** once **v** is known. In electromagnetism the electric current and electric field are primary variables, with the current driven by electric fields, and then the magnetic field is a secondary variable produced by the currents. However, in MHD the basic physics is quite different since the plasma velocity (**v**) and magnetic field (**B**) are the primary variables, determined by the induction equation and the equation of motion, while the resulting current (**j**) and electric field (**E**) are secondary and may be deduced from (7) and (8) if required.

If V_0, l_0 are typical values for the velocity and length-scale over which the quantities are varying, the ratio of the first to the second term on the right-hand side of (9) is, in order of magnitude, the *magnetic Reynolds number*

$$R_m = \frac{l_0 V_0}{\eta}. \tag{12.10}$$

Thus, for example, in an active region where $\eta \approx 1 m^2 s^{-1}$, $l_0 \approx 10^5 m$, $V_0 \approx 10^4 m s^{-1}$, we find $R_m \approx 10^9$ and so the second term on the right of (9) is completely negligible. This is the case in almost all of the solar atmosphere - the only exception is in regions (such as current sheets) where the length-scale is extremely small, so small that $R_m \overset{<}{\sim} 1$ and the second term on the right of (9) becomes important.

If $R_m \ll 1$, the induction equation reduces to

$$\frac{\partial \mathbf{B}}{\partial t} = \eta \nabla^2 \mathbf{B}, \tag{12.11}$$

and so **B** is governed by a diffusion equation, which implies that field variations (irregularities) on a scale L_0 diffuse away on a time-scale of

$$\tau_d = \frac{L_0^2}{\eta}, \tag{12.12}$$

which is obtained simply by equating the orders of magnitude of both sides of (11). The corresponding speed at which they diffuse is

$$v_d = \frac{L_0}{\tau_d} = \frac{\eta}{L_0}. \tag{12.13}$$

With $\eta \approx 1 m^2 s^{-1}$, the decay time for a sunspot is (with $L_0 = 10^6 m$) 10^{12} sec = 30,000 years, so that the process whereby sunspots are observed to disappear in a few years, cannot just be diffusion. Similarily, the diffusion time for a magnetic field pervading the Sun as a whole (with $L_0 = 7 \times 10^8 m$) is 5×10^{17} sec = 10^{10} years. This is of the order of the age of the Sun, so a magnetic field that was present at its formation has not had time to diffuse much.

The main reason for variations in R_m from one phenomenon to another is variations in the appropriate length-scale L_0. If $R_m \gg 1$, the induction equation becomes

$$\frac{\partial \mathbf{B}}{\partial t} = \nabla \times (\mathbf{v} \times \mathbf{B}), \tag{12.14}$$

and Ohm's Law reduces to $\mathbf{E} + \mathbf{v} \times \mathbf{B} = 0$ so that the total electric field vanishes. Consider a curve $\mathcal{C}$ (bounding a surface $\mathcal{S}$) which is moving with the plasma, in a time dt an element $\mathbf{ds}$ of $\mathcal{C}$ sweeps out an element of area $\mathbf{v}dt \times \mathbf{ds}$. The rate of change $d/dt \int \mathbf{B} \cdot \mathbf{dS}$ of magnetic flux through $\mathcal{C}$ then consists of two parts, namely

$$\int_{\mathcal{S}} \frac{\partial \mathbf{B}}{\partial t} \cdot \mathbf{dS} + \int_{\mathcal{C}} \mathbf{B} \cdot \mathbf{v} \times \mathbf{ds}.$$

As $\mathcal{C}$ moves, so the flux changes partly because the magnetic field is changing with time (the first term) and partly because of the motion of the boundary (the second term). Then, by putting $\mathbf{B} \cdot \mathbf{v} \times \mathbf{ds} = -\mathbf{v} \times \mathbf{B} \cdot \mathbf{ds}$ and applying Stokes theorem to the second term, we obtain

$$\frac{d}{dt} \int \mathbf{B} \cdot \mathbf{dS} = \int_{\mathcal{S}} \left(\frac{\partial \mathbf{B}}{dt} - \nabla \times (\mathbf{v} \times \mathbf{B}) \right) \cdot \mathbf{dS},$$

which vanishes in the present approximation.

Thus the total magnetic flux through C remains constant as it moves with the plasma. In other words, we have proved *magnetic flux conservation*, namely, that plasma elements that form a flux tube initially do so at all later times (Figure 12.1). There is also *magnetic field line conservation*, namely that, if two plasma elements lie on a field line initially, they will always do so (Figure 12.2). At $t = t_1$, say, suppose elements P_1 and P_2 lie on a field line, which may be defined as the intersection of two flux tubes. Then at some later time, $t = t_2$, by magnetic flux conservation P_1

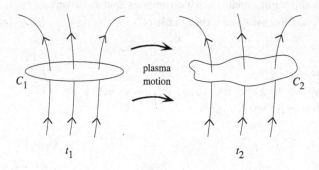

Figure 12.1. Magnetic flux conservation - if a curve C_1 is distorted into C_2 by a plasma motion, the flux through C_1 at t_1 equals the flux through C_2 at t_2.

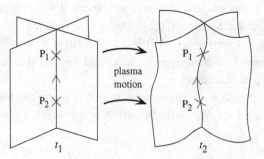

Figure 12.2. Magnetic field line conservation - if plasma elements P_1 and P_2 lie on a field line at t_1, then they will lie also on a field line at t_2.

and P_2 will still lie on both tubes, and so they will lie on the field line defined by their intersection. We interpret the above results by saying that the magnetic field lines move with the plasma - we say that they are *frozen into the plasma* - plasma can move freely along field lines, but in motion perpendicular to them they are dragged with the plasma or vice versa.

12.2.4 The Lorentz force

The induction equation is one equation relating our fundamental variables **v** and **B**; the other is the equation of motion

$$\rho \frac{d\mathbf{v}}{dt} = -\nabla p + \mathbf{j} \times \mathbf{B}. \tag{12.15}$$

The term $-\nabla p$ is the plasma pressure gradient: it acts from regions of high plasma pressure to low pressure and is perpendicular to the isobars (curves of constant pressure). The magnetic force ($\mathbf{j} \times \mathbf{B}$) is perpendicular to the magnetic field, and so any plasma acceleration parallel to the magnetic field must be caused by other forces. By substituting for $\mathbf{j}$ from Ampère's law (12.7a) and using an identity for the triple vector product, we can write it as

$$\mathbf{j} \times \mathbf{B} = (\nabla \times \mathbf{B}) \times \mathbf{B}/\mu = (\mathbf{B} \cdot \nabla)\frac{\mathbf{B}}{\mu} - \nabla \left(\frac{B^2}{2\mu} \right). \tag{12.16}$$

The two terms on the right-hand side have important physical interpretations. Since the second term has the same form as $-\nabla p$, we can say that it represents the effect of a *magnetic pressure* of magnitude $B^2/(2\mu)$. It gives a force when B^2 varies with position, and the direction of the force is from regions of high magnetic pressure to low magnetic pressure.

The first term represents the effect of a *magnetic tension* parallel to the magnetic field of magnitude B^2/μ per unit area. It produces a force when the field lines are curved, just like an elastic band or rope. By writing $\mathbf{B} = B\hat{\mathbf{s}}$ in terms of the unit vector $\hat{\mathbf{s}}$ along the magnetic field, the tension term $(\mathbf{B} \cdot \nabla)\mathbf{B}/\mu$ may be written

$$\frac{B}{\mu}\frac{d}{ds}(B\hat{\mathbf{s}}) = \frac{B^2}{\mu}\frac{d\hat{\mathbf{s}}}{ds} + \frac{B}{\mu}\frac{dB}{ds}\hat{\mathbf{s}} = \frac{B^2}{\mu}\frac{\hat{\mathbf{n}}}{\mathcal{R}} + \frac{d}{ds}\left(\frac{B^2}{2\mu} \right)\hat{\mathbf{s}},$$

where $\hat{\mathbf{n}}$ is the principal normal and $\mathcal{R}$ is the radius of curvature. The second term on the right of this equation is irrelevant, since it cancels with the component of $-\nabla(B^2/(2\mu))$ parallel to $\mathbf{B}$, as it must since $\mathbf{j} \times \mathbf{B}$ is perpendicular to $\mathbf{B}$. However, the first term on the right is the magnetic tension term, which shows that when the radius of curvature is small the tension force is large.

Thus we have shown that the magnetic force has two distinct effects: the magnetic pressure tries to compress the plasma and produces a net force if B varies with position; the magnetic tension tends to make the field lines shorten themselves like elastic bands and gives a net force if the field lines are curved.

12.3 Magnetohydrostatics

12.3.1 Introduction

With gravity included, the equation of motion is

$$\rho \frac{d\mathbf{v}}{dt} = -\nabla p + \mathbf{j} \times \mathbf{B} + \rho \mathbf{g}, \tag{12.17}$$

where

$$\mathbf{j} = \nabla \times \mathbf{B}/\mu, \tag{12.18}$$

and, if L, v_0, L/v_0 are typical values for the length-scale, plasma velocity and time-scale, respectively, the orders of magnitude of the current from (12.18) is $j_0 = B_0/(\mu L_0)$. Then, in terms of the typical density (ρ_0), pressure (p_0) and magnetic field (B_0), the sizes of each term in equation (12.17) are $\rho_0 v_0^2/L$, p_0/L, $B_0^2/(\mu L)$, $\rho_0 g$, respectively.

Now, provided the magnetic term is of the same order as the largest force term, we have *force balance* if the first term is much smaller than the third, namely

$$v_0^2 \ll \frac{B_0^2}{\mu \rho_0} \equiv v_A^2, \tag{12.19}$$

where v_A is known as the Alfvén speed. Thus, provided magnetic structures in the solar atmosphere are moving or evolving much more slowly than the Alfvén speed (typically 1000 km s^{-1} in the corona), they may be regarded as passing through a series of equilibria. Then (3.1) reduces to the equation for *magnetohydrostatic force balance*

$$0 = -\nabla p + \mathbf{j} \times \mathbf{B} + \rho \mathbf{g}, \tag{12.20}$$

where $\mathbf{j} = \nabla \times \mathbf{B}/\mu$, $\nabla \cdot \mathbf{B} = 0$ and $\rho = p/(\mathcal{R}T)$.

When gravity is negligible in this equation it reduces to *magnetostatic balance*

$$0 = -\nabla p + \mathbf{j} \times \mathbf{B}. \tag{12.21}$$

This arises when the fourth term in (3.1) is much smaller than the third, namely $l \ll B_0^2/(\mu \rho_0 g) = 2H/\beta$, where

$$H = \frac{p_0}{\rho_0 g} = \frac{\mathcal{R}T_0}{g} \tag{12.22}$$

is the *pressure scale height* and

$$\beta = \frac{p_0}{B_0^2/(2\mu)} \tag{12.23}$$

is the *plasma beta*, namely the ratio of plasma to magnetic pressure.

In turn, in (12.21) the magnetic force dominates if $\frac{1}{2}\beta \ll 1$, and then it reduces further to the equation

$$\mathbf{j} \times \mathbf{B} = \mathbf{0}$$

for a *force-free field*, in which the magnetic field is in equilibrium with itself under a balance between magnetic pressure and magnetic tension forces.

Suppose now that gravity is directed vertically downwards in the negative z-direction $\mathbf{g} = -g\hat{\mathbf{z}}$. Then the component of (12.20) parallel to a particular magnetic field line is $0 = -dp/ds - \rho g \cos\theta$, where s is measured along the field, or, since $ds \cos\theta = dz$ we have $0 = -dp/dz - \rho g$. After putting $\rho = p/(\mathcal{R}T)$ this becomes

$$\frac{dp}{dz} = -p\frac{g}{\mathcal{R}T},$$

which may be integrated to give

$$p = p_0 \exp\left(-\int_0^z \frac{g}{\mathcal{R}T}dz\right), \tag{12.24}$$

where p_0 is the pressure at the base ($z = 0$) of the field line. If the variation $T(z)$ of the temperature with height is known, (12.24) determines the pressure and therefore the density.

If in particular the temperature is uniform ($T = T_0$) then

$$p = p_0 e^{-z/H}, \tag{12.25}$$

so that the pressure (and density) decrease exponentially with height, with H being the vertical distance over which the pressure falls off by a factor e. Down in the photosphere where $T_0 = 5000$K, the scale height is about 150 km or less. By contrast, in the corona where $T_0 = 2 \times 10^6$K, say, the scale height is about 100 Mm, and so the density falls off much more slowly. Indeed for many purposes in the corona we can neglect the effect of gravity - i.e., when the vertical scales of interest are less than 100 Mm.

12.3.2 Potential fields

If the pressure gradients and gravitational force are negligible, a particular case of a force-free field is when the current vanishes, so that

$$\nabla \times \mathbf{B} = \mathbf{0} \tag{12.26}$$

where

$$\nabla \cdot \mathbf{B} = 0. \tag{12.27}$$

Equation (12.26) may be satisfied identically by putting $\mathbf{B} = \nabla \psi$ and then (12.27) gives Laplace's equation

$$\nabla^2 \psi = 0. \tag{12.28}$$

Thus many of the general results of potential theory may be applied. For example: *If the normal field component (B_n) is imposed on the boundary S of a volume (V), then the potential solution inside V is unique.*

Another useful theorem is: *If B_n is prescribed on S, then the field with the minimum magnetic energy is the potential field.*

For example, it is known that during a solar flare the normal field component at the solar surface does not change appreciably, and so the magnetic energy source for such a flare must come from a sheared force-free field with energy in excess of potential.

Several types of potential solution are of interest.

(a)   We may calculate the field at a point $P(x, y)$ due to an imposed $B_n(x)$ on the x-axis. It is

$$\mathbf{B} = \frac{1}{\pi} \int \frac{B_n(x')}{R} \hat{\mathbf{R}} dx'. \tag{12.29}$$

A similar technique may be used to calculate the field above a plane or the surface of a sphere rather than a line.

(b)   Separable solutions in rectangular cartesian coordinates or in spherical or cylindrical polar coordinates may also be found.

12.3.3   Force-free fields

If pressure gradients and gravity are negligible we have

$$\mathbf{j} \times \mathbf{B} = \mathbf{0}, \tag{12.30}$$

so that $\mathbf{j} = \nabla \times \mathbf{B}/\mu$ is parallel to $\mathbf{B}$. In other words

$$\nabla \times \mathbf{B} = \alpha \mathbf{B}, \tag{12.31}$$

where α is a function of position. Such a magnetic field is called *force-free*. Equation (12.30) looks disarmingly simple, but very little has been done so far in understanding the nature of its solutions in general.

Taking the divergence of (12.31) we find $\nabla \cdot \nabla \times \mathbf{B} = \nabla \cdot (\alpha \mathbf{B})$, or

$$0 = \alpha \nabla \cdot \mathbf{B} + \mathbf{B} \cdot \nabla \alpha.$$

Thus, since $\nabla \cdot \mathbf{B}$ vanishes, $\mathbf{B} \cdot \nabla \alpha = 0$, which implies that, although in general α varies in space, it has a constant value along each field line.

If instead we take the curl of (12.31)

$$\nabla \times (\nabla \times \mathbf{B}) = \nabla \times (\alpha \mathbf{B}),$$

or

$$\nabla(\nabla \cdot \mathbf{B}) - \nabla^2 \mathbf{B} = \alpha \nabla \times \mathbf{B} + \mathbf{B} \times \nabla \alpha.$$

Thus, after putting $\nabla \cdot \mathbf{B} = 0$ and substituting for $\nabla \times \mathbf{B}$ from (12.31), we have

$$-\nabla^2 \mathbf{B} = \alpha^2 \mathbf{B} + \mathbf{B} \times \nabla \alpha. \tag{12.32}$$

However, in general this is no simpler than the original equation (12.32) since it represents three coupled equations for the three coefficients of $\mathbf{B}$.

In the particular case when α is uniform and so possesses the same constant value on all field lines, (12.32) reduces to the simple equation

$$(\nabla^2 + \alpha^2)\mathbf{B} = 0. \tag{12.33}$$

Its solutions give so-called "constant-α" or linear force-free fields and may be found by a variety of techniques, generalising many of the techniques for solving Laplace's equation. The beauty of (12.33) is that it is a linear equation so that the solutions may be superposed, but the disadvantage is that often the solutions have undesirable properties such as unphysical field reversals at large distances.

There are several general theorems which lend an air of mystery to the structure of force-free fields, since very little is known of their general properties:

 (i) If the flux and topological connections on the surface S of a simply connected volume V are given and the field possesses a minimum energy, then it is force-free. However, the converse is not true, so that a force-free field does not necessarily produce a minimum energy.

 (ii) If $\mathbf{j} \times \mathbf{B}$ vanishes everywhere within V and on S, then the magnetic field $\mathbf{B}$ is identically zero. Thus, a nontrivial (i.e., non-zero) field that is force-free inside V must be stressed somewhere on S. In other words, force-free fields are possible but they must be anchored down somewhere on the boundary. You cannot build a force-free field from currents enclosed entirely within a volume.

 (iii) No magnetic field having a finite energy can be force-free everywhere. Thus a nontrivial field that is force-free everywhere must possess a singularity somewhere: this is also true in particular for a potential field where one is familiar with fields produced by, for instance, point monopoles or dipoles.

 (iv) An axisymmetric, force-free, poloidal magnetic field must be current-free.

 (v) For a perfectly conducting plasma in a closed volume V, the *magnetic helicity* is

$$K = \int_V \mathbf{A} \cdot \mathbf{B} \, dV, \tag{12.34}$$

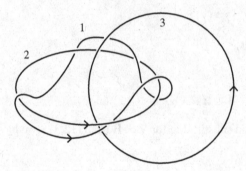

Figure 12.3. An example of three linked tubes with linkage numbers $L_{12} = 3$, $L_{13} = -1$ and $L_{23} = -1$.

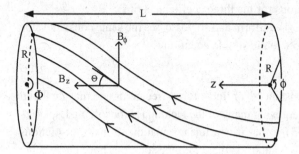

Figure 12.4. Notation for a straight magnetic flux tube.

where $\mathbf{B} = \nabla \times A$ is invariant, and the state of minimum energy is a linear force-free field. K is a measure of the sum of the twist and the linkage of flux tubes. Thus, for two linked tubes

$$K = \Theta_1 F_1^2 + \Theta_2 F_2^2 + 2LF_1F_2, \qquad (12.35)$$

where Θ_1 and Θ_2 are the twists of the two tubes, F_1, F_2 their fluxes and L the linking number (Figure 12.3). Reconnection can convert linkage helicity to twist helicity, but it preserves the total helicity.

12.3.4   Magnetic flux tubes

Straight tubes with cylindrical symmetry (Figure 12.4) have field lines and current lines lying on cylindrical surfaces and magnetic components in cylindrical polars of

$$\mathbf{B} = (0, B_\phi(R), B_z(R)), \qquad (12.36)$$

The amount Φ (the *twist*) by which a field line is twisted about the z-axis in going from one end of the tube to another is

$$\Phi = \frac{LB_\phi}{RB_z}. \qquad (12.37)$$

In general Φ varies with R and so field lines on different surfaces have different inclinations. The radial component of the magnetostatic force balance

$(\nabla p = \mathbf{j} \times \mathbf{B})$ is

$$\frac{dp}{dR} + \frac{d}{dR}\left(\frac{B_\phi^2 + B_z^2}{2\mu}\right) + \frac{B_\phi^2}{\mu R} = 0, \tag{12.38}$$

where the terms simply represent the plasma pressure gradient, the magnetic pressure gradient and the magnetic tension force, respectively. We are allowed to prescribe two of the four functions p, B_ϕ, B_z, Φ of R and deduce the other two from (12.37) and (12.38): which ones are prescribed depends on the extra physics of the particular problem being considered.

12.4   Magnetohydrodynamic waves

In a gas there are sound waves, which propagate equally in all directions at the sound speed. In a plasma there are also waves, but they are of several types. Waves are very important in the solar atmosphere and throughout the cosmos. For example, they may be seen propagating out of sunspots or away from large solar flares. They are also a prime candidate for heating the solar atmosphere.

For an elastic string one may derive the equations for a small displacement from equilibrium and show that they have the form of a wave equation with a wave speed equal to the square root of the tension divided by the mass density. The approach is similar for a plasma, where one sets up the equations for small disturbances of the plasma and finds that they too satisfy a wave equation.

12.4.1   Sound waves

Consider a uniform plasma at rest with no magnetic field, a pressure p_0 and density ρ_0. Suppose there is a small disturbance with velocity $\mathbf{v}'$, pressure $p_0 + p'$, density $\rho_0 + \rho'$, satisfying the equations

$$\rho\frac{d\mathbf{v}}{dt} = -\nabla p, \qquad \frac{\partial \rho}{\partial t} + \nabla \cdot (\rho\mathbf{v}) = 0, \qquad \frac{p}{\rho^\gamma} = K = \frac{p_0}{\rho_0^\gamma},$$

where motions are so rapid that there is no heat exchange and γ is the ratio of specific heats (5/3 for a fully ionised hydrogen plasma). These may be linearised, neglecting squares and products of small quantities, to give

$$\rho_0\frac{\partial \mathbf{v}'}{\partial t} = -\nabla p', \qquad \frac{\partial \rho'}{\partial t} + \rho_0\nabla \cdot \mathbf{v}' = 0, \qquad p' = c_s^2\rho', \tag{12.39}$$

where $c_s^2 = \gamma p_0/\rho_0$.

Suppose first of all that $\rho' = \rho'(x, t)$ and $\mathbf{v}' = v'(x, t)\hat{\mathbf{x}}$, so that (12.39) reduce to

$$\rho_0\frac{\partial v'}{\partial t} = -\frac{\partial p'}{\partial x}, \qquad \frac{\partial \rho'}{\partial t} + \rho_0\frac{\partial v'}{\partial x} = 0, \qquad p' = c_s^2\rho'.$$

If $\mathbf{v}'$ and p' are eliminated between these three equations, we obtain

$$\frac{\partial^2 \rho'}{\partial t^2} = c_s^2 \frac{\partial^2 \rho'}{\partial x^2}, \tag{12.40}$$

so that ρ' satisfies a wave equation and disturbances propagate along the x-axis with speed c_s (the *sound speed*). Formally, we may Fourier analyse an arbitrary disturbance into components, since this is a linear equation, and consider each component separately, namely

$$\rho' = const \times e^{i(kx - \omega t)}, \tag{12.41}$$

where ω is the wave frequency, k the wave number, $2\pi/k$ the wavelength and $2\pi/\omega$ the period of the wave. Then $\partial \rho'/\partial t = -i\omega \rho'$ and $\partial \rho'/\partial x = ik\rho'$, so that (12.40) gives

$$\omega^2 = k^2 c_s^2. \tag{12.42}$$

This is the *dispersion relation* for sound waves, relating ω and k. The solution (12.41) represents a wave moving without change of shape at constant speed $\omega/k = c_s$, which is known as the *phase speed* of the wave.

More generally, for an arbitrary linear disturbance $\rho'(x, y, z, t)$, Equation (12.39) implies, after eliminating $\mathbf{v}'$ and p', that

$$\frac{\partial^2 \rho'}{\partial t^2} = c_s^2 \nabla^2 \rho'. \tag{12.43}$$

Instead of (12.41) we now consider a general Fourier component

$$\rho' = \text{constant} \times e^{i(\mathbf{k}\cdot\mathbf{r} - \omega t)}. \tag{12.44}$$

Then $\partial \rho'/\partial x = ik_x \rho'$, $\partial \rho'/\partial y = ik_y \rho'$, $\partial \rho'/\partial z = ik_z \rho'$ or, in compact notation, $\nabla \rho' = i\mathbf{k}\rho'$, and, as before, $\partial \rho'/\partial t = -i\omega \rho'$.

Equation (12.43) therefore implies the dispersion relation

$$\omega^2 = k^2 c_s^2, \tag{12.45}$$

so that the waves propagate equally in all directions with phase speed c_s. In addition, Equation (12.39a) becomes $\rho_0 \omega \mathbf{v}' = \mathbf{k} p'$, so that $\mathbf{v}'$ is parallel to the direction $\mathbf{k}$ of propagation. In other words, we have a longitudinal wave, with *phase velocity* $(\omega/k)\mathbf{k} = c_s \mathbf{k}$.

The overall procedure therefore for considering simple linear waves is to start with the MHD equations for variations from equilibrium of a uniform plasma. Then, by neglecting squares and products of small quantities, these are reduced to a set of linear partial differential equations. By looking for plane-wave solutions of the form (12.44), these are in turn reduced to a set of homogeneous algebraic equations for the amplitudes of the variables, which leads to the dispersion relation.

12.4.2 Alfvén waves

On disturbing a uniform magnetic field, one would expect, by analogy with an elastic band, the magnetic tension to make a wave propagate along the field lines with speed

$$v_A = \sqrt{\left(\frac{\text{tension}}{\rho}\right)} = \frac{B}{\sqrt{(\mu\rho)}}, \qquad (12.46)$$

known as the *Alfvén speed*, since the tension is B^2/μ. Consider, therefore, an ideal plasma (with negligible dissipation), initially at rest with a uniform field $\mathbf{B}_0 = B_0\hat{\mathbf{z}}$ and density ρ_0. The effect of a disturbance is to introduce a velocity $\mathbf{v}'$ and to make the other variables $\mathbf{B}_0 + \mathbf{B}'$, $\rho_0 + \rho'$. Suppose it is so small that squares and products of v', B', ρ' can be neglected. Then, the pressureless MHD equations become

$$\frac{\partial \mathbf{B}'}{\partial t} = \nabla \times (\mathbf{v}' \times \mathbf{B}_0), \qquad (12.47)$$

$$\rho_0 \frac{\partial \mathbf{v}'}{\partial t} = (\nabla \times \mathbf{B}') \times \frac{\mathbf{B}_0}{\mu}, \qquad (12.48)$$

where $\nabla \cdot \mathbf{B}' = 0$ and ρ' is given by

$$\frac{\partial \rho'}{\partial t} + \nabla \cdot (\rho_0 \mathbf{v}') = 0, \qquad (12.49)$$

and p' by $p' = c_s^2 \rho'$.

Look for wave-like solutions by supposing the perturbation quantities behave like $\exp[i(\mathbf{k} \cdot \mathbf{r} - \omega t)]$ so that (12.47) and (12.48) reduce to

$$-\omega \mathbf{B}' = \mathbf{k} \times (\mathbf{v}' \times \mathbf{B}_0) = (\mathbf{B}_0 \cdot \mathbf{k})\mathbf{v}' - \mathbf{B}_0(\mathbf{k} \cdot \mathbf{v}'), \qquad (12.50)$$

$$-\mu\rho_0\omega \mathbf{v}' = (\mathbf{k} \times \mathbf{B}') \times \mathbf{B}_0 = \mathbf{B}'(\mathbf{B}_0 \cdot \mathbf{k}) - \mathbf{k}(\mathbf{B}' \cdot \mathbf{B}_0), \qquad (12.51)$$

where $\mathbf{k} \cdot \mathbf{B}' = 0$, $p' = c_s^2 \rho'$ and

$$-\omega\rho' + \rho_0 \mathbf{k} \cdot \mathbf{v}' = 0. \qquad (12.52)$$

For *Alfvén waves* propagating along the magnetic field $\mathbf{B}_0$, we make $\mathbf{k}$ parallel to $\mathbf{B}_0$ and assume $\mathbf{v}'$ is normal to $\mathbf{k}$. Then (12.50) and (12.51) reduce to

$$-\omega \mathbf{B}' = B_0 k \mathbf{v}', \qquad (12.53)$$

$$-\mu\rho_0\omega \mathbf{v}' = B_0 k \mathbf{B}' - \mathbf{k}(\mathbf{B}' \cdot \mathbf{B}_0), \qquad (12.54)$$

where the last term vanishes, since by (12.53) $\mathbf{B}'$ is parallel to $\mathbf{v}'$, which is normal to $\mathbf{k}$ by assumption, and $\mathbf{B}_0$ is parallel to $\mathbf{k}$ so that $\mathbf{B}' \cdot \mathbf{B}_0 = 0$. Thus (12.53) and (12.54) imply that $\omega^2 = k^2 v_A^2$.

More generally, the Alfvén waves may also propagate at some angle θ to $\mathbf{B}_0$, in which case we still assume $\mathbf{k} \cdot \mathbf{v}' = 0$. Then (12.50) and (12.51) this time become

$$-\omega \mathbf{B}' = (\mathbf{B}_0 \cdot \mathbf{k})\mathbf{v}'$$

$$-\mu\rho\omega \mathbf{v}' = (\mathbf{B}_0 \cdot \mathbf{k})\mathbf{B}' - \mathbf{k}(\mathbf{B}' \cdot \mathbf{B}_0),$$

where again $\mathbf{B}' \cdot \mathbf{B_0} = 0$ (which may be seen most easily by taking $\mathbf{k}\cdot$ the second equation and remembering that $\mathbf{k} \cdot \mathbf{v}' = \mathbf{k} \cdot \mathbf{B}' = 0$). Thus these two equations give the dispersion relation for Alfvén waves as

$$\omega^2 = k^2 v_A^2 \cos^2 \theta, \tag{12.55}$$

since $\mathbf{B_0} \cdot \mathbf{k} = B_0 k \cos \theta$ and $v_A^2 = B_0^2/(\mu \rho_0)$.

By comparison with sound waves, which are longitudinal and have non-zero density perturbation (ρ'), Alfvén waves are transverse in the sense that $\mathbf{v}'$ is perpendicular to the direction $\mathbf{k}$ of propagation and so by (12.52) ρ' (and therefore also p') vanishes. The phase velocity is $\omega/k = \pm v_A \cos \theta$ in the direction of $\mathbf{k}$, so that the speed of propagation depends on the direction and there is no propagation perpendicular to $\mathbf{B_0}$ ($\theta = \frac{1}{2}\pi$).

12.4.3 Compressional Alfvén waves

Consider now the case when $\mathbf{k} \cdot \mathbf{v}' \neq 0$. Substitute for $\mathbf{B}'$ from (12.50) in (12.51), so that

$$\mu \rho_0 \omega^2 \mathbf{v}' = [(\mathbf{B_0} \cdot \mathbf{k})\mathbf{v}' - \mathbf{B_0}(\mathbf{k} \cdot \mathbf{v}')](\mathbf{B_0} \cdot \mathbf{k}) - \mathbf{k}[(\mathbf{B_0} \cdot \mathbf{k})(\mathbf{v}' \cdot \mathbf{B_0}) - B_0^2(\mathbf{k} \cdot \mathbf{v}')]. \tag{12.56}$$

This represents three linear homogeneous equations for three unknowns (v_x', v_y', v_z') and so in principle the determinant of coefficients would give a relation between the coefficients, namely the dispersion relation. But since $\mathbf{v}'$ only appears in the forms $\mathbf{v}'$, $\mathbf{k} \cdot \mathbf{v}'$ and $\mathbf{B_0} \cdot \mathbf{v}'$, we may take $\mathbf{k}\cdot$ and $\mathbf{B_0}\cdot$ (12.56) to give two equations for $\mathbf{k} \cdot \mathbf{v}'$ and $\mathbf{B_0} \cdot \mathbf{v}'$ namely,

$$\mu \rho_0 \omega^2 (\mathbf{B_0} \cdot \mathbf{v}') = 0, \tag{12.57}$$

and

$$\mu \rho_0 \omega^2 (\mathbf{k} \cdot \mathbf{v}') = k^2 B_0^2 (\mathbf{k} \cdot \mathbf{v}'). \tag{12.58}$$

Thus, from (12.58), either $\mathbf{k} \cdot \mathbf{v}' = 0$ or

$$\omega^2 = k^2 v_A^2, \tag{12.59}$$

which is the dispersion relation for compressional Alfvén waves. These waves propagate equally in all directions, like sound waves, and, since $\mathbf{k} \cdot \mathbf{v}' \neq 0$, (12.52) implies that ρ' and p' are in general non-zero. For propagation across the field ($\mathbf{k} \cdot \mathbf{B_0} = 0$) it can easily be seen from (12.56) that $\mathbf{v}'$ is parallel to $\mathbf{k}$ and therefore the mode is longitudinal.

12.4.4 Magnetoacoustic waves

We have found that there are two waves when the pressure vanishes, namely the Alfvén and compressional Alfvén waves, and one wave when the magnetic field vanishes, namely the sound wave. If pressure fluctuations are included in the MHD equations by adding a term $-\nabla p'$ to the right of (12.48), the effect is to add a term $-\mu \mathbf{k} p'$ to the right of (12.51) and $\mathbf{k} c_s^2 \mu \rho_0 (\mathbf{k} \cdot \mathbf{v}')$ to the right of (12.56). The Alfvén waves (for which $\mathbf{k} \cdot \mathbf{v}'$ vanishes) are unaltered since ρ' and p' vanish. However, the sound and compressional Alfvén waves are coupled together to give two *magneto-acoustic waves*.

Equations (12.57) and (12.58) become

$$-\omega^2 (\mathbf{B_0} \cdot \mathbf{v}') = -(\mathbf{B_0} \cdot \mathbf{k}) c_s^2 (\mathbf{k} \cdot \mathbf{v}'), \tag{12.60}$$

$$(\mu \rho_0 \omega^2 - k^2 c_s^2 \mu \rho_0 - k^2 B_0^2)(\mathbf{k} \cdot \mathbf{v}') = -k^2 (\mathbf{B_0} \cdot \mathbf{k})(\mathbf{B_0} \cdot \mathbf{v}'). \tag{12.61}$$

Thus, either $\mathbf{k} \cdot \mathbf{v}'$ and $\mathbf{B_0} \cdot \mathbf{v}'$ both vanish, when (12.56) gives the dispersion relation (12.55) for Alfvén waves, or (12.60) and (12.61) imply

$$\omega^4 - \omega^2 k^2 (c_s^2 + v_A^2) + c_s^2 v_A^2 k^4 \cos^2 \theta = 0. \tag{12.62}$$

This is the dispersion relation for slow and fast magnetoacoustic waves. The smallest root for ω^2 / k^2 gives the slow mode and the largest root the fast mode. The particular cases $p_0 = 0$ (i.e., $c_s^2 = 0$) and $B_0 = 0$ (i.e., $v_A^2 = 0$) reduce to the dispersion relations ($\omega^2 = k^2 v_A^2$ and $\omega^2 = k^2 c_s^2$) for compressional Alfvén and sound waves, respectively, as expected.

12.4.5 Shock waves

(i)   *Hydrodynamic shock*

Small-amplitude sound waves propagate without change of shape, but when the amplitude is finite the crest can move faster than its trough, causing a progressive steepening. Ultimately, the gradients become so large that dissipation becomes important, and a steady shock wave shape may be attained with a balance between the steepening effect of the nonlinear convective term and the broadening effect of dissipation. The dissipation inside the shock front converts the energy being carried by the wave gradually into heat. The effect of the passage of the shock is to compress and heat the gas.

We model a shock as a plane discontinuity, although in reality it is a very thin transition region (Figure 12.5). The shock travels at speed U, say, into a gas at rest and accelerates the shocked gas to a speed U_2. In a frame of reference moving with the shock the fluid ahead has speed $v_1 = U$, density ρ_1 and pressure p_1, while the corresponding variables behind the shock are $v_2 = U - U_2$, ρ_2 and p_2, say.

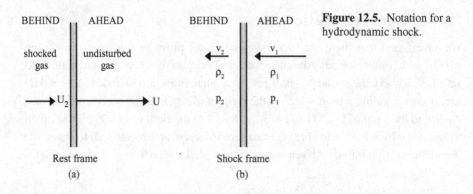

Figure 12.5. Notation for a hydrodynamic shock.

Conservation of mass, momentum and energy then gives

$$\rho_2 v_2 = \rho_1 v_1, \tag{12.63}$$

$$p_2 + \rho_2 v_2^2 = p_1 + \rho_1 v_1^2 \tag{12.64}$$

$$\rho_2 v_2 + \left(\rho_2 e_2 + \tfrac{1}{2}\rho_2 v_2^2\right) v_2 = p_1 v_1 + \left(\rho_1 e_1 + \tfrac{1}{2}\rho_1 v_1^2\right)v_1, \tag{12.65}$$

where $e = p/[(\gamma - 1)\rho]$ is the internal energy. These may be solved to give the *Rankine-Hugoniot jump relations*:

$$\frac{\rho_2}{\rho_1} = \frac{(\gamma + 1)M_1^2}{2 + (\gamma - 1)M_1^2}, \frac{v_2}{v_1} = \frac{2 + (\gamma - 1)M_1^2}{(\gamma + 1)M_1^2}, \frac{p_2}{p_1} = \frac{2\gamma M_1^2 - (\gamma - 1)}{\gamma + 1}, \tag{12.66}$$

which are supplemented by the condition for an isolated system from the second law of thermodynamics that the entropy $s = c_v \log(p/\rho^\gamma)$ must increase, namely, $s_2 \geq s_1$, where c_v is the specific heat at constant volume and $M_1 = v_1/c_{s1}$ is the *Mach number*, namely the ratio of the shock speed to the sound speed ($c_{s1} = (\gamma p_1/\rho_1)^{1/2}$).

Consequences of these equations are:

(i) $M_1 \geq 1$, so that the shock speed (v_1) exceeds the sound speed ahead of the shock;

(ii) $v_2 \leq c_{s2}$, so that in the shock frame the flow is subsonic behind the shock (and supersonic ahead of it);

(iii) $p_2 \geq p_1$ and $\rho_2 \geq \rho_1$, so that the shock is compressive;

(iv) $v_2 \leq v_1$ and $T_2 \geq T_1$, so that the maximum density ratio is $(\gamma + 1)/(\gamma - 1)$, whereas the pressure ratio increases with M_1, like M_1^2.

(ii)   *Perpendicular magnetic shock*

In the presence of a magnetic field, we now have three wave modes, and when their amplitudes are large the Alfvén wave can propagate without steepening, whereas the slow and fast magnetoacoustic modes steepen to form shocks. Derivation of the jump relations is more complicated, since: there is an extra variable (**B**); **B** and **v** may be inclined away from the shock normal; and the entropy condition is replaced by an "evolutionary condition".

The jump relations for mass, momentum, energy and magnetic flux of a perpendicular shock now imply that

$$\frac{v_2}{v_1} = \frac{1}{X}, \qquad \frac{B_2}{B_1} = X, \qquad \frac{p_2}{p_1} = \gamma M_1^2\left(1 - \frac{1}{X}\right) - \frac{1 - X^2}{\beta_1},$$

where $\beta_1 = 2\mu p_1/B_1^2$ and $X = \rho_2/\rho_1$ is the density ratio, which is the positive solution of

$$2(2 - X)X^2 + [2\beta_1 + (\gamma - 1)\beta_1 M_1^2 + 2]\gamma X - \gamma(\gamma + 1)\beta_1 M_1^2 = 0. \quad (12.67)$$

Consequences of these jump relations together with the evolutionary condition (that the perturbation caused by a small disturbance be small and unique) are:

  (i) Equation (12.67) has only one root since $1 < \gamma < 2$;
 (ii) the effect of the magnetic field is to reduce X below its hydrodynamic value;
(iii) the shock is compressive with $X \geq 1$;
 (iv) the shock speed (v_1) must exceed the fast magnetoacoustic speed $\sqrt{(c_{s1}^2 + v_A^2)}$ ahead of the shock;
  (v) magnetic compression is limited to the range $1 < B_2/B_1 < (\gamma + 1)/(\gamma - 1)$, where for $\gamma = 5/3$ the upper limit is 4.

(iii)   *Oblique magnetic shocks*

Set up axes in a frame moving with the shock, as before, and assume **v** and **B** lie in the xy-plane. If we choose axes moving parallel to the shock at such a speed that **v** is parallel to **B**, the jump relations simplify greatly and may be solved to give

$$\frac{v_{2x}}{v_{1x}} = \frac{1}{X}, \qquad \frac{v_{2y}}{v_{1y}} = \frac{v_1^2 - v_{A1}^2}{v_1^2 - X v_{A1}}, \qquad \frac{B_{2x}}{B_{1x}} = 1,$$

$$\frac{B_{2y}}{B_{1y}} = \frac{(v_1^2 - v_{A1}^2)X}{v_1^2 - X v_{A1}^2}, \qquad \frac{p_2}{p_1} = X + \frac{(\gamma - 1)X v_1^2}{2c_{s1}^2}\left(1 - \frac{v_2^2}{v_1^2}\right), \qquad (12.68)$$

where the compression ratio is

$$X = \frac{\rho_2}{\rho_1}, \qquad c_{s1}^2 = \frac{\gamma p_1}{\rho_1}, \qquad v_{A1}^2 = \frac{B_1^2}{\mu\rho_1}, \qquad \cos\theta = \frac{v_{1x}}{v_1}, \qquad (12.69)$$

and X is a solution of

$$(v_1^2 - X v_{A1}^2)^2\{X c_{sl}^2 + \tfrac{1}{2}v_1^2 + \tfrac{1}{2}v_1^2\cos^2\theta[X(\gamma - 1) - (\gamma + 1)]\}$$
$$+ \tfrac{1}{2}v_{A1}^2 v_1^2\sin^2\theta X\{[\gamma + X(2 - \gamma)]v_1^2 - X v_{A1}^2\{[(\gamma + 1) - X(\gamma - 1)]\} = 0.$$
$$(12.70)$$

Equation (12.70) has three solutions, which give the slow shock, Alfvén wave and fast shock, the forms of the resulting field lines being shown in Figure 12.6.

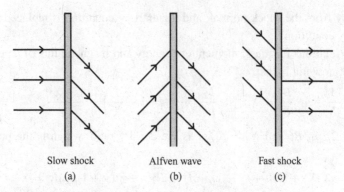

$$\text{Slow shock} \qquad \text{Alfven wave} \qquad \text{Fast shock}$$
$$\text{(a)} \qquad\qquad \text{(b)} \qquad\qquad \text{(c)}$$

Figure 12.6. Magnetic field lines for special oblique waves.

The slow and fast shocks have the following properties:

(i) they are compressive with $X > 1$ and $p_2 > p_1$;

(ii) they conserve the sign of B_y so that $B_{2y}/B_{1y} > 0$,

(iii) for the slow shock $B_2 < B_1$, so that **B** refracts towards the shock normal and B decreases as the shock passes;

(iv) for the fast shock $B_2 > B_1$, so that the shock makes **B** refract away from the normal and B increase;

(v) the flow speed (v_{1x}) ahead of the shock exceeds the appropriate wave speed, while the speed (v_{2x}) behind it is smaller than the wave speed;

(vi) the flow normal to the shock is slowed down ($v_{2x} < v_{1x}$);

(vii) in the limit as $B_x \to 0$, the fast shock becomes a perpendicular shock, while the slow shock becomes a *tangential discontinuity*, for which v_x and B_x vanish and there are arbitrary jumps in v_y and B_y, subject only to total pressure balance ($p_2 + B_2^2/(2\mu) = p_1 + B_1^2/(2\mu)$).

In the limit when $v_1 = v_{A1}$ and $X \neq 1$, (12.68) implies that $B_{2y} = 0$ and we have a *switch-off shock*. Since $\mathbf{v_1}$ and $\mathbf{B_1}$ are parallel, this implies that

$$v_{1x} = \frac{B_{1x}}{\sqrt{(\mu\rho)}}, \tag{12.71}$$

so that the shock propagates at the Alfvén speed based on the normal component of the magnetic field. For a shock propagating along $\mathbf{B_1}$, the fast-shock solution is $X = v_1^2/v_{A1}^2$ for a *switch-on shock*.

When $v_1 = v_{A1}$, the conservation of flux implies that $v_{2y}/v_{1y} = B_{2y}/B_{1y}$, while the equations for conservation of x-momentum and energy reduce to $p_2 = p_1$ and $B_{2y}^2 = B_{1y}^2$. Thus, in addition to the trivial solution ($\mathbf{B_2} = \mathbf{B_1}$, $\mathbf{v_2} = \mathbf{v_1}$), we also have

$$B_{2y} = -B_{1y} \quad , \quad B_{2x} = B_{1x},$$
$$v_{2y} = -v_{1y} \quad , \quad v_{2x} = v_{1x},$$

for a finite-amplitude *Alfvén wave*, sometimes called an *intermediate wave* or *rotational discontinuity*. The tangential component of magnetic field is reversed by the wave, but there is no change in pressure or density.

12.5   Concluding comment

We have here given a brief description of the basic processes involved in solar MHD, including an introduction to the MHD equations (Section 12.2) and to the subjects of magnetohydrostatics (Section 12.3) and MHD waves (Section 12.4). The ways in which these processes are at work in the Sun are described throughout the other chapters of this book and elsewhere.

The magnetic field of the Sun is generated in the solar interior by dynamo action (Chapter 6), as illuminated recently both by helioseismology results (Chapters 3–5) and by detailed observations of the magnetic properties of the solar surface (Chapters 7–8). Furthermore, many dynamic processes that both heat the plasma and drive motions in the chromosphere and transition region are probably caused by MHD effects (Chapters 9–11 and 17). Solar activity itself is essentially an MHD process driven in a cyclic manner (Chapter 13) and it involves sunspots and solar prominences, whose structure and behaviour is created by the magnetic field. In an eruptive solar flare, the onset of a lack of magnetic equilibrium causes an eruption and drives energy release by reconnection. Furthermore, coronal mass ejections are large-scale eruptions of the corona driven magnetically, as described by MHD. Again, MHD provides the crucial environment within which particle acceleration (Chapter 14) and radio processes (Chapter 15) take place. Finally, both coronal oscillations and the driving of the solar wind are also inherently magnetic processes.

References

Bateman, G (1978). *MHD Instabilities*, MIT Press, Cambridge, USA.

Benz, A.O. (1993). *Plasma Astrophysics*, Kluwer, Dordrecht, Netherlands.

Choudhuri, A.R. (1998). *Physics of Fluids and Plasmas*, Cambridge University Press, Cambridge.

Priest, E.R. (1982). *Solar MHD*, Reidel, Dordrecht, Netherlands.

Priest, E.R. and Forbes, T.G. (2000). *Magnetic Reconnection: MHD Theory and Applications*, Cambridge University Press, Cambridge.

13

Solar activity

Z. Švestka

CASS UCSD, La Jolla, California, USA and SRON Utrecht, The Netherlands

13.1 Solar cycles

Although the Sun illuminating our Earth looks like a steadily shining celestial body, its surface is actually the seat of continuous changes and powerful activity. As the *Solar and Heliospheric Observatory* (SOHO, Domingo, Fleck, and Poland, 1995) and the *Transition Layer and Coronal Explorer* (TRACE, Handy *et al.*, 1999) spacecraft in recent years revealed, in extreme ultraviolet lines the solar surface and solar atmospheric layers look highly turbulent: all the time and everywhere one can see brightness variations, loop formations and decays, plasma flows, and ejections of gas indicating permanent changes of the structures that we see in the solar atmosphere (e.g., Schrijver *et al.*, 1999).

However, this is not what we call 'solar activity' — all these changes are still considered to occur on the 'quiet Sun'. The real processes called *solar activity*, which have their impacts also on the Earth environment, appear only in limited parts of the solar surface, and their occurrence varies quasi-periodically with time, creating 11-year *cycles of solar activity*. Each new solar cycle is born close to the solar poles and its activity then slowly propagates to lower heliographic latitudes. The real length of one cycle is actually about 22 years, with magnetic polarities reversed on the northern and southern hemispheres during the two 11-year parts of the cycle. (For more about solar cycles see Harvey, 1992.)

When looking at the Sun in white light — which was the only possible way to observe the Sun before the early years of this century — we see the lowest level of the solar atmosphere which is called the *photosphere*. In the photosphere solar activity manifests itself as sunspots and groups of sunspots (Figure 13.1). Therefore,

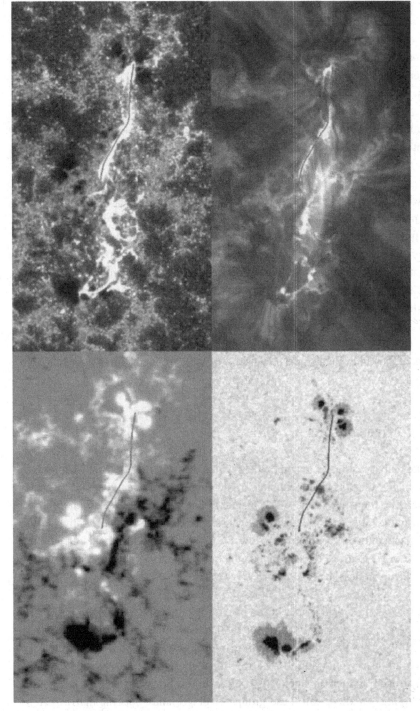

Figure 13.1. *Left*: a sunspot group in the photosphere (*below*) and the corresponding photospheric magnetic field (*above*). *Right*: images of the solar transition layer taken with TRACE. All images were taken during a flare at 09:12 UT on 11 December 1999. In the magnetogram black and white denote the opposite magnetic polarities. A line marks the position of a flare ribbon and helps the reader to align the images.

solar cycles were long characterized (and still are) by the so-called *Wolf numbers* or *relative sunspot numbers R*. A daily *R* is the sum of ten-times the number of sunspot groups on the Sun plus the number of individual sunspots in all of them. A monthly or yearly *R* is the average daily *R* during a month or a year.

Although we talk generally about '11-year solar cycle', individual solar cycles differ very much as their lengths are concerned: during the past three centuries their duration varied between 9 and 16 years. Also the strengths (heights) of the cycles greatly vary: between the early 18th century, when frequent sunspot observations began, and the present days the maxima of smoothed yearly means of *R* during 25 different cycles were observed as low as 46 and as high as 190; prior to that, in the 17th century, there are reliable indications that a period of extremely low cycles occurred, the so-called *Maunder Minimum* of solar activity (Eddy (1976), cf., Chapter 6 of this book), when the maxima of yearly *R* of at least 7 cycles did not exceed the value of 20.

When the active (visible) part of a new solar cycle begins, rare sunspot groups appear first at high latitudes (between 40 and 50 heliographic degrees) and as the frequency of their occurrence increases, their positions move progressively closer to the equator. A few years after the onset of the active cycle, when the spots appear mostly at latitudes below 25°, the cycle reaches its *maximum*. Thereafter the activity slowly declines, with spotgroups gradually approaching the equator, and eventually reaches a *minimum* when sunspot appearance becomes very rare and for many days the Sun can stay without any sunspot at all. However, usually still before the last sunspots of the old cycle disappear, sunspots of the new one begin to appear at high latitudes.

13.2   Active regions

This solar activity is due to magnetic field which exists in the Sun (see Chapter 6 of this book) and sunspots become visible in the photosphere when ropes of magnetic flux emerge to the solar surface. The magnetic flux in the central dark *umbra* of a spot is usually between 1500 and 3000 gauss and as a consequence of this strong field the temperature in a spot is much lower (4000 K or less) than in the surrounding photosphere. That causes the dark appearance of sunspots in which the dark umbra is surrounded by a less dark *penumbra* (Figure 13.1).

However, as Figures 13.1–13.3 show, the situation looks quite differently if we observe sunspot groups in higher layers of the solar atmosphere, the *chromosphere*, *transition layer*, and *corona*. From about 1930, the chromosphere could be regularly observed in a spectrohelioscope and in more recent years through filters which make it possible to observe the Sun in a narrow monochromatic band, usually centered on the hydrogen Balmer Hα line (Figure 13.2). In this line, sunspots are surrounded by *bright plages* and we call these groups of plages and spots the *active regions* on the Sun. And since 1973, when *Skylab* orbited the Earth and imaged the Sun in

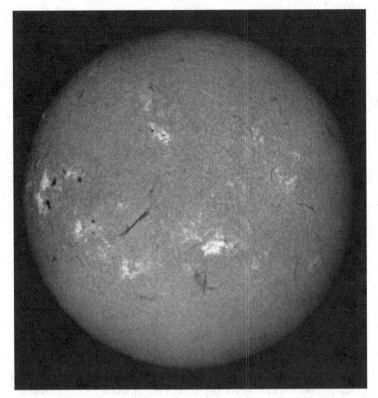

Figure 13.2. Image of the active Sun in the Hα line (made at Big Bear Solar Observatory) on 21 May 1980. One can see here several active regions with bright plages and dark spots, dark filaments (prominences in projection on the disk), and a two-ribbon flare south-west of the disk center. North is up and east is to the left.

X-rays, we can also see the highest layer of the solar atmosphere, the *solar corona* in which these active regions look very extensive and bright (Figure 13.3). In the most recent years, TRACE has revealed high-resolution images of the solar corona and the transition layer between the chromosphere and the corona, of which examples are shown in Figure 13.1.

Active regions appear where magnetic flux emerges from subphotospheric layers to the solar chromosphere and corona (see the comparison of photospheric and magnetic images in Figure 13.1). There are many flux emergences on the solar surface, all the time creating *ephemeral active regions*, which in X-rays are seen as *bright points* on the Sun (some examples – bright dots – can be seen in Figure 13.3). However, only few of them grow further (with newly emerging flux subsequently added to them) and eventually develop into much larger *active regions*. Several of them are seen in Figures 13.2 in the Hα line and in Figure 13.3 in X-rays.

Most active regions are *bipolar*, with two main spots and surrounding plages of opposite magnetic polarities (compare Figure 13.1, left.) The two polarities are

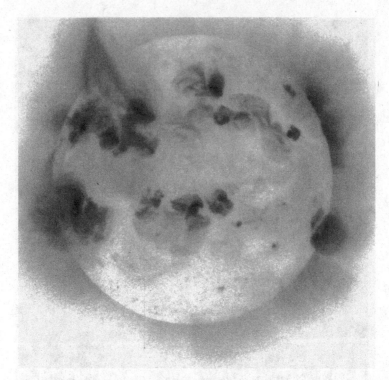

Figure 13.3. Image of the active Sun in soft X-rays (made on 23 April 2000 on board *Yohkoh*. Note the sets of bright loops that cross the neutral lines in active regions, the enormous loop with a 'cusp' that interconnects two active regions on the north-east, and several bright points on the disk. North is up and east is to the left.

connected by chromospheric and coronal *loops*, particularly well seen in X-rays (Figure 13.3), and are separated by a *polarity inversion line* below the tops of these loops where the longitudinal magnetic field is zero. (Compare the schematic drawing in Figure 13.9a,b.) Solar atmospheric gas slowly accumulates along these lines and cools. In the Hα line we see then along these inversion lines *dark filaments* (many are seen in Figure 13.2) which on the solar limb look like bright *prominences* (see an example in Figure 13.8). Dark filaments often survive long after the active region decays and they can be formed also outside active regions (like the one to the south-east of the Sun's center in Figure 13.2). Remember this fact when we later talk, in Section 13.6, about coronal mass ejections.

In some cases irregular patches of magnetic flux of different polarities emerge in an active region and cause irregularities in its magnetic structure. Then several different polarity inversion lines exist in the region and more dark filaments can form there. While all active regions are seats of various kinds of active processes, the most powerful events of solar activity occur in these magnetically complex active regions, where magnetic polarities are mixed. Most active is the so-called δ-*configuration*, when umbrae of opposite polarity are embedded in one common penumbra.

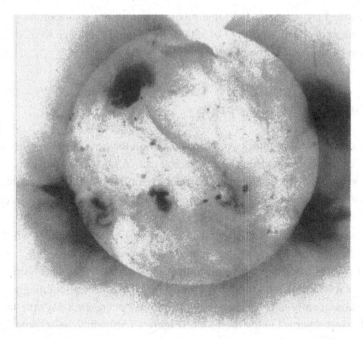

Figure 13.4. The longest transequatorial interconnecting loop of the new solar cycle connecting active regions over a distance of 700 000 km. (*Yohkoh* soft X-ray image of the Sun, after Fárník *et al.*, 1999.)

13.3   Complexes of activity and interconnecting loops

Solar activity seems to prefer, sometimes for a period of many months, selected *active longitudes* where active regions form more frequently than elsewhere on the Sun, creating so-called *complexes of activity* (Howard and Švestka, 1977; Gaizauskas *et al.*, 1983). This seems to be due to some irregularities in the distribution of the subphotospheric magnetic flux that emerges through the photosphere, and once such an irregularity is formed, it takes a long time before effects of solar rotation, which slightly varies with the latitude, remove it (Ruzmaikin, 1998).

Many active regions in such a complex of activity are connected by coronal loops that very often extend across the equator. We call them *interconnecting loops*. They can be best observed in soft X-rays, as was first done on *Skylab* in 1973 (Chase *et al.*, 1973) and presently by *Yohkoh* (Figure 13.4). Due to magnetic field variations in interconnected active regions, these connections vary all the time in their shape and brightness. In a new solar cycle, when active regions begin to emerge at high latitudes, it has been found that these interconnections can be extremely long, up to 60 heliographic degrees (700 000 km, i.e., 55 diameters of the Earth, Figure 13.4).

Because of convective motions on the Sun, it seems impossible that such long connections could exist below the photosphere and survive an emergence through it into the higher atmospheric layers. Therefore, the most plausible explanation is that they are formed by magnetic field-line reconnection in the solar corona (Tsuneta, 1996; Fárnik *et al.*,1999; Bagalá *et al.*, 2000; for general reference to reconnection processes see Priest and Forbes, 2000).

13.4   Surges, jets, and sprays

In most active regions brightness variations occur all the time, reflecting new emergence of magnetic flux, old-flux decay, or interactions between individual active-region loops. In many active regions, particularly in those which are magnetically complex, also many ejections of material appear, mostly rooted in tiny patches of magnetic polarities embedded inside, or penetrating into, a region of opposite polarity. At their footpoints one often observes small short-lived brightenings which have been called *Ellerman bombs* or *moustaches* (e.g., Bruzek, 1972).

In the Hα line the most common ejection is a bright or dark *surge* (Figure 13.5a), in which solar material is first moving upward into the corona along magnetic field lines and subsequently falls back again to the chromosphere. In X-rays the Japanese satellite *Yohkoh* observed many bright *jets* (Figure 13.5b) which have some features in common with surges, but apparently are not exactly the same phenomena (Shibata *et al.*, 1992). Most probably, as the satellite TRACE recently revealed, both have a similar cause, but the hot jets and cold surges move along different trajectories through the corona (Schrijver *et al.*, 1999). Some X-ray jets are injected along interconnecting

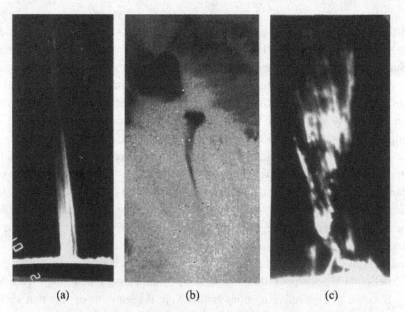

(a)   (b)   (c)

Figure 13.5. (a) Bright surge in the Hα line, observed at Wroclaw Observatory. (b) X-ray jet observed by *Yohkoh*. (c) A spray, observed at Wroclaw Observatory.

loops (Fárník and Švestka, 2002). In the last years SOHO and TRACE could detect some jets even in polar coronal holes (Wang *et al.*, 1998) and in quiet Sun regions, which reflects the complexity of magnetic field structure also outside active regions.

A much more powerful phenomenon seen in Hα images is a *spray* (Figure 13.5c), in which large amounts of active-region plasma are abruptly ejected into the corona and often escape into interplanetary space, possibly being one of the sources of coronal mass ejections (cf., Section 13.8).

13.5   Solar flares

However, the most powerful brightening in an active region is a *solar flare* (Švestka, 1976; Tandberg-Hanssen and Emslie, 1988). In the optical range almost all flares can be seen only in monochromatic light (most observations are made in the Hα line), but some major flares also emit in the white light: the first flare ever detected was discovered by Carrington on 1 September 1859 when he observed a large sunspot group looking at the photosphere (Carrington, 1859).

After Hale's invention of the spectrohelioscope, which made it possible to observe continuously the whole solar surface in the Hα line, from the early thirties flares have been observed regularly at many solar observatories throughout the world and listed in monthly reports. As the resolving power of solar instruments improved, smaller and smaller flares and flare-like phenomena could be detected in active regions on the Sun. Thus first the category of *subflares* was added to the original flares, and later the categories of *microflares* and still smaller *nanoflares*. Obviously, flare-like processes can occur in an active region on all scales and an idea - originally due to Parker (1988) - is that nanoflares occur everywhere on the Sun and may be the actual source of coronal heating.

Generally, flares are of two different kinds (Pallavicini *et al.*, 1977). There are *confined flares*, in which preexisting loops in an active region suddenly brighten and thereafter slowly decay, and *eruptive flares*, in which the whole configuration of loops crossing an inversion line in an active region is disrupted and must be newly rebuilt.

As polarity inversion lines are very often marked by dark filaments (quiescent prominences projected on the solar disk), many eruptive flares begin with a filament activation (Figures 13.6 and 13.8). Most flares, and essentially all small flares are confined flares which often originate through an interaction of two active-region loops which magnetically reconnect (Machado *et al.*, 1988). But the most powerful and energetic phenomena are the eruptive flares (see Švestka *et al.*, 1992) which are also closely related to *coronal mass ejections* (cf., Sections 13.6 and 13.7).

Figure 13.9 shows schematically the development of an eruptive flare, following the original suggestion of Kopp and Pneuman (1976), later improved and further developed by many other authors. The originally closed magnetic field in an active region, in which a filament (prominence) is embedded, suddenly opens. Reasons for it

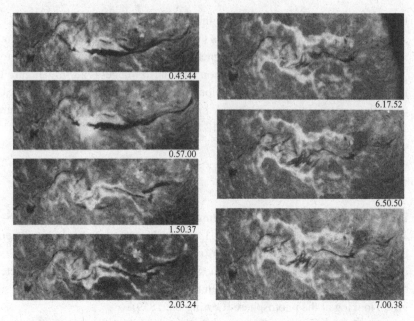

Figure 13.6. Development of an eruptive flare on 3–4 September 1982. A dark filament disappears, two bright ribbons are formed, their distance increases, and 'post'-flare loops connect the ribbons. (Photographed in the Hα line, after Morishita, 1987). Time is given in UT.

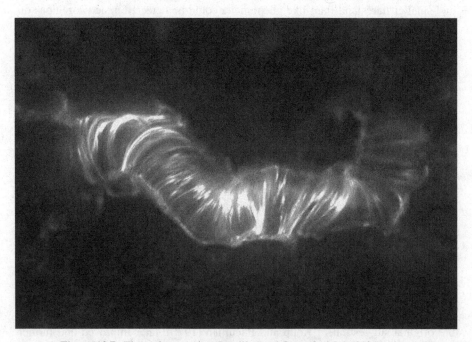

Figure 13.7. The major eruptive 'Bastille Day' flare of 14 July 2000 as imaged by TRACE in the 195 Å EUV line of FeIX/x. Very narrow ribbons are connected by a rich set of loops.

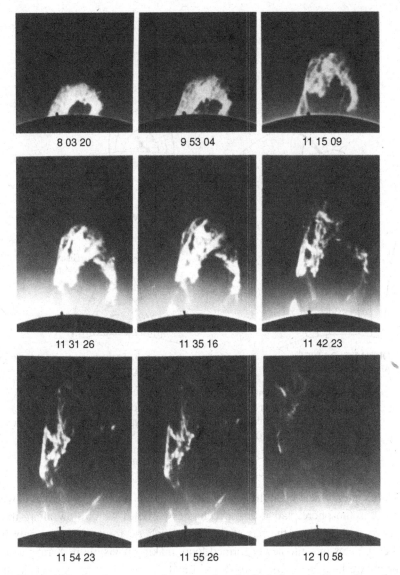

Figure 13.8. A quiescent prominence on the solar limb, its activation and eruption. Photographed at the Astronomical Observatory in Wroclaw, Poland on 18 August 1980. Time is given in UT.

can be a newly emerging magnetic flux, a confined flare nearby, a wave disturbance coming along the solar surface from another source of activity, or some internal instability (e.g., a shear of field lines which exceeds a certain critical limit). As field lines open (Figure 13.9c) plasma begins to flow from the dense chromosphere upward to the corona so that gas pressure decreases and magnetic pressure begins to prevail. That leads to sequential reconnections of the open field lines which begin to create new loops in the active region (Figure 13.9d). The reconnection process

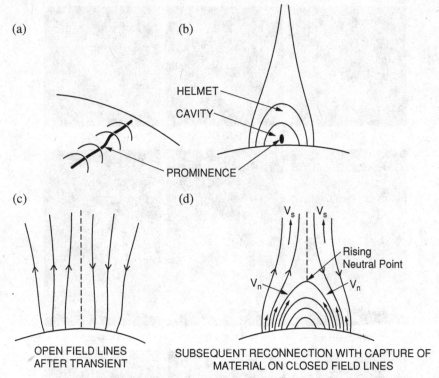

Figure 13.9. The interpretation of eruptive flares. (a,b) Two different views of the preflare situation when a dark filament (prominence) extends along a polarity inversion line and is embedded in a system of loops forming a coronal helmet structure. (c) Opening of magnetic field lines. (d) Subsequent closing of field lines, creating the flare loops. (After Kopp and Pneuman, 1976.)

produces intense heating at the top of each new loop which is conducted downward to the chromosphere, and it also accelerates particles which flow along the loop to its footpoints. Thus the gas at the chromospheric footpoint is strongly heated and evaporates into the newly formed loop, making it visible in X-rays and high-temperature lines as a *flare loop* (Cargill and Priest, 1983; Forbes and Malherbe, 1986). The loop then cools, but in between other loops are formed above it through reconnection of other field lines and the whole process is repeated in each of them. Thus the loop system gradually grows. After some time the lowest loops cool to about 10000 K and begin to be visible in the Hα line. This takes some time, so that earlier, when no X-ray observations of eruptive flares were available, these cooled structures, although formed earlier during the flare process, were called *post-flare loops*.

While in compact flares energy is suddenly released and thereafter the flare structure cools and decays, in eruptive flares energy is released during each reconnection and thus this process of energy release, though decreasing in efficiency, can continue for many hours. Because each reconnection produces new X-ray flux, one can see enhanced emission in X-rays during the whole time of the repeated reconnections. Therefore, based on these X-ray records see example in Figure 13.3, eruptive flares are also often called *long-decay events* (Kahler, 1977).

When looking at the chromospheric image of an eruptive flare in the Hα line, one observes at the footpoints of newly formed coronal loops the heated chromosphere, in the form of two bright ribbons, which slowly separate as the flare loop system grows (Figure 13.6). Therefore, these phenomena were earlier called (and often still are) *two-ribbon flares*. Inner edges of the two ribbons are connected by bright or dark loops. Figure 13.2 shows an example of such a flare (to the southwest of the center) and the bright patches between the bright ribbons are tops of the loops that connect them. And Figure 13.6 shows a system of dark loops connecting the ribbons. Above them are higher loops which still have too high temperature to be visible in the Hα line, but one can see them in EUV lines and in X-rays. (An example is in Figure 13.7.) Both in the Hα line and in high-temperature spectral regions one can see how in all eruptive flares the loop system, while decaying, slowly grows, sometimes for many hours.

13.6   Coronal mass ejections and coronal storms

The opening of magnetic field lines which initiates an eruptive flare is connected with ejection of material. This was first recognized long ago on metric radio waves, when strong outbursts of radio emission were seen high in the corona, some continuously moving upward. (For radio observations of the Sun see Chapter 15 of this book.) However, only spacecraft observations, first by OSO-7 and *Skylab* in early seventies (Tousey, 1973, Hildner *et al.*, 1976, and references therein) showed plasma ejections from the Sun – *coronal transients*, now called *coronal mass ejections* (CMEs) – which moved with high speeds through the corona into interplanetary space. (See examples of recent observations by LASCO in Figure 13.10.) But eruptive flares are only one of the sources of these CMEs which, as we know now, play the most important role in solar–terrestrial relations. Everywhere on the Sun magnetic field lines, closed across a polarity inversion line, can be disrupted, open, and eject solar plasma into space. Eruptive flares are only one special – and apparently the most energetic – case of the field-line opening, when strong magnetic field inside an active region is involved in the process (St. Cyr and Webb, 1991; Švestka, 2001). Harrison (1996) suggested we call all processes of this kind, irrespective where they occur, *coronal storms*.

Because in Hα images of the Sun polarity inversion lines inside active regions are usually marked by dark filaments, the opening of magnetic field is often first made apparent in Hα images by a filament activation. The filament structure begins to change, parts of the filament slowly rise into the corona, the speed of the rise accelerates, and finally the whole filament (or prominence, when seen on the limb)

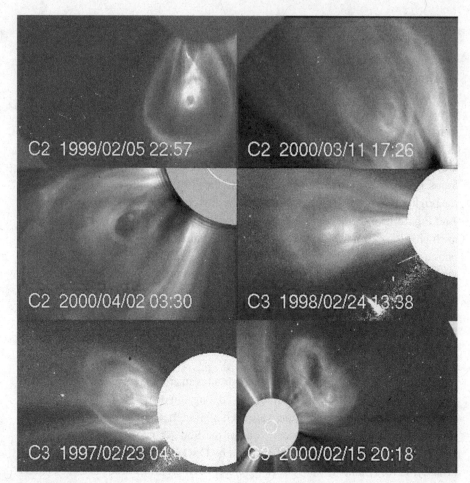

Figure 13.10. Examples of different coronal mass ejections observed by LASCO on board SOHO in 1997–2000. The Sun is hidden behind the black or white disk.

erupts (see examples in Figures 13.6 and 13.8.) Only later, when the open field lines begin to reconnect, do bright flare ribbons and loops begin to appear, but at that time the ejected material (often with the filament embedded in it) already propagates with a speed of a few hundred km/s high in the corona into interplanetary space.

However, one can very often see a similar process also far from any active region on the quiet solar surface. A quiescent filament (several are seen in Figure 13.2) becomes activated and eventually erupts. Only, on the quiet Sun, we do not see in Hα any eruptive flare, because magnetic field there is not strong enough to produce all the flare effects which we see in active regions. These activations of quiescent filaments have been known for some 50 years and called *disparition brusques*, but only in the seventies did space observations in X-rays reveal that these disruptions can have equally important effects in interplanetary space and at the Earth as major flares in active regions (see, e.g, Kahler, 1977). From *Skylab* data Munro *et al.* (1979) found

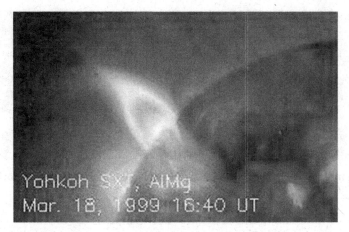

Figure 13.11. A coronal helmet imaged in X-rays by the Soft X-ray Telescope on board *Yohkoh*.

that at least 50% of coronal mass ejections were associated with filament eruptions without chromospheric flares, and this was later confirmed by observations made on board other spacecraft; from *Solar Maximum Mission* data, St. Cyr and Webb (1991) found that only 34% of CME's were associated with flares.

When the source of a CME is close to the solar limb, we find in X-rays that it arises predominantly from *coronal helmet streamers* – closed large-scale structures extending above channels of polarity inversion lines with a cusp at their tops (an example is shown in Figure 13.11). In coronal images one can first see a slow growth or brightening of the structure (Hundhausen, 1993) which subsequently erupts into the high corona and interplanetary space, with both legs still connected to the Sun. Often CMEs form expanding 'bubbles' (see examples at the left-hand side of Figure 13.10) and many have the shape of twisted ropes (cf., Section 13.7).

CMEs have very large dimensions, their average angular size being about 45 degrees. Thus they represent the disruption of closed magnetic structures along a huge part of the solar surface. The CME mass is between 10^{15} and 10^{16} grams. Their fronts propagate through interplanetary space with speeds between 100 and 2000 km/s, with an average about 400 km/s. CMEs with high velocities drive a shock ahead (Gosling *et al.*, 1991) which can accelerate particles of the ambient solar wind, as well as particles preaccelerated in eruptive flare regions, to high energies (compare Section 13.10 and Chapter 14 of this book).

As we discuss in more detail in the last section of this article, CMEs are the sources of geomagnetic storms at the Earth. However, in order to hit the Earth at least a part of the CME must propagate along the Sun–Earth line. Thus the most geoeffective CMEs must originate in central parts of the solar disk and we look at them head-on, while CMEs are best seen when their source is close to the solar limb where we see them from the side. Howard *et al.* (1982) were the first to demonstrate that Earth-directed

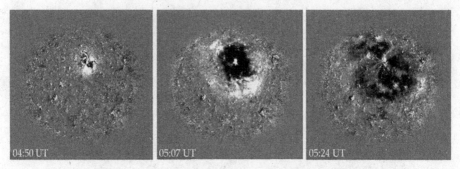

Figure 13.12. The source of a CME on the solar disk observed by the EIT onboard SOHO.

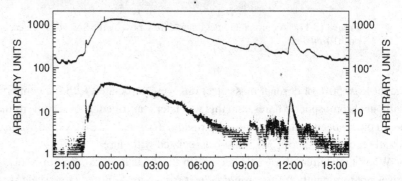

Figure 13.13. A long-decay X-ray burst. Note the impulsive burst at its onset. Time is in UT hours.

CMEs appear as a diffuse ring or half-ring of enhanced brightness surrounding the Sun and called them *halos*. Our knowledge of CME halos greatly increased after the launch of LASCO coronagraphs on board SOHO in 1996, as LASCO coronagraphs can better detect faint coronal structures. Also the EIT experiment on SOHO helps in the identification of geoeffective CMEs, because it can observe the early stages of CMEs on the solar disk (Figure 13.12). (Also see Plunkett and Wu, 2000, and references therein).

13.7   Relation between CMEs and flares

Sheeley *et al.* (1975) were the first to show that sources of CMEs (they called them coronal transients at that time) are characterized by unusually long soft-X-ray bursts (Figure 13.13), and Kahler (1977) introduced for these events the term *Long Decay Event* (LDE). He found this characteristic X-ray behaviour common to both flare-associated and non-flare-associated CMEs. Thus in X-rays one can see as an atmospheric response to the field-line opening in all cases, but chromospheric images are very misleading: when looking at the Sun in the Hα line, one can see as the source of a

CME either a two-ribbon flare, if the opening occurred in the strong field of an active region, or an activated filament without any chromospheric flare, if the opening took place in weak fields surrounding a quiescent filament, or simply nothing, if field lines opened along a polarity inversion line where no dark filament was embedded before.

This variety of chromospheric situations at the sites of CME origins (still complicated by the fact that many CMEs have their sources hidden behind the solar limb) has led to a lot of confusion about the origin of CME's for more than two decades. Many authors tried to find, and discussed, differences between CMEs associated with flares and those without flares (Gosling *et al.*, 1976; MacQueen and Fisher, 1983; Dryer, 1996; Sheeley *et al.*, 1999, Andrews and Howard, 2001). Some authors (Kahler, 1992; Gosling, 1993; Webb, 1996) even suggested that flares have essentially no direct relation to CMEs. Some facts seemed to support their conclusion:

First, there are many CMEs without flares. We have explained reasons for it above: a coronal storm can appear everywhere on the Sun.

Second, the opening of loops along a polarity inversion line often comprises a much larger extent than that of a flare inside an active region. Indeed, sometimes, in X-rays on *Yohkoh* or in EUV-line images on SOHO and TRACE one observes very extensive arcades associated with CMEs (see an example in Figure 13.14). Thus,

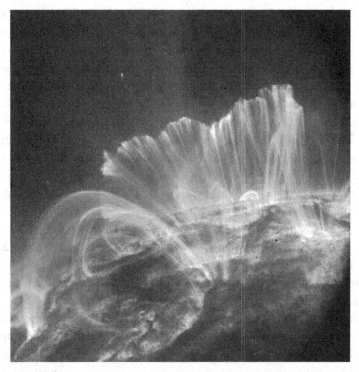

Figure 13.14. An arcade near the solar limb observed by TRACE on 9 November 2000.

obviously, an opening at the base of a CME can involve a very extensive region on the Sun. When opening avoids any active region, we do not see any flare. However, if it extends into an active region, the subsequently closing field lines there are seen as 'post'-flare loops of an eruptive flare.

Third, one would expect to see first a CME, when the field opens, and only afterwards a flare (as the field begins to close), and this is not always the case (e.g., Harrison, 1995). However, the instability that starts the opening needs a trigger, and in many cases this trigger is a confined flare (see, e.g., Harrison *et al.*, 1983; Jordan *et al.*, 1997; Innes *et al.*, 1999) – one can recognize it by a short-lived impulsive peak in the X-ray burst at the onset of the event, as in Figure 13.13. Thus observers may see first a confined flare, then the beginning of a CME, and then an eruptive flare, not recognizing that the first and the second flare brightenings are of a completely different kind.

Fourth, flare-associated CMEs propagate with constant speed, while those without flares tend to show acceleration during their rise (MacQueen and Fischer, 1983; Gosling, 1993; St. Cyr *et al.*, 1999; Sheeley *et al.*, 1999). Also, those with constant speed are generally brighter and faster (Tappin and Simnett, 1997; Andrews and Howard, 2001). But these differences in speed and acceleration can be explained by different energy inputs into the CME during its origin in a flaring or non-flaring region (Švestka, 2001).

Thus, it looks likely that the concept of a coronal storm, described above, is basically correct and that the differences in shapes and speeds depend mainly on the strength of the magnetic field at the site of the opening; and, in consequence of it, on the energy released. Of course, even when we accept this simple explanation, individual CMEs can differ one from another in very significant ways. In particular now, when LASCO on board SOHO provides CME images of excellent quality (see Figure 13.10), observed CME events reveal a large complexity and diversity of forms.

One reason for this is, as shown by Gibson and Low (1998), that we see these three-dimensional structures only in two dimensions and their orientation relative to the line of sight gives rise to a variety of geometrical appearances. Many CMEs also have the form of a magnetic flux rope, a helical current-carrying coil extending from the Sun but attached at its footpoints to its surface (see, e.g., Rust and Kumar, 1994, 1996; Chen *et al.*, 1997; Sterling and Hudson, 1997). These twisted magnetic ropes are often S-shaped, forming sigmoids, and sigmoid shapes are good indicators of non-potentiality and thus large free potential energy (Canfield *et al.*, 1999).

However, the main reason for the great variety of CME shapes and structures is the complexity of the underlying sources on the Sun. The polarity inversion lines of the longitudinal magnetic field are seldom straight, and often, in particular in active regions, several such lines can be involved in an eruption. Eruptions from these configurations have little resemblance to an eruption along a straight or sigmoidal zero line. In some cases, multiple ejections follow one after another (Lyons and Simnett, 1999).

13.8   Other sources of CMEs

Although most CMEs seem to originate in coronal storms, some obviously do not originate in this way. We have several kinds of evidence for this.

Several authors have found (first Kahler *et al.*, 1989) that some confined flares can also be sources of CMEs. This implies that one encounters here another kind of CME, not associated with the field-line openings that are present in coronal storms. However, according to Kahler *et al.* the CMEs associated with confined flares are narrow ejections, while those with LDEs are broad and extensive. Thus the sources of these narrow CMEs seem to be 'ejections of material along magnetic field lines, predominantly in sprays which are defined as 'ejections with speeds that are in excess of the escape velocity from the Sun' (see Figure 13.5c). Whether also other ejections (surges and jets - compare Section 13.4) can be associated with CMEs is doubtful. Shibata *et al.* (1995) reported several events of jets which were associated with X-ray plasma ejections that 'looked like miniature versions of CMEs'. But obviously these jet-associated events were quite different from coronal-storm CMEs.

Another source of CMEs, different from coronal storms, are activated interconnected loops (cf., Section 13.3). Already Kahler (1991) reported CMEs which apparently did not arise from activations along polarity inversion line channels, like coronal storms, but corresponded to magnetic connections between high-latitude regions of a new solar cycle and low-latitude regions of the old cycle, or to connections between active regions in opposite hemispheres. At that time, there were no spacecraft capable of seeing loops that interconnected those regions. But many such connections were later seen in X-rays by *Yohkoh* (Fárnik *et al.*, 1999), and Khan and Hudson (2000) really found disappearances of several transequatorial interconnecting loops associated with major flares and followed by CMEs. In these cases no system of loops crossing a polarity inversion line erupts, but instead an extensive loop that connects distant active regions expands into the corona and through it into interplanetary space.

Recently Lewis and Simnett (2000) found sources of CMEs coinciding with a coronal hole. They suggest that magnetic reorganization at the hole boundaries could act as a trigger for the destabilization of other structures. It may be that we encounter here also coronal storms of a very similar kind as described above, but subsequent collapses of borders of coronal holes could also create another kind of instability and these CMEs may differ from the coronal-storm events.

13.9   Causes of instabilities

We still do not know well what destabilizes closed structures and leads to their opening, thus producing CMEs anywhere on the Sun and eruptive flares in active regions. Several triggers have been suggested, and all of them can alternatively cause the instabilities observed:

(i)   Emerging new magnetic flux (e.g., Feyman and Martin, 1995; Plunkett *et al.*, 1997; Tang *et al.*, 1999; Wang and Sheeley, 1999). Such a drastic change of magnetic configuration can easily cause a field opening.

(ii) Slow-mode wave propagating from another solar disturbance along the solar surface (Bruzek, 1952; Rust and Švestka, 1979; Lyons and Simnett, 1999; Khan and Hudson, 2000). The filament channels are sensitive to lateral disturbances, in particular in weaker fields outside active regions.

(iii) Excessive sheer in an arcade of loops and a coronal helmet streamer (e.g., Antiochos *et al.*, 1999 and references therein). The larger is the angle under which loops cross the neutral line, the more energy is stored in the structure, and the less stable is the configuration.

(iv) Priest *et al.* (1989) showed that solar prominences (filaments) must be supported by a large-scale curved and twisted magnetic flux tube. When the flux tube is not twisted, it cannot support dense plasma against gravity. Apparently, any relaxation of the twist can lead to an instability. Many CMEs indeed have the form of a magnetic flux rope, as we mentioned earlier.

(v)   A catastrophic loss of mechanical equilibrium (Lin *et al.*, 1998 and references therein).

As many of these (and other possible) processes involve magnetic field reconnection, readers are referred to the recent book by Priest and Forbes (2000) about the reconnection process.

Essentially all these processes can occur equally well in the quiet Sun as in an active region, and there are more polarity inversion lines separating weak fields than strong fields. Thus it is not surprising to find that a higher number of CMEs originates outside active regions.

13.10   Accelerated particles

Many active processes on the Sun are apparently due to magnetic field-line reconnection (see the book by Priest and Forbes, 2000), and every reconnection process can accelerate electrons and atomic nuclei to higher energies. In addition to that, eruptions in the chromosphere and corona incite wave motions which propagate both along the solar surface (these waves are often called *Moreton waves*) and upward through the corona into interplanetary space. Some of these disturbances develop into shock waves which can be another source of particle acceleration, or can produce second-step acceleration of particles preaccelerated earlier elsewhere on the Sun. (For details, the reader is referred to Chapter 14 of this book).

Therefore, the active Sun is a rich source of energetic particles, mainly electrons, protons, and helium nuclei, which in some events can reach energies of hundreds of MeV, and exceptionally particles are recorded even in the BeV range. The most energetic source of accelerated particles are eruptive flares and CME-associated shocks (cf., Section 13.6), but also confined flares are sometimes sources of intense flows of accelerated particles, in particular electrons.

Particles accelerated in the solar atmosphere are responsible for various kinds of radio emission recorded from the Sun. Clouds of particles, captured in magnetic traps above active regions, cause radio noise storms on metric waves which can last for many days. They have been called *Type I* radio bursts. Shock waves from flares produce travelling radio disturbances, called *Type II* bursts. Very frequent radio events are short-lasting *Type III* bursts which are caused by accelerated electrons streaming along open magnetic field lines high into the corona and in particularly intense events deep into interplanetary space, up to the Earth distance. And particles trapped in magnetic clouds associated with CMEs produce powerful *Type IV* bursts, partly stationary and partly moving upward. (For more details, see Chapter 15 in this book.)

Accelerated electrons also give rise to *hard X-rays* in solar active regions by bremsstrahlung, and particles accelerated to high energies produce in some flares γ *rays* which are partly continuum, originating through bremsstrahlung of relativistic electrons, and partly lines excited through electron−positron annihilation, neutron capture by protons and helium nuclei in the photosphere, or by transitions in excited nuclei of heavier atoms (see Chapter 14 of this book).

Particles propagating through interplanetary space produce disturbances in the Earth enviroment, about which we will talk more in the last section of this review.

13.11   Impacts of solar activity at the Earth

The Sun, as the central body of our planetary system, has very serious impacts, of various kinds, on interplanetary space and the environments of planets. Near the solar poles the magnetic field lines are open and solar plasma flows continuously into space creating there the *fast solar wind* blowing around the Earth deep into outer regions of the planetary system. At lower latitudes, coronal helmet streamers (like the one shown in Figure 13.11) and possibly active regions during periods of field-line openings are sources of *slow solar wind*. (Also see Chapter 19 of this book.) Streams of accelerated particles, both electrons and atomic nuclei, propagate at various places through interplanetary space. And in addition to these streams of plasma and particles, coronal mass ejections send plasma clouds and shock waves in various directions through interplanetary space and eventually cause other particle accelerations there. All this creates highly variable and very complex conditions in the space between the Sun and the Earth and in the last decade we began to speak about, and regulary study, the *space weather*. In particular during periods of high solar activity weather in space is very stormy and the situations extremely complex. For us, of course, most important is the impact at the Earth itself.

When a flare appears on the Sun, it is the source of X-rays which influence Earth's ionosphere and thus cause disturbances in radio communications around the Earth. A major eruptive (long-decay) flare can disturb radio contacts for many hours. This disturbance is the same for flares occurring anywhere on the visible solar surface.

If a flare (or another active phenomenon on the Sun) accelerates particles, they also can arrive at the Earth, but not from all positions on the solar surface, because they are guided by interplanetary magnetic field lines which – due to the solar rotation – are curved into Archimedean spirals. The most intense particle events come from about 45° west from the solar central meridian, and generally from the western solar hemisphere. Such particle events are usually characterized by an impulsive onset. On the other hand, particles accelerated by CME-associated shocks in interplanetary space do not show any such dependence, and are characterized by a gradual rise (Reames, 1992). Often we see a combination of an impulsive event (accelerated on the Sun) followed by a gradual phase (accelerated in interplanetary space).

The most energetic flares emit protons with energy exceeding 500 MeV which arrive at the Earth some 15 minutes after the flare onset, produce streams of neutrons in Earth's atmosphere, and cause the so-called *ground level effects*. Flares that produce protons of such high energies are sometimes called *cosmic-ray flares*. Flares that emit protons with energies higher than 10 MeV are often called *proton flares*. Particles of lower energy are guided by the Earth magnetic field to the polar regions and cause there absorption of radio waves (*polar cap absorption*) and intense *aurorae*. All these effects are delayed by tens of minutes to several hours after the flare onset, depending on the energy of the propagating particles.

Thereafter, moving with much slower speeds, usually a few hundred km/s, a coronal mass ejection, often with a shock wave, arrives at the Earth, if it propagates in the right direction toward us. This arrival – two or three days after its origin on the Sun – has a strong impact at the Earth's magnetosphere and causes a *geomagnetic storm* which sometimes can last for several days and has serious impact on communications all around the Earth.

Before the discovery of coronal mass ejections (CMEs) in the seventies, all effects of the Sun on the magnetosphere were ascribed to major solar flares. Now we know that the real agent that causes geomagnetic storms are CMEs, which can originate also in quiet parts of the Sun, without any observed chromospheric flare. But it is misleading to jump to a conclusion, as some authors did (cf., Section 13.7) that flares are not important any more in solar terrestrial relations. Flares are excellent indicators of coronal storms and actually indicate the strongest, fastest, and most energetic disturbances coming from the Sun. According to Webb (1995), the largest geomagnetic storms are caused by fast CMEs which usually are associated with flares, while moderate or small storms mostly have no association with flares. Flares are also sources of short-wave radiation that affects the ionosphere, and produce a significant fraction of accelerated particles that cause disturbances in space and at the Earth.

Without any doubt, active processes on the Sun also influence the weather at the Earth, but these effects are indirect − depending on the behavior of the magnetosphere and ionosphere and on the meteorological situation at the time of the disturbance arrival − so that they are very complex: the same effect on the Sun can have quite different consequences at different places of the Earth. Therefore, we still know very little about it. But the active Sun is surely a very important factor in our life.

Acknowledgements

Thanks are due to Prof. Eric Priest who read the manuscript and helped to improve its contents. Illustrations used in this review were obtained at the Big Bear Solar Observatory, California, USA (*courtesy* Dr H. Zirin), University Observatory Wroclaw, Poland (*courtesy* Dr B. Rompolt), *Yohkoh* satellite (*courtesy* Dr H.S. Hudson and the *Yohkoh* SXT Team), TRACE spacecraft (*courtesy* Drs. C.J. Schrijver and A.M. Title), and experiments onboard the SOHO spacecraft: Large-Angle Spectroscopic Coronagraph (LASCO, *courtesy* Dr K.P. Dere and the LASCO consortium) and Extreme Ultraviolet Imaging Telescope (EIT, *courtesy* Dr J.B. Gurman). Most illustrations were reprinted from the journal *Solar Physics* and from the CD-ROM enclosure to *Solar Physics* **200** published by Kluwer Academic Publishers in Dordrecht, Holland.

References

Andrews, M.D. and Howard, R.A. (2001). *Space Sci. Rev.*, **95**, 147.

Antiochos, S.K., DeVore, C.R., and Klimchuk, J.A. (1999). *Astrophys. J.*, **510**, 485.

Bagalá, L.G., Mandrini, C.H., Rovira, M.G., and Démoulin, P. (2000). *Astron. Astrophys.*, **363**, 779.

Bruzek, A. (1952). *Z. Astrophys.*, **31**, 111.

Bruzek, A. (1972). *Solar Phys.*, **26**, 94.

Canfield, R.C., Hudson, H.S., and McKenzie, D.E. (1999). *Geophys. Res. Lett.*, **26**, 6.

Cargill, P.J. and Priest, E.R. (1983). *Astrophys. J.*, **266**, 383.

Carrington, R.C. (1859). *Monthly Notices Royal Astron. Soc.*, **20**, 13.

Chase, R.C., Krieger, A.S., Švestka, Z., and Vaiana, G.S. (1976). *Space Res.*, **XVI**, 917.

Chen, J. *et al.* (1997). *Astrophys. J.*, **490**, L191.

Domingo, V., Fleck, B., and Poland, A.I. (1995). *Solar Phys.*, **162**, 1.

Dryer, M. (1996). *Solar Phys.*, **169**, 421.

Eddy, J.A. (1976). *Science*, **192**, 1189.

Fárnik, F. and Švestka, Z. (2002). *Solar Phys.*, In press.

Fárnik, F., Karlický, M., and Švestka, Z. (1999). *Solar Phys.*, **187**, 33.

Feyman, J. and Martin, S.F. (1995). *J. Geophys. Res.*, **100**, 3355.

Forbes, T.G. and Malherbe, J.M. (1986). *Astrophys. J.*, **302**, L67.

Gaizauskas, V., Harvey, K.L., Harvey, J.W., and Zwaan, C. (1983). *Astrophys. J.*, **265**, 1056.

Gibson, S.E. and Low, B.C. (1998). *Astrophys. J.*, **493**, 460.

Gosling, J.T. (1993). *J. Geophys. Res.*, **98**, 18937.

Gosling, J.T., Hildner, E., MacQueen, R.M., Munro, R.H., Poland, A.J., and Ross, C.L. (1976). *Solar Phys.*, **48**, 389.

Gosling, J.T., McComas, D.J., Philips, J.L., and Bame, S.J. (1991). *J. Geophys. Res.*, **96**, 7831.

Handy, B.N. *et al.* (1999). *Solar Phys.*, **187**, 229.

Harrison, R. (1995). *Astron. Astrophys.*, **304**, 585.

Harrison, R. (1996). *Solar Phys.*, **166**, 441.

Harrison, R.A., Simnett, G.M., Hoyng, P., LaFleur, H., and van Beek, H.F. (1983). *Solar Phys.*, **84**, 237.

Harvey, K.L. (ed.) (1992). 'The Solar Cycle', *ASP Conf. Ser.*, **27**.

Hildner, E., Gosling, J.T., MacQueen, R.M., Munro, R.H., Poland, A.I., and Ross, C.L. (1976). *Solar Phys.*, **48**, 127.

Howard, R.A., Sheeley, N.R., Koomen, M.J., and Michels, D.J. (1982). *Astrophys. J.*, **263**, L101.

Howard, R.F. and Švestka, Z. (1977). *Solar Phys.*, **54**, 65.

Hundhausen, A.J. (1993). *J. Geophys. Res.*, **98**, 13177.

Innes, D.E. *et al.* (1999). *Solar Phys.*, **186**, 337.

Jordan, S., Garcia, A., and Bumba, V. (1997). *Solar Phys.*, **173**, 359.

Kahler, S.W. (1977). *Astrophys. J.*, **214**, 891.

Kahler, S.W. (1991). *Astrophys. J.*, **378**, 398.

Kahler, S.W. (1992). *Ann. Rev. Astron. Astrophys.*, **30**, 113.

Kahler, S.W., Sheeley, N.R. Jr, and Liggett, M. (1989). *Astrophys. J.*, **344**, 1026.

Khan, J.I. and Hudson, H.S. (2000). *Geophys. Res. Lett.*, **27**, 1083.

Kopp, R.A. and Pneuman, G.W. (1976). *Solar Phys.*, **50**, 85.

Lewis, D.J. and Simnett, G.M. (2000). *Solar Phys.*, **191**, 185.

Lin, J., Forbes, T.G., Isenberg, P.A., and Démoulin, P. (1998). *Astrophys. J.*, **540**, 1006.

Lyons, M.A. and Simnett, G.M. (1999). *Solar Phys.*, **186**, 363.

MacQueen, R.M. and Fisher, R.R. (1983). *Solar Phys.*, **89**, 89.

Machado, M.E., Moore, R.L., Hernandez, A.M., Rovira, M.G., Hagyard, M.J., and Smith J.B. Jr. (1988). *Astrophys. J.*, **326**, 425.

Morishita, H. (1987). 'Selected Solar Hα Photographs', Norikura Solar Observatory Publication.

Munro, R.H., Gosling, J.T., Hildner, E., MacQueen, R.M., Poland, R.M., and Ross, C.L. (1979). *Solar Phys.*, **61**, 201.

Pallavicini, R., Serio, S., and Vaiana, G.S. (1977). *Astrophys. J.*, **216**, 108.

Parker, E.N. (1988). *Astrophys. J.*, **330**, 474.

Plunkett, S.P. and Wu, S.T. (2000). *IEEE Trans. Plasma Sci.*, **28**, 1807.

Plunkett, S.P. *et al.* (1997). *Solar Phys.*, **175**, 699.

Priest, E. and Forbes, T. (2000). *Magnetic Reconnection*, Cambridge University Press.

Priest, E.R., Hood, A.W., and Anzer, U. (1989). *Astrophys. J.*, **344**, 1010.

Reames, D.V. (1992). *AIP Conf. Proc.* **264**, 213.

Rust, D.M. and Kumar, A. (1994).
Proceedings of the Third SOHO Workshop,
ESTEC, Noordwijk, p. 39.

Rust, D.M. and Kumar, A. (1996).
Astrophys. J., **464**, L199.

Rust, D.M. and Švestka, Z. (1979). *Solar Phys.*, **63**, 279.

Ruzmaikin, A. (1998). *Solar Phys.*, **181**, 1.

Schrijver, C.J. *et al.* (1999), *Solar Phys.*, **187**, 261.

Sheeley, N.R., Jr. *et al.* (1975). *Solar Phys.*, **45**, 377.

Sheeley, N.R., Jr., Watters, J.H., Wang, Y.-M., and Howard, R.A. (1999).
J. Geophys. Res., **104**, 24739.

Shibata, K. *et al.* (1992). *Publ. Astron. Soc. Japan*, **44**, L173.

Shibata *et al.* (1995). *Astrophys. J.*, **451**, L83.

Solar Physics (2001). Topical Issue on the 2000 Bastille Day Flare. *Solar Phys.*, **204**.

St. Cyr, O.C. and Webb, D.F. (1991). *Solar Phys.*, **136**, 379.

St. Cyr, O.C., Burkepile, J.T., Hundhausen, A.J., and Lecinski, A.H. (1999).
J. Geophys. Res., **104**, 12493.

Sterling, A.C. and Hudson, H.S. (1997).
Astrophys. J., **491**, L55.

Švestka, Z. (1976). *Solar Flares*, D.Reidel Publishing House, Dordrecht, Holland.

Švestka, Z. (2001). *Space Sci. Rev.*, **95**, 135.

Švestka, Z., Jackson, B.V., and Machado, M.E. (eds.) (1992). 'Eruptive Solar Flares', *Lecture Notes in Physics*, **399**.

Tandberg-Hanssen, E. and Emslie, A.G. (1988). *The Physics of Solar Flares*, Cambridge University Press, UK.

Tang, Y.H., Mouradian, Z., Schmieder, B., Fang, C., and Sakurai, T. (1999). *Solar Phys.*, **185**, 143.

Tappin, S.J. and Simnett, G.M. (1997). *ESA*, **SP-415**, 117.

Tousey, R. (1973). *Adv. Space Res.*, **13**, 713.

Tsuneta, S. (1996). *Astrophys. J.*, **456**, L63.

Wang, Y.-M. and Sheeley, N.R. Jr. (1999). *Astrophys. J.*, **510**, L157.

Wang, Y.-M. *et al.* (1998). *Astrophys. J.*, **508**, 899.

Webb, D.F. (1995). *Rev. Geophys.*, Suppl. July, 577.

Webb, D.F. (1996). *ASP Conf. Ser.*, **95**, 219.

14

Particle acceleration

A.G. Emslie, J.A. Miller
School of Graduate Studies, UAH, Huntsville, AL 35899, USA

14.1 Introduction

The acceleration of particles to high energies is a ubiquitous phenomenon at sites throughout the Universe. Impulsive solar flares offer one of the most impressive examples anywhere, releasing up to 10^{32} ergs of energy over timescales of several tens of seconds to several tens of minutes. Much of this energy is expended on energizing the ambient particles, which can then remain either trapped at the Sun or escape into interplanetary space. The radiation from trapped particles (Ramaty and Murphy 1987) consists in general of (1) continuum emission, which ranges from radio and microwave wavelengths to soft ($\sim$ 1–20 keV) X-rays, hard ($\sim$ 20–300 keV) X-rays, and finally gamma rays (above $\sim$ 300 keV), which may have energies in excess of 1 GeV; (2) narrow gamma-ray nuclear deexcitation lines between $\approx$ 4 and 8 MeV; (3) broad nuclear deexcitation lines in about the same energy range, and (4) high-energy neutrons. The particles that escape into space (e.g., Reames 1990) can be detected directly, and often have compositions quite different than that of the ambient solar atmosphere.

It is generally accepted that the basic source of this released energy lies in stressed, current-carrying magnetic fields, if for no other reason than that this is the only energy reservoir of sufficient magnitude (see, e.g., discussion in Tandberg-Hanssen and Emslie 1988). As we shall summarize below, most of the available magnetic energy reservoir is in fact released in the form of accelerated charged particles, both electrons and ions. The Sun is therefore a very efficient particle accelerator, although the mechanisms through which magnetic energy is transformd into accelerated suprathermal particles has yet to be satisfactorily elucidated.

Solar flares are one of two general types, *gradual* and *impulsive* flares (but see Cliver 1996 for a refinement). Gradual events are large, occur high in the corona, have long-duration soft and hard X-rays and gamma rays, are electron poor, are associated with Type II radio emission and coronal mass ejections (CMEs), and produce interplanetary energetic ions with coronal abundance ratios. Impulsive events are more compact, occur lower in the corona, produce short-duration radiation, and exhibit dramatic abundance enhancements in the escaping and trapped energetic ions (Mandzhavidze and Ramaty 1993). Their ^{3}He/^{4}He ratio is ~ 1, which is a huge increase over the coronal value of about 5×10^{-4}, and they also posses other ion abundance enhancements as well (see below). The general scenario that emerged from these observations is that energetic particles in gradual events are accelerated by a CME-driven shock, while those particles in impulsive events are accelerated by another mechanism(s). In this paper, we review candidates for this mechanism.

When trying to probe the acceleration mechanism in impulsive flares, one is immediately confronted by the great wealth of diagnostic data, which is unequaled in other areas of astrophysics. Furthermore, impulsive flares exhibit quite a bit of variation. The approach we see as most fruitful is to distill the results obtained from these many and varied observations into a few central requirements that any theory must account for simultaneously (if it is to be considered a promising candidate for the actual mechanism). While glossing over many of the details, this approach should ensure that at least the major aspects of flares are explained. We believe that theories which focus exclusively on one or two observations should be avoided, since they might not account for any other and eventually lead to a collection of unrelated and possibly contradictory proposed mechanisms and no overall scheme. Furthermore, theories that invoke *ad hoc* accelerating agents, such as large-scale hydrodynamic shocks, are not considered here because of the difficulty in reducing them to the basic (magnetic reconnection) mechanism of flare energy release.

After summarizing the order-of-magnitude observational constraints that a viable acceleration model must account for, we then consider the inherently simplest acceleration model – a coherent electric field. We point out significant problems with this simple picture, in both the small-scale (super-Dreicer) and large-scale (sub-Dreicer) regimes. We then turn our attention to *stochastic* acceleration models, which avoid most of these problems and also account rather nicely for a wide variety of solar flare observational characteristics.

14.2   Observational constraints

Evidence for particle acceleration in solar flares comes from a variety of observations. First, there are direct observations of accelerated "solar energetic particles" (SEPs) injected into interplanetary space and observed by spacecraft in Earth orbit and throughout the solar system (heliosphere). Second, there are the radiation signatures produced by the particles as they interact with the solar environment. These range from radio wave signatures (such as gyrosychrotron radiation and coherent plasma radiation), through thermal signatures (in optical lines and continua, ultraviolet and extreme ultraviolet, and soft X-rays), to nonthermal signatures in the hard X-ray and gamma-ray regions of the electromagnetic spectrum. Space here does not permit a comprehensive review of these observations. However, in order to characterize the extent to which particle acceleration plays a role in the overall flare process, some order-of-magnitude estimates of fluxes, etc., will be useful. For a detailed discussion of how these constraints are derived, see Miller *et al.* (1997) or Vestrand and Miller (1998). We briefly note here that most of the information on the number of interacting electrons and their acceleration timescale is provided by fitting hard X-ray bremsstrahlung emissions, while similar information for the interacting ions is obtained by modeling the nuclear deexcitation line and pion radiation emission. Furthermore, while all of the following features may not be present in a given individual flare, they are routinely observed and are essential aspects that should be accounted for by a successful model.

14.2.1 Electrons

One of the most robust indicators of electron acceleration in solar flares is the hard X-ray signature produced by bremsstrahlung of the accelerated electrons on ambient protons and heavy ions. The cross-section for this process is well-known, and is a function of the incident electron energy, the emergent photon energy, the direction of the emitted photon relative to the incident electron velocity vector, and the polarization state of the emitted photon. Observed hard X-ray spectra are typically fairly steep (e.g., power-laws with a spectral index $\gamma > 3$) and so the bulk of the emitted bremsstrahlung is at low (deka-keV) energies. At such energies the bremsstrahlung cross-section is fairly isotropic and independent of polarization; it is a function $\sigma(\epsilon, E)$ (cm^2 keV^{-1}) only of the electron (E) and photon (ϵ) energies involved and so rather well determined.

The bremsstrahlung emissivity at photon energy ϵ of an electron of initial energy E_o as it traverses a target is given by

$$i(\epsilon, E_o) = \int_{E=\epsilon}^{E_o} \sigma(\epsilon, E)\, n(\mathbf{r}[t])\, v(E) \frac{dE}{|dE/dt|}, \tag{14.1}$$

where $n(\mathbf{r}[t])$ (cm^{-3}) is the target density, $v(E)$ (cm s^{-1}) is the velocity and dE/dt (keV s^{-1}) is the energy loss rate of the electron in the target. This last quantity can be written as

$$\frac{dE}{dt} = \sigma_E(E) n v \tag{14.2}$$

so that the formula for the bremsstrahlung emissivity becomes

$$i(\epsilon, E_o) = \int_{E=\epsilon}^{E_o} \frac{\sigma(\epsilon, E)}{\sigma_E(E)} \, dE. \tag{14.3}$$

This equation basically states that the *bremsstrahlung efficiency* η (ergs of bremsstrahlung produced per erg of electron energy required) is a weighted mean of the ratio of the cross-section for bremsstrahlung production to that for other energy loss processes. In the solar atmosphere, the dominant energy loss process in the relevant energy range is Coulomb collisions on ambient electrons (losses on ambient ions are several orders of magnitude less important – Emslie 1978), and the ratio of the cross-sections is of order 10^{-5}. It is therefore a relatively straightforward matter to estimate the rate of production of energy in accelerated electrons – one simply multiplies the hard X-ray emissivity by $\sim 10^5$. This results in energy input rates of order 10^{28} to 10^{30} erg s^{-1} in large events.

This energy production rate corresponds to the acceleration of some 10^{36} to 10^{37} electrons s^{-1}, and to an associated current of some 10^{17} to 10^{18} Amps. To illustrate the severe constraints imposed by such values, we note the following:

- The number of electrons in the corona of a typical flaring region is of order $nV \sim 10^{10} \times 10^{27} \simeq 10^{37}$. Thus, to accelerate the required rate of electrons requires not only that all of the available electrons are accelerated in a few seconds, but also that we replenish the acceleration site with fresh electrons on a continual basis;
- The steady-state magnetic field associated with a current of 10^{18} Amps in a region of transverse dimension 10^9 cm is, by Ampère's Law, of order 10^8 Gauss. Other than being orders of magnitude larger than solar magnetic fields, such a magnetic field would contain an energy $B^2/8\pi V \simeq 10^{42}$ ergs, several orders of magnitude larger than the electrons that supposedly produce it;
- The voltage required to intitiate a current I in a region of length ℓ (and corresponding inductance $\sim \mu_o \ell$) within a time τ is $V \simeq \mu_o \ell I / \tau$. With $I = 10^{18}$ Amps, $\ell = 10^7$ m, and $\tau = 1$ s, this gives $V \simeq 10^{19}$ Volts and a corresponding field-aligned electric field of 10^{10} V cm^{-1}, an absurdly large number.

These factors, taken together, imply that if electron acceleration takes place through large-scale processes (such as the action of a coherent electric field), then the current pattern must be highly fragmented (Holman 1985), with current closure and acceleration replenishment accomplished through the collective interaction of many such "sub-regions" (Emslie and Hénoux 1995). Such problems are alleviated, however, if the acceleration takes place through stochastic processes such that the current pattern associated with the acceleration is not in one predominant direction.

To summarize:

 (i)   Approximately 10^{36} to 10^{37} electrons s^{-1} are accelerated above the hard X-ray producing threshold energy of 20 keV for $\sim$ 100 s;

 (ii) the total > 20 keV electron energy content is thus $\sim 10^{31}$ ergs;

 (iii) the maximum electron energy is $\sim$ 100 MeV;

 (iv) they are accelerated out of the thermal or quasithermal background;

 (v)   they are accelerated simultaneously (to within a second or so) with the ions;

 (vi) real-time electron replenishment of the acceleration region must occur.

14.2.2   Ions

Similar arguments apply to the determination of the population of accelerated ions, except that here the predominant emission process is not bremsstrahlung, but nuclear reactions on ambient ions. For an excellent review of the characteristics of the accelerated ion population, see Ramaty (1986). Here we simply summarize the salient results:

 (i)   Approximately 10^{35} protons s^{-1} are accelerated above the nuclear deexcitation line producing threshold energy of $\approx$ 1 MeV for $\sim$ 100 s;

 (ii) the total > 1 MeV proton energy content is thus also $\sim 10^{31}$ ergs;

 (iii) the maximum proton energy is $\sim$ 1 GeV;

 (iv) they are accelerated out of the thermal or quasithermal background;

 (v)   they are accelerated simultaneously with the electrons; and

 (vi) they possess significant abundance enhancements (see below).

14.3   Direct electric field acceleration

The basic physics behind electron aceleration by an electric field is straightforward: the equation of motion for an electron in a collisional plasma is

$$m\frac{dv}{dt} = e\,\mathcal{E} - \nu(v)mv, \tag{14.4}$$

where m and e are the magnitude of the electron mass and charge, respectively, $v(t)$ is the speed of the electron, $\mathcal{E}$ is the applied electric field and $\nu(v)$ is the collision frequency at speed v. Most of the electrons (i.e., those in the bulk of the Maxwellian distribution) satisfy Equation (1) in steady-state by drifting at velocity $v_D = e\mathcal{E}/m\nu_c$, where ν_c is the collision frequency representative of the thermal speed of the bulk population; this gives rise to a steady current density $j = nev_d$ that is proportional to $\mathcal{E}$, i.e., the Ohm's law $j = \sigma \mathcal{E}$. For electrons with velocities significantly greater than the thermal velocity, however, one must recognize the explicit (inverse cube) dependence of $\nu(v)$: $\nu = \nu_c(v_t/v)^3$, where v_t is the thermal velocity. This requires that we write the equation of motion as

$$m\frac{dv}{dt} = e\mathcal{E} - \frac{K}{v^2},$$

(14.5)

where $K = m\nu_c v_t^3$. Electrons in the high-energy tail of the distribution, viz. those with initial velocities $v > v_{crit} = \sqrt{K/e\mathcal{E}} = v_t\sqrt{\mathcal{E}_D/\mathcal{E}}$ (where $\mathcal{E}_D = m\nu_c v_t/e$ is the *Dreicer field*), suffer a net positive force and hence an initial increase in velocity. Because of the inverse velocity dependence of the collisional drag term K/v^2, the collisional drag term is reduced even further, leading to an even greater net force and an unstable acceleration profile. This ultimately causes runaway acceleration of such electrons out of the initial Maxwellian distribution.

Acceleration by *super*-Dreicer ($\mathcal{E} \gg \mathcal{E}_D$) fields has been considered by Martens and Young (1990) and Litvinenko (1996). This model invokes very large electric field strengths (or order 1 V cm^{-1}), produced as a result of rapid inflow of reconnecting field lines ($\mathbf{E} = \mathbf{v} \times \mathbf{B}$) and consequently very short acceleration lengths $\ell \ll 1$ km. Because of the very short acceleration length compared to the collisional mean free path, the acceleration can be treated as collisionless. To avoid acceleration of particles to unreasonably high energies, these authors suppose that the accelerated particles are efficiently removed from the acceleration region once they reach reasonable energy values. This occurs due to the increase in gyroradius with energy, which causes the particle orbit to leave the region of intense electric field.

Martens and Young (1990) considered an acceleration region in the form of a very thin reconnection current sheet. The typical dimensions of the sheet are 10^9 cm $\times 10^9$ cm, and the width of the sheet is 10^2 cm. Both the influx of "seed" particles and outflux of accelerated particles occur over the large $x - y$ planes bounding the acceleration region. The magnetic field vectors lie in the $x - z$ plane (with the B_x component suffering the reversal across the central current sheet lying in the $x - y$ plane) and the electric field $\mathbf{E}$ is in the y direction perpendicular to the current sheet. While this model is capable of accelerating *protons* to energies of order 1 MeV, which drag along electrons to create a neutral beam, it is limited in its ability to directly accelerate electrons to hard X-ray-producing energies because of the tendency of the particles to escape the sheet. In the Martens and Young model, both electrons and protons are accelerated to comparable *velocities*, and as a result the typical maximum electron energy is ~ 1 keV. Litvinenko (1996) showed that the inclusion of a magnetic field component B_z aligned parallel to the electric field vector $\mathbf{E}$ would more effectively magnetize the particles along the direction of the acceleration and hence allow electrons to be accelerated to energies much larger than the ~ 1 keV appropriate to the Martens and Young (1990) model.

Litvinenko estimates that electron energies of order 100 keV can be attained in such a model.

Super-Dreicer acceleration models represent an efficient way of accelerating particles to high energies. However, there are several issues that remain to be clarified. First, the very small acceleration volume ($\sim 10^{20}$ cm^3) implies a specific energy release rate of some 10^{10} erg cm^{-3} s^{-1}. This corresponds to the annihilation of a 1000 gauss field every microsecond. This is approximately the timescale corresponding to the passage of an Alfvénic disturbance across the 10^2 cm wide sheet. The small acceleration region must produce some $10^{36} - 10^{37}$ accelerated particles per second, which, for a density of 10^{10} cm^{-3}, implies a recycling of all the particles in the sheet on similar microsecond timescales. The model, then, invokes a very rapid reconnection of inflowing magnetic field lines, and efficient acceleration of all the particles initially frozen-in to the reconnecting field. Second, the treatments to date all utilize a "test particle" approach, with no inclusion of the large electric and magnetic fields that are created by the huge electric currents associated with the accelerated particles themselves. Third, it is not at all clear how the stability of a sheet with an aspect ratio (width to thickness) of 10^7 can be maintained for the duration of the impulsive phase of a flare.

For typical coronal parameters (e.g., density $n = 10^{10}$ cm^3, $T = 10^7$ K), $v_t \simeq 10^9$ cm s^{-1} and $v_c \simeq 200$ s^{-1}, so that the Dreicer field $\mathcal{E}_D \simeq 4 \times 10^7$ statvolt cm$^{-1} \sim 10^{-4}$ V cm^{-1}. To accelerate an electron to 10 keV over a distance of 10^9 cm requires an electric field of only 10^{-5} V cm^{-1}, which is in the sub-Dreicer ($\mathcal{E}/\mathcal{E}_D \simeq 0.1$) regime. Hence it is possible that electrons in the high-energy tail of the ambient Maxwellian distribution are energized by a sub-Dreicer electric field which acts over a distance comparable to the size of the flaring magnetic structure.

The original theory of sub-Dreicer electron runaway (Dreicer 1959) was aimed at understanding acceleration in terrestrial tokamak reactors, which have a geometry fundamentally different from the solar flare. In the tokamak case, the accelerated runaway electrons continuously circle the reactor and thereby maintain a homogeneous distribution at all velocities and times. Since the electrons never leave the "acceleration region," the length of the acceleration region is formally infinite; the electrons see the applied electric field $\mathcal{E}$ and continue to gain velocity at all times. This leads to the formation of an accelerated electron spectrum that extends to formally infinite energies (cf. Fuchs et al. 1986, 1988). By contrast, in the solar flare case, the runaway electrons are accelerated out of the acceleration region into the ambient flare plasma in a relatively short time (typically of order 0.1 seconds for flare loops of length $L \sim 10^9$ cm and electron velocities $\sim 10^{10}$ cm s^{-1}). This raises several important issues. First, after this time, the electrons leave the acceleration region, causing the disappearance of the driving term $e\mathcal{E}$. The electrons are now subject only to the Coulomb drag force $-K/v^2$, as in the standard collisional thick-target model (e.g., Brown 1972; Emslie 1978). Second, the accelerated electrons do not naturally recycle through the acceleration region, and so the region must be replenished by some ancillary mechanism. Thus, application of results from tokamak research to the solar flare scenario (e.g., Holman and Benka 1992) give highly implausible results, such as a maximum electron energy that exceeds the total potential drop across the acceleration region. Let us therefore briefly consider the application of sub-Dreicer acceleration to the finite acceleration regions found within a solar flare.

Consider, then, a one-dimensional acceleration region which extends from $x = 0$ to $x = L$, with the applied electric field $\mathcal{E}$ acting to the left, so that electrons are accelerated to the right, i.e., toward increasing values of x. These electrons will emit thin-target bremsstrahlung through collisions with ambient ions that are present in the accleration region itself, and thick-target bremsstrahlung when they emerge from the acceleration region into the collisionally dominated thick-target chromosphere at and beyond $x = L$. For sufficiently low coronal column densities $nL < 10^{17} E^2$, where E is the electron energy in keV, the thick-target emission dominates. With $n = 10^{10}$ cm^{-3} and $L = 10^9$ cm, this corresponds to all energies greater than about 10 keV. Therefore, for now we consider mainly the electron spectrum that emerges from the region at $x = L$. At early times, this spectrum corresponds to electrons that started at $x_o \simeq L$, which are accelerated to only relatively low energies. Later, electrons from lower values of x_o appear. These electrons take longer to emerge but, having been acted upon by the applied force for a larger distance, do so with higher energies. The hard X-ray signature of this initial transient can therefore be represented by an instantaneously narrow spectrum, the mean energy of which increases with time. If we neglect entirely Coulomb collisions in the acceleration region, the emergent spectrum $F(E, t)$ can be represented by the simple delta-function analytic form

$$F(E) \sim \delta(\, E - e\mathcal{E}[L - x_o(t)]), \quad 0 < t < \sqrt{\frac{2mL}{e\mathcal{E}}}, \tag{14.6}$$

where $x_o(t) = L - \frac{1}{2}(e\mathcal{E}/m)t^2$ is the position at which the electrons that emerge at time t originated. The presence of Coulomb collisions complicates this somewhat, especially at low energies, but this can be treated through a simple numerical solution of the basic equation of motion. The results show that the above analytic approximation to the emergent spectrum is quite accurate. For typical flare parameters, this corresponds to a very brief ($t < 0.1$ s) burst of hard X-rays at steadily increasing energies, up to a maximum energy of $\epsilon \simeq e\mathcal{E}L$. This is nothing like the hard X-ray signature (either spectrally or temporally) that is observed in flares.

A more realistic analysis of the hard X-ray yield must, however, take into account the replenishment of the acceleration region with fresh electrons during (and after) the transient phase represented by Equation (3). It is by now well established (e.g., Miller et al. 1997) that the number of electrons accelerated during a solar flare, and the total associated current, is so large that any acceleration region of reasonable dimension ($\sim 10^{27}$ cm^3) and density ($\sim 10^{10}$ cm^{-3}) would be depleted in a second or less, so that some sort or replenishment or "recycling" of the acceleration region is a strict requirement of *any* acceleration model. It has also been established (e.g., Holman 1985; Miller *et al.* 1997) that any model invoking large-scale direct electric field acceleration must consist of a very large ($\gg 10^6$) separate current channels, in order to keep the magnetic fields associated with the electrical currents carried by the accelerated electrons to acceptable levels.

Emslie and Hénoux (1995) have incorporated this feature of a large number of current channels to explain how the replenishment of the acceleration region could in fact occur. In their model, the oppositely-directed currents in two adjacent current channels is closed through cross-field drift of *protons* at the chromospheric ends of the channels. Ionization of hydrogen atoms, and electron-proton recombination at the same rate provides the necessary sources, and sinks, for the charged particle species. It has been shown by Emslie and Hénoux (1995) that this mechanism cannot only qualitatively, but also *quantitatively*, account for the current closure process in solar flares. Specifically, it is shown that the rate of recombinations and ionizations in the partially-ionized chromosphere is sufficient to absorb the downward electron flux at one loop footpoint, and provide the fresh electrons at an adjacent loop footpoint, respectively. In addition, the off-diagonal components of the (anisotropic) conductivity tensor in the collisional, magnetized plasma, are sufficiently large to permit cross-field drift of protons at the required rate.

While this model accounts for current closure through the supply of fresh electrons at the end of the acceleration region, it is important to note that the real issue is the rate of replenishment of "seed" electrons with $v > v_{crit}$, i.e., those that will suffer runaway acceleration. This "seed" population is replenished not by the high-energy tail of the fresh electrons introduced at the end of the loop ($x = 0$), but rather by *in situ* collisional repopulation of the high-energy tail at all points along the loop, as the local electron distribution strives to regain a Maxwellian distribution at a frequency $\nu(v)$. In steady-state, we expect, then, the introduction of fresh electrons at the end of the loop. These drift along the acceleration region, maintaining the number (and current) densities along its length. Collisional repopulation of the high-energy tail throughout the region constitute the "seed" population for subsequent Dreicer runaway acceleration.

Since diffusion of bulk electrons into the high-energy tail by local collisional processes is the dominant "seed" population replenishment mechanism, then the emergent flux spectrum $F(E)$ (electrons cm^{-2} s^{-1}) will be flat. To see this, consider the following argument:

For a uniform density acceleration region each spatial interval $x_o < x < x_o + \Delta x$ of the region will produce the same number of accelerated electrons $\Delta N = \dot{n} A \Delta x$ per unit time, where $\dot{n}$ is the volumetric rate of addition of fresh "seed" particles and A is the loop cross-sectional area. Furthermore, these electrons emerge from the acceleration region with energies in the range $e\mathcal{E}(L - x_o - \Delta x) < E < e\mathcal{E}(L - x_o)$, which has a width $\Delta E = e\mathcal{E}\Delta x$ at all energies. The emergent flux spectrum

$F(E) = \Delta N/(A\,\Delta E)$ is therefore flat: $F(E) = \dot{n}/e\mathcal{E}$. If we set $\dot{n} = v_{crit}\,n$, where v_{crit} is the effective repopulation frequency of the high-energy tail of "seed" particles (i.e., those with initial velocities $v > v_{crit}$), then, we obtain the flat flux spectrum

$$F(E) = \frac{n}{e\mathcal{E}}\,v_{crit}. \tag{14.7}$$

This electron flux spectrum has an associated hard X-ray spectrum

$$i(\epsilon) \sim \int_{\epsilon}^{e\mathcal{E}L} \sigma(\epsilon, E)F(E)\,dE, \tag{14.8}$$

where $\sigma(\epsilon, E)$ is the bremsstrahlung cross-section. If we use the Kramers cross-section $\sigma(\epsilon, E) = \sigma_o/\epsilon E$, then the resulting form of $i(\epsilon)$ is

$$i(\epsilon) = \frac{\sigma_o}{\epsilon}\,\frac{n}{e\mathcal{E}}\,v_{crit}\,\ln\left(\frac{e\mathcal{E}L}{\epsilon}\right). \tag{14.9}$$

This form, which depends only on the parameters n, $\mathcal{E}$, $\mathcal{E}_D$ and L (the thermal velocity v_t can be derived from the values of n and $\mathcal{E}_D$), has a form totally inconsistent with observed hard X-ray spectra in solar flares, which typically show a steeply decreasing distribution over a very wide energy range, frequently with "breaks" consistent with multiple-power-law fits.

To produce a hard X-ray spectrum consistent with those observed in solar flares therefore requires the superposition of a large number of separate "kernels," each with a different set of parameters $\mathcal{E}, \mathcal{E}_D, n, L$:

$$I(\epsilon) = \int_0^{\infty} i(\epsilon; \mathcal{E}, \mathcal{E}_D, n, L)\,n_{\mathcal{E}}(\mathcal{E})\,n_{\mathcal{E}_D}(\mathcal{E}_D)\,n_n(n)\,n_L(L)\,d\mathcal{E}\,d\mathcal{E}_D\,dn\,dL, \tag{14.10}$$

where the various subscripted n's are the distributions of the various parameters throughout the entire flare volume and $i(\epsilon; \mathcal{E}, \mathcal{E}_D, n, L)$ is the "kernel" hard X-ray spectrum produced by a single accleration region with specified values of the parameters. This assumption of multiple "kernel" spectra is not entirely *ad hoc*: it is consistent with, or even required by, the need for a large number of acceleration channels due to electrodynamic constraints (Holman 1985; Emslie and Hénoux 1995; Miller *et al.* 1997). Indeed, it would be very surprising if the hard X-ray spectrum from a single acceleration region *did* correspond closely to observed spectra. However, considering the freedom introduced by the various distribution functions above, we see that construction of just about *any* observed spectrum is possible, making it difficult to ascertain the validity of a coherent electric field acceleration model.

14.4   Stochastic acceleration

The problems that arise from a coherent electric field acceleration model are significant, and cause us to seriously question whether any viable coherent acceleration mechanism can indeed be responsible for particle acceleration in flares. Fortunately, there is an alternative – a stochastic acceleration mechanism that accelerates particles quasi-isotropically and so avoids the global electrodynamic issues addressed above.

Of course, a more persuasive argument for stochastic acceleration should address its positive features, rather than simply dwelling on the problems associated with the alternatives. Hence, it is worthwhile to first ask whether or not there is good reason to believe that stochastic acceleration is occurring in flares at all. The answer is "yes", and this answer is supported strongly by the observed ^{3}He enhancement. This enhancement is as spectacular as it is strange. The ^{3}He cyclotron frequency is located between the H and ^{4}He cyclotron frequencies, and these ions are not enhanced or preferentially accelerated. The only known way this could occur is if ^{3}He were accelerated by waves that were excited in a narrow frequency range just around its cyclotron frequency. This basic idea was originally proposed by Fisk (1978), and subsequent models just differ in the specific wave mode that is employed (e.g., cf. Fisk with Temerin and Roth 1992 or Miller and Viñas 1993). However, regardless of the specific model, ^{3}He should be considered essentially as proof that stochastic acceleration at least operates in impulsive flares.

Stochastic acceleration may be broadly defined as any process in which a particle can either gain or lose energy in a short interval of time, but where the particles systematically gain energy over longer times. There are two varieties of stochastic acceleration: Fermi (or nonresonant) and resonant. We consider briefly Fermi acceleration first. Fermi (1949) pointed out that collisions with randomly moving magnetic scattering centers (e.g., large-amplitude waves) will lead to a systematic increase in particle energy, and this process was further investigated by Davis (1956), who emphasized its diffusive or stochastic nature, Parker and Tidman (1958), who first applied it to flares, Tverskoi (1967), and Ramaty (1979). In the scattering-center rest frame, the particle suffers an elastic collision, typically as a result of conservation of the first adiabatic invariant and magnetic mirroring. Head-on collisions in the lab frame thus lead to an increase in the particle energy, while following collisions lead to a decrease. However, since the relative velocity is higher for head-on collisions, their frequency is greater as well, and so there will be a net gain of energy over long timescales.

This idea has recently received further attention for electron acceleration. Gisler and Lemons (1990) and Gisler (1992) have shown through Monte Carlo simulations that, in certain instances, Fermi acceleration can be efficient for accelerating electrons out of the background distribution. LaRosa and Moore (1993) and LaRosa et al. (1994) have applied Fermi acceleration to flares and have argued that it can account for the energization of a large fraction of the ambient electrons to ≈ 25 keV. They assume that during the flare tens or hundreds of elementary flux tubes with radii of order 10^8 cm undergo reconnection and proposed that a shear flow instability in the jets resulting from reconnection produces fast mode waves at similar scales. The wave energy at these large scales then cascades to smaller scales and ultimately to the electrons (LaRosa *et al.* 1996) through the Fermi mechanism. This model is similar to that of Miller *et al.* (1996), except that it assumes high levels of turbulence in many small ($\approx 10^{24}\,\mathrm{cm}^3$) regions, as opposed to the injection of many packets of low-amplitude turbulence in a single large ($\approx 10^{27}\,\mathrm{cm}^3$) region.

The rate of Fermi acceleration of electrons from 0.1 to 20 keV under flare conditions is of order a few tenths of a second once the waves reach wavelengths of about 1 km. While this model is encouraging for some aspects of the electron acceleration problem, it cannot address ion acceleration in a self-consistent manner. That is, parameters which permit the efficient energization of electrons almost by definition do not permit the acceleration of ions. This mechanism should thus be viewed as, at best, very incomplete. A more promising variant of stochastic acceleration is resonant acceleration. While often thought to be rather complicated, this mechanism is in fact quite simple, since it involves only two ingredients: resonance and resonance overlap.

Resonance occurs when the condition $x \equiv \omega - k_\parallel v_\parallel - \ell |\Omega|/\gamma = 0$ is satisfied. Here, $v_\parallel$, γ, and Ω are the particle's parallel speed (with respect to the ambient magnetic field B_0), Lorentz factor, and cyclotron frequency; while ω and $k_\parallel$ are the wave frequency and parallel wavenumber. The quantity x is the frequency mismatch parameter. If the harmonic number $\ell \neq 0$, its sign depends upon the sense of rotation of the electric field and the particle in the plasma frame: if both rotate in the same sense (right or left handed) relative to B_0, then $\ell > 0$; if not, then $\ell < 0$. When $x = 0$, the frequency of rotation of the wave electric field is an integer multiple of the frequency of gyration of the particle in its guiding center frame, and the sense of rotation is the same. The particle thus sees an electric field for a sustained length of time and will either be strongly accelerated or decelerated, depending upon the relative phase of the field and the gyromotion. If $\ell = 0$, there is matching between the parallel motion of the particle and the wave parallel electric or magnetic field.

When a particle is in resonance with a single small-amplitude wave, $v_\parallel$ executes approximate simple harmonic motion about that parallel velocity which exactly satisfies the resonance condition (Karimabadi *et al.* 1992). There is no energy gain on average. The frequency ω_b of oscillation is proportional to the square root of the wave amplitude, and if $|x| \leq 2\omega_b$, the particle and wave effectively are in resonance. Hence, the exact resonance condition $x = 0$ does not have to be satisfied in order for a strong wave-particle interaction to occur.

This brings us to the second ingredient: resonance overlap, which is what yields large average energy gains. To understand overlap, consider two neighboring waves, i and $i + 1$, where $i + 1$ will resonate with a particle of higher energy than i will. A particle initially resonant with wave i will periodically gain and lose a small amount of $v_\parallel$. If the gain at some time is large enough to allow it to satisfy $|x| \leq 2\omega_{b,i+1}$, where $\omega_{b,i+1}$ is the bounce frequency for wave $i + 1$, then the particle will resonate with that wave next. After "jumping" from one wave to the next in this manner, the particle will have achieved a net gain in energy. If other waves are present that will resonate with even higher energy particles, the particle will continue jumping from resonance to resonance and achieve a maximum energy corresponding to the last resonance present. If the wave spectrum is discrete, then the spacing of waves is critical; however, if the spectrum is continuous (as is almost certainly the case in flares, and is the case in our model below), then resonance overlap will automatically occur. Of course, the particle can also move down the resonance ladder, but over long timescales, there is a net gain in energy and stochastic acceleration is the result.

A number of different forms of electron acceleration by electromagnetic waves have been considered for flares. The most familiar of these is gyroresonant stochastic acceleration by turbulence with frequencies below the electron cyclotron frequency Ω_e. (For our purposes, turbulence refers simply to a continuous spectrum of randomly phased monochromatic waves.) Alfvén, fast mode, and whistler waves were among the first to be considered (Melrose 1974). Electrons can gyroresonate with the first two waves via $\ell = \pm 1$, due to the presence of both right- and left-handed electric field components, but $\ell = +1$ is most important for whistlers.

Since $\omega \leq \Omega_H$ for both Alfvén and fast mode waves, we see from the resonance condition that $\gamma |v_\parallel|$ must be greater than about $(m_p/m_e)v_A$ for electrons to resonate, where v_A is the Alfvén speed. For a v_A of about $2000 \, \mathrm{km \, s^{-1}}$, this requires electron energies of $\approx 6 \, \mathrm{MeV}$. While possibly important for the acceleration of ultrarelativistic electrons, these waves cannot accelerate electrons out of the thermal distribution or through hard X ray producing energies. Whistlers, with $\Omega_H \ll \omega \ll \Omega_e$, yield a resonance requirement of $\gamma |v_\parallel| \gg (m_p/m_e)^{1/2} v_A$. The threshold condition for whistlers is then $20 \, \mathrm{keV}$, so that these waves could accelerate hard X ray producing electrons. However, since the threshold is still well above the thermal energy, these waves cannot accelerate electrons directly from the thermal plasma either. Whistlers and Alfvén waves have been used to accelerate deka-keV "seed" electrons to ultra-relativistic energies (Miller and Ramaty 1987).

However, Steinacker and Miller (1992) and Hamilton and Petrosian (1992) point out that relaxation of the $\omega \ll \Omega_e$ requirement and the inclusion of higher-frequency waves reduces the energy threshold to values inside the electron distribution. Hamilton and Petrosian (1992) have calculated electron and X ray spectra, the latter of which compare favorably with SMM and Hinotori observations, as well as with the high-resolution spectra of Lin et al. (1981). Steinacker and Miller (1992) showed that the acceleration times could be reproduced if the whistler turbulence energy density was about 10% of the magnetic field energy density and that acceleration to the highest observed energies could occur if lower-frequency waves on the branch were also included.

The acceleration of electrons from the thermal distribution by gyroresonance thus requires the transfer of spectral energy up the fast mode branch into the whistler and electron cyclotron regimes, since it would appear likely that the initial turbulence exists at low frequencies ($\omega \ll \Omega_H$). A cascade of power is one way to achieve this, but such a process has not been investigated and so is speculative at present (for $\omega > \Omega_H$; in the MHD regime, it is well established that cascading occurs). These whistler acceleration models also have important implications for the overall flare energetics. For a power law spectral density, Steinacker and Miller (1992) showed that for wavelengths shorter than $\approx 10^6$ cm, the wave energy density needs to be about 10% the ambient magnetic field energy density in order for the electron acceleration time to be consistent with observations. Hence, if the cascading produces a power law spectral density (which is the case where cascading has been investigated), and if the low-wavenumber cutoff corresponds to about one to one tenth the scale size of the flare ($\approx 10^9$ cm), then the total wave energy density exceeds the estimated ambient magnetic field energy density by a significant amount, the exact value depending on the slope of the turbulence spectrum.

One way to avoid the issue of cascading over a large frequency range, and the possibly exorbitant energy requirements, is to simply use the long-wavelength MHD waves (specifically, the shear Alfvén and fast mode waves – Swanson 1989) directly. The Alfvén waves posses a transverse left-hand polarized electric field, which can resonate with the ions via the $\ell = +1$ (or cyclotron) resonance. We see from the resonance condition and the Alfvén wave dispersion relation that in order for a wave to cyclotron resonate with a low-energy ion (say near the thermal speed), ω must be near Ω. On the other hand, as the ion gains energy, ω can become much less than Ω. Hence, a broad-band spectrum of Alfvén waves extending up to Ω can stochastically cyclotron accelerate ions from the tail of the thermal distribution to high energies (see also Barbosa 1979; Eichler 1979). In general, a shear Alfvén wave also has a right-hand polarized electric field which could cyclotron resonate with electrons. However, the disparity between the wave and electron cyclotron frequency requires that the electrons be ultrarelativistic. Similarly, electrons could resonate with the left-hand component via $\ell = -1$, but this again requires very high-energy electrons. Hence, Alfvén waves are not useful for accelerating electrons from the background.

The fast mode waves posses a compressive magnetic field, which can couple with either ions or electrons via the $\ell = 0$ (or Landau) resonance. We see from the resonance condition and the fast mode dispersion relation that (i) a wave of any ω will only Landau resonate with a particle having speed v greater than the Alfvén speed v_A, and (ii) as the energy increases the wave propagation angle must approach $90°$. Hence, fast mode waves having a distribution of propagation directions can resonate with particles from Alfvénic to relativistic energies, and this leads specifically to transit-time acceleration (Miller 1997). Now, initially in a flare plasma of temperature $T \approx 3$ MK, only electrons will be present in large numbers above v_A, and so these particles will be preferentially accelerated by the fast mode waves. However, should the ions become superAlfvénic (e.g., as a result of cyclotron acceleration by Alfvén waves), then they too will be transit-time accelerated. Lastly, even though fast mode waves of any ω can resonate with superAlfvénic particles, the acceleration rate is proportional to their wavenumber or frequency.

14.4.1 The cascading turbulence model

An important aspect of stochastic acceleration in general is that it is not directed, as with DC electric fields. This allows cospatial return currents to form, which draw particles up from the denser and cooler chromosphere, ensure charge neutrality, and provide the replenishment for the acceleration region that is necessary to sustain the large fluxes mentioned above. This is incorporated in the model of Miller (2000), which consists of just a few elements:

(i)   During the primary flare energy release phase, long wavelength MHD Alfvén and fast mode waves are excited. This assumption is consistent with the simulations of Yokoyama (1998; 1999), which demonstrate that tens of percent of the released energy is in the form of such waves. It is also consistent with the results of fast reconnection, where, in addition to whistlers, long-wavelength waves are also copiously produced (Drake 2000; see also Shay and Drake 1998).

(ii) These waves then cascade in a Kolmogorov-like fashion to smaller wavelengths (e.g., Verma *et al.* 1996), forming a power-law spectral density.

(iii) When the mean wavenumber of the fast mode waves has increased sufficiently, the transit-time acceleration rate for superAlfvénic electrons can overcome Coulomb energy losses, and these electrons are accelerated out of the thermal distribution and to relativistic energies (Miller *et al.* 1996). As the Alfvén waves cascade to higher wavenumbers, they can cyclotron resonate with progressively lower energy ions. Eventually, they will resonate with ions in the tail of the thermal distribution, which will then be accelerated to relativistic energies as well (Miller and Roberts 1995).

(iv) When the ions become superAlfvénic (above ≈ 1 MeV nucleon^{-1}), they too can suffer transit-time acceleration by the fast mode waves and will receive an extra acceleration "kick."

Acceleration occurrs in the coronal portion of the flare loop, which is modeled by an ionized homogeneous plasma extending a distance L along a constant magnetic field $\boldsymbol{B}_0$. For simplicity, consider only a H plasma (the effect of the heavier ions will be addressed below). The evolution of the electron distribution function N_e [MeV^{-1} cm^{-3}] is governed by a Fokker-Planck equation in kinetic energy E space,

$$\frac{\partial N_e}{\partial t} = -\frac{\partial}{\partial E} \left\{ \left[A_e + \left(\frac{dE}{dt}\right)_{Ce} \right] N_e \right\}$$

$$+ \frac{1}{2} \frac{\partial^2}{\partial E^2} [(D_e + D_{Ce})N_e] - \frac{N_e}{T_e} + S_e, \tag{14.11}$$

where D_e and A_e are the diffusion and convection coefficients resulting from transit-time acceleration by the fast mode waves, and $(dE/dt)_{Ce}$ and D_{Ce} are the convection and diffusion coefficients for electron-electron Coulomb collisions (Huba 1994).

Escape from the acceleration region is taken into account with a leaky-box loss term, where the characteristic escape time T_e is just the typical transit time $2L/v$, where v is the electron speed and the factor of 2 comes from the average pitch angle cosine being $1/2$ along a given direction. The source term S_e takes into account replenishment by a cospatial return current, and is taken to be a drifting Maxwellian of temperature 10^5 K. The drift speed is equal to the average speed of the escaping particles and its normalization is set so that the electron number in the acceleration region remains constant.

The fast mode waves are taken to be isotropic, and their spectral density W_{TFM} [total (field plus plasma motion) wave energy density per unit wavenumber k] is determined by a wave diffusion equation (Zhou and Matthaeus 1990)

$$\frac{\partial W_{TFM}}{\partial t} = \frac{\partial}{\partial k}\left[k^2 D_{FM} \frac{\partial}{\partial k}\left(k^{-2} W_{TFM}\right)\right] - \gamma_{FM} W_{TFM} + S_{FM}, \qquad (14.12)$$

where the diffusion coefficient D_{FM} describes Kolmogorov-like cascading and depends on v_A and W_{TFM}. Since the fast mode waves can interact with both protons and electrons, the damping rate γ_{FM} is a function of both the electron and proton distribution functions. The source term for the waves S_{FM} takes into account injection at some low wavenumber.

The proton distribution function N_p [MeV^{-1} cm^{-3}] is governed by a Fokker-Planck equation just like equation (1), except that the acceleration convection and diffusion coefficients A_p and D_p are now determined by the Alfvén wave spectral density W_{TA} along with the fast mode spectral density W_{TFM}. The coefficients for the Coulomb collisions take into account both proton-proton and proton-electron collisions; however, as with the electrons, Coulomb collisions are not an important effect in the cases we consider, and could actually be ignored altogether to a good approximation. The source term is also a drifting Maxwellian, but now the normalization and drift speed are such that the proton number remains constant. We assume that the Alfvén waves propagate only parallel and antiparallel to B_0, in which case their spectral density W_{TA} evolves according to

$$\frac{\partial W_{TA}}{\partial t} = \frac{\partial}{\partial k_\parallel}\left(D_{\parallel\parallel} \frac{\partial W_{TA}}{\partial k_\parallel}\right) - \gamma_A W_{TA} + S_A, \qquad (14.13)$$

where the $D_{\parallel\parallel}$ is the wave diffusion coefficient describing the cascade of energy of energy in $k_\parallel$ space; γ_A is the damping rate, due to only the protons; and S_A is a source term for the injection of the waves at some low wavenumber.

The diffusion and convection coefficients for transit-time electron acceleration by fast mode waves and cyclotron proton acceleration by Alfvén waves, along with the corresponding wave damping rates, can be found in Miller *et al.* (1996) and Miller and Reames (1996), respectively, and we will not repeat them for the sake of brevity. The coefficients for transit-time proton acceleration can be easily derived from Miller *et al.* (1996). The wave diffusion coefficients that describe cascading can be found in Miller *et al.* (1996) and Miller and Roberts (1995). Here it is sufficient to note the highly coupled and nonlinear nature of this system: the electrons are coupled to the fast mode waves; the protons are coupled to both the fast mode and Alfvén waves; and the cascading of both wave species depends upon the wave spectral density characteristics. After transforming to normalized variables, these four coupled nonlinear partial differential equations are solved numerically using Chang-Cooper (1970) finite differencing, which is a very stable variant of fully-implicit differencing. The resulting quasilinear code is a totally self-consistent treatment of particle acceleration and wave evolution and damping, and conserves particle number and energy to within one part in a million.

We take $B_0 = 500\,\mathrm{G}$, $v_A = 0.036c$, the temperature of the plasma to be initially 3 MK, and the H density to be $10^{10}\,\mathrm{cm}^{-3}$, which remains constant due to the return current. The waves are injected at the wavenumber $k_0 = 3.9 \times 10^{-3}\Omega_H/c$, which corresponds to a wavelength of 10^7 cm. However, as discussed in Miller and Roberts (1995), the nature of the particle acceleration is independent of the specific value of k_0. Both wave species are injected at a constant rate of $8 \times 10^{-9}U_B\Omega_H = 380\,\mathrm{ergs\,cm}^{-3}\,\mathrm{s}^{-1}$, where U_B is the energy density of B_0 and Ω_H is the H cyclotron frequency. The only remaining parameter is L.

14.4.2   Baseline case

Here we take $L = 1.6 \times 10^4 c/\Omega_H = 10^8$ cm. The particle distributions and wave spectral densities between $t = 0$ to $5 \times 10^5 T_H$ are given in Figure 14.1, where $T_H = \Omega_H^{-1} \approx 2 \times 10^{-7}$ s. During this 0.1 s initial time interval, the waves cascade rapidly to higher k through the inertial range and encounter the dissipation range at high k, where their energy is expended on particle acceleration. The Alfvén wave spectral density turns over sharply at $k_\parallel \approx 27\Omega_H/c$, since this wavenumber corresponds to the proton cyclotron frequency in the linear dispersion relation approximation $\omega = v_A|k_\parallel|$ that was used, and the Alfvén branch cannot exist at higher frequencies. As the waves enter the dissipation range at high k, they draw particles out of the tail of the thermal distribution and accelerate them to higher energies, which is evident in the left-hand panels. At $t = 5 \times 10^5 T_H$, the Alfvén U_A and fast mode U_{FM} wave energy densities are ≈ 31 and 15 ergs cm^{-3}, respectively, where they approximately remain for the duration of the process. Note that both of these are much less than U_B, and correspond to a total $\delta B/B$ of about 0.015. Hence, the turbulence is very low amplitude at all times, even though the injection rate is relatively high.

The particle distributions and wave spectral densities in the next 0.1 s interval are shown in Figure 14.2. During this time, all quantities are relatively unchanged, except N_p, which reflects the fact that protons are still being accelerated to higher energies since their acceleration rate is less than that for electrons. At $t = 10^6 T_H$, the proton U_p and electron U_e energy densities are 35 and 13 ergs cm^{-3}, respectively, and are both higher than the initial value of about 6.2 ergs cm^{-3}.

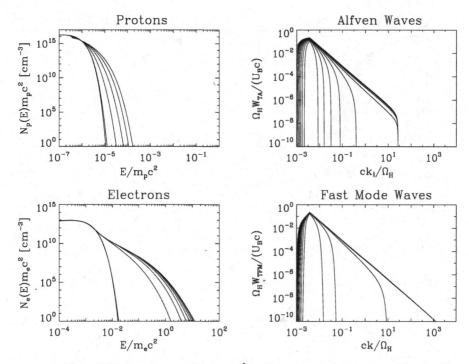

Figure 14.1. For time $t = 0$ to $5 \times 10^5 T_H$: The proton and electron distributions N_p and N_e (*left-hand panels*) at 10 equally-spaced times in the above interval, with kinetic energy E normalized to the corresponding particle rest mass energy. The Alfvén and fast mode wave spectral densities W_{TA} and W_{TFM} (*right-hand panels*) at 10 equally-spaced times in the above interval, with wavenumber normalized to Ω_H/c and energy density normalized to U_B. The particle distributions at progressively later times move out toward the right, as with the spectral densities. The proton cyclotron period $T_H \approx 2 \times 10^{-7}$ s in this all and subsequent figures.

For $t = 10^6$ to $5 \times 10^6 T_H$, the distributions and spectral densities are shown in Figure 14.3. During this next 0.8 s time interval, a significant number of protons are pushed above the Alfvén speed by cyclotron acceleration, and are thus able to interact with the fast mode waves and suffer significant transit-time acceleration. This results in a marked hardening of the proton distribution at energies above ≈ 1 MeV and a dramatic increase in U_p. Specifically, U_p grows to about 160 ergs cm^{-3} at the end of 1 s, with U_e remaining relatively constant.

The next 1 s interval is not noteworthy, and instead we show in Figure 14.4 the distributions and spectral densities for $t = 10^7$ to $2 \times 10^7 T_H$. At the beginning of this interval, equilibrium has been achieved, and the escape of energetic particles is balanced by the acceleration of colder particles that flowed into the acceleration region to offset the decrease in total number. During equilibrium, $U_p \approx 190$ ergs cm^{-3}, while $U_e \approx 11$ ergs cm^{-3}, and the drift speeds of both the inflowing electrons and protons are subAlfvénic. Between ≈ 50 and 500 keV, the electron distribution is approximately a power law of spectral index -2, but softens considerably at higher

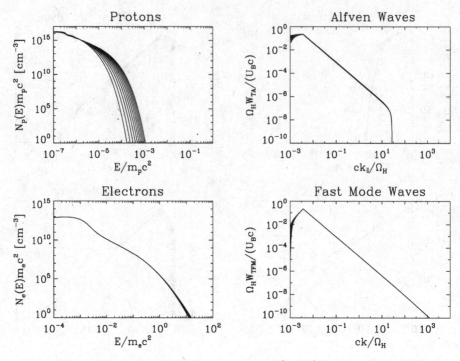

Figure 14.2. Same as Fig. 1, but for $t = 5 \times 10^5$ to $10^6 T_H$. While progressively later proton distributions move to the right, the electron distribution falls back slightly to the left.

energies. Above 1 MeV, the proton distribution is also a power law, but remains one with spectral index -2.4 all the way above 1 GeV.

In Figure 14.5, we summarize the results of this simulation by showing the total number density of escaped particles and their volumetric escape rate as a function of time. At equilibrium, the escape rate of > 1 MeV protons $F_p(> 1 \text{ MeV})$ is $\approx 10^{34} V_{27} \text{ s}^{-1}$, while the escape rate of > 20 keV electrons $N_e(> 20 \text{ keV})$ is $\approx 1.8 \times 10^{36} V_{27} \text{ s}^{-1}$, where V_{27} is the acceleration region volume in units of a typical volume of 10^{27} cm^3. This electron flux is consistent with the large-flare requirement in §2, while the proton flux is slightly low.

The effect of the loop length can be seen by considering a couple of additional examples. For a shorter region, with $L = 7 \times 10^3 c / \Omega_H = 4.4 \times 10^7$ cm, the qualitative behavior of the particle distributions and the wave spectral densities are very similar to those above, except that now more protons escape before becoming superAlfvénic. Basically, they do not remain in the acceleration region long enough to reach 1 MeV in appreciable numbers. The proton distribution above 1 MeV is thus steeper ($\propto E^{-2.8}$), and the normalization is much smaller. The proton energy density at equilibrium thus also decreases to ≈ 113 ergs cm^{-3}, but it is still much higher than the ≈ 10 ergs cm^{-3} in the electrons. At equilibrium, the volumetric escape rate of > 1 MeV protons

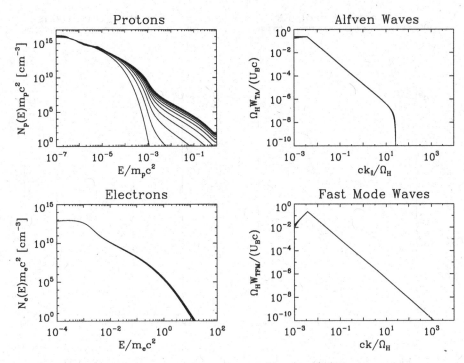

Figure 14.3. Same as Fig. 1, but for $t = 10^6$ to $5 \times 10^6 T_\mathrm{H}$.

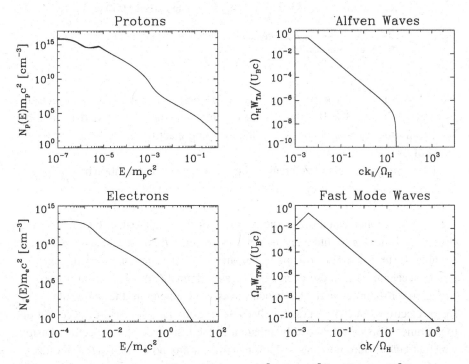

Figure 14.4. Same as Fig. 1, but for $t = 10^7$ to $2 \times 10^7 T_\mathrm{H}$. After $t \approx 10^7 T_\mathrm{H}$, equilibrium has been established.

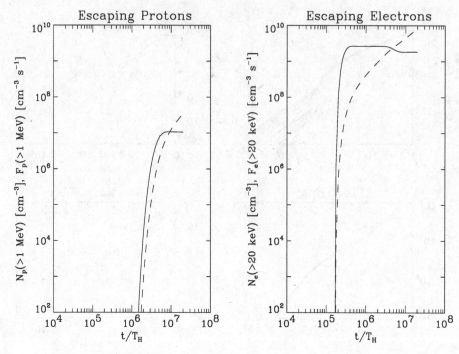

Figure 14.5. Summary of $L = 1.6 \times 10^4 c/\Omega_H$ case. *Left panel*: The total number density $N_p(> 1\,\text{MeV})$ of $> 1\,\text{MeV}$ protons that have escaped from the acceleration region (*dashed line*), and the volumetric escape rate $F_p(> 1\,\text{MeV})$ of $> 1\,\text{MeV}$ protons (*solid line*). *Right panel*: The total number density $N_e(> 20\,\text{keV})$ of escaped $> 20\,\text{keV}$ electrons (*dashed line*), and the volumetric escape rate $F_e(> 20\,\text{keV})$ of $> 20\,\text{keV}$ electrons (*solid line*).

$F_p(> 1\,\text{MeV})$ is $\approx 1.4 \times 10^{33} V_{27}\,\text{s}^{-1}$, while the escape rate of $> 20\,\text{keV}$ electrons $F_e(> 20\,\text{keV})$ is $\approx 3 \times 10^{36} V_{27}\,\text{s}^{-1}$. Hence, the escaping proton flux decreases by almost an order of magnitude, while the escaping electron flux goes up almost a factor of two.

If shortening the acceleration region suppresses proton acceleration by forcing the protons to escape before they receive an extra boost from the fast mode waves, then lengthening it should have the opposite effect. For a length L of $2 \times 10^5 c/\Omega_H = 1.2 \times 10^9\,\text{cm}$. Once again, the basic behavior of the particle distributions and the spectral densities are qualitatively the same. However, in this case, many more protons are able to survive against escape to become superAlfvénic, and the efficiency of proton acceleration at these energies increases dramatically. This in turn leads to strong proton damping of the fast mode waves, which then saps the source of energy for the electrons, and which then finally causes the efficiency of electron acceleration to plummet. This effect is readily apparent in the fast mode wave spectral density, which at equilibrium now has a sharp turnover at around $k = 5\Omega_H/c$ (and lies in the MHD regime). In the baseline case, it extended as a power law to much higher wavenumbers (bordering on the whistler regime). However, the wave energy densities are not dissimilar from the baseline case, with the Alfvén and fast mode values being ≈ 40 and $13\,\mathrm{ergs\,cm^{-3}}$, respectively.

As far as the particles are concerned, at equilibrium the proton distribution above $1\,\mathrm{MeV}$ reverts to a power law of spectral index ≈ -2.4 (as in the baseline case), albeit with a higher normalization; the electron distribution between 50 and $500\,\mathrm{keV}$ becomes much softer and assumes a power law form of spectral index ≈ -3.6. The total proton and electron energy densities are now widely disparate, and are about 830 and $6.3\,\mathrm{ergs\,cm^{-3}}$, respectively. As in all these cases, the drift speeds of the inflowing electrons and protons remain subAlfvénic at all times.

There is a precipitous decline in the escaping electron flux at about 1 s that coincides with substantial numbers of protons reaching $\approx 1\,\mathrm{MeV}$ and then absorbing large amounts of fast mode wave energy, which in turn quenches electron acceleration. For almost the preceding second, however, the escape rate of $> 20\,\mathrm{keV}$ electrons exceeds $\approx 3 \times 10^{35} V_{27}\,\mathrm{s^{-1}}$. At equilibrium, $F_p(> 1\,\mathrm{MeV})$ is $\approx 6.3 \times 10^{34} V_{27}\,\mathrm{s^{-1}}$, while $F_e(> 20\,\mathrm{keV})$ is $\approx 7 \times 10^{31} V_{27}\,\mathrm{s^{-1}}$. This proton flux is consistent with the observational requirement, while the electron flux falls short of the requirement.

In summary, then, this model is characterized by a very small number of free parameters. Since any deviations from the above typical values for acceleration region parameters like density, B_0, and temperature can be offset by small changes in length and turbulence energy density injection rate, only these two parameters are actually free. While conceptually simple, this model is able is able to account for a wealth of energetic particle data; specifically,

(i)   The acceleration region length L determines the relative proportion of energetic escaping protons to energetic escaping electrons: specifically, the longer the acceleration region, the greater the number of protons. This offers a novel explanation for the so-called "electron dominated" flares (e.g., Petrosian et al. 1994), which are rich in relativistic electrons but show no evidence of gamma-ray lines and thus no evidence for significant proton acceleration. In our model, such flares are a consequence of the flare acceleration region scale size being smaller than $\sim 10^8$ cm. In other words, "electron dominated" flares are a consequence of small acceleration regions (and robust turbulence injection). With HESSI, at least the geometrical dependence could be explored.

(ii) Protons can be easily accelerated beyond 1 GeV and electrons beyond 1 MeV in about 1 s. The time over which particles are accelerated is as long as the generation of the turbulence. At equilibrium, the total energy contained in the protons dwarfs that in the electrons.

(iii) Our baseline case comes close to accounting for all the major aspects of particle acceleration in a large impulsive flare, with only the $> 1\,\mathrm{MeV}$ proton flux being somewhat too low. This can be rectified by assuming a larger turbulence energy injection rate ($>$ 400 ergs cm^{-3} s^{-1} for both Alfvén and fast mode waves), which will then also place the electron flux near the upper limit of its range.

(iv) Hence, an injection of $>$ 400 ergs cm^{-3} s^{-1} of both Alfvén and fast mode waves at any wavelength throughout an acceleration region of scale size not exceeding several times 10^8 cm will account for all the major features of the energetic particles from the largest impulsive solar flares.

We thus submit that stochastic acceleration is by far the most promising candidate for acceleration in impulsive flares. One item that was not addressed is the issue of heavy ion abundance enhancements, and that of ^{3}He. First, ^{3}He is, as far as we can determine, preferentially accelerated (at least initially) by a totally separate mechanism, such as that proposed in Temerin and Roth (1992). At higher energies, it can certainly be accelerated as above, but its initial acceleration out of the thermal distribution must have another component. Second, calculations have shown that the heavy ions (^{4}He and above) can be incorporated successfully into the cascading turbulence framework (e.g., Mason et al. 2002).

Observed abundance enhancements are illustrated in Table 1, which gives the typical ion abundance ratios for particles $\gtrsim$ 1 MeV nucleon^{-1} (data from Great Debate [1995] and Reames et al. references therein; Share and Murphy [1997]). The abundances for the gradual events are essentially the same as those in the ambient corona, and is the basic evidence for the shock accelerated nature of these particles. The abundances for the impulsive events show enhancements of ^{3}He, Ne, Mg, Si, and Fe relative to C, N, O, and ^{4}He, which in turn are not enhanced relative to one another.

We have included these heavy ions and have obtained enhancements quite similar to those in Table 14.1. Qualitatively, in a multiion flare plasma, the Alfvén waves will encounter Fe first. Iron will be strongly accelerated but is not abundant enough to damp the waves. Thus, some wave energy will cascade to higher frequencies where

Table 14.1. *Ion abundance ratios*

Ratio	Impulsive Flares	Gradual Flares (Corona)
^{3}He/^{4}He	$\sim$1 ($\times$2000 increase)	$\sim$ 0.0005
^{4}He/O	$\approx$46	$\approx$55
^{4}He/H	$\approx$0.5	$\approx$0.1
C/O	$\approx$0.436	$\approx$ 0.471
N/O	$\approx$0.153	$\approx$ 0.128
Ne/O	$\approx$0.416 ($\times$2.8 increase)	$\approx$0.151
Mg/O	$\approx$0.413 ($\times$2.0 increase)	$\approx$0.203
Si/O	$\approx$0.405 ($\times$2.6 increase)	$\approx$0.155
Fe/O	$\approx$1.234 ($\times$8.0 increase)	$\approx$0.155

it encounters Ne, Mg, and Si. The same way, these ions suffer strong acceleration but the wave dissipation is not complete. Some wave energy then cascades to reach ^{4}He, C, N, and O. Thus, iron will resonate with the most powerful waves; Ne, Mg, and Si will resonate with waves having less power; and the other heavies will resonate with even less powerful waves. Hence, Fe should be enhanced more than the Ne group relative to the He group. Since ^{4}He, C, N, and O all have the same cyclotron frequency and behave similarly, they should not be enhanced relative to each other. This qualitative argument has been verified by quasilinear simulations (Miller 2002), which are the generalization of the simpler case discussed above. These simulations have been found to yield an Fe/O ratio of about 2.1, and Ne/O, Mg/O, and Si/O ratios of about 0.9, for a wide range of parameters. The ^{4}He, C, N, and O ions are not enhanced relative to each other. This is quite good agreement with Table 1. More notably, perhaps, the heavy ion spectra fit amazing well the interplanetary spectra observed by ULEIS and SIS on the Advanced Compostion Explorer (Mason *et al.* 2002).

14.5 Conclusions

Despite decades of observations in X-rays and gamma-rays, the mechanism for particle acceleration in solar flares remains an enigma. However, in the opinion of these authors, the electrodynamic problems peculiar to models that invoke acceleration by a coherent electric field (either small-scale, super-Dreicer; or large-scale, sub-Dreicer) are sufficiently troubling that their validity must be questioned at a fairly basic level. On the other hand, acceleration through some form of stochastic process, such as interaction with a cascade of magnetohydrodynamic waves energized by the primary (magnetic) energy dissipation mechanism, shows great promise at accounting for most of the salient features of accelerated particle distributions (both electrons and ions) in flares.

References

Barbosa, D. D. (1979). *Astrophys. J.*, **233**, 383.

Benka, S.G. and Holman, G.D. (1994). *Astrophys. J.*, **435**, 469.

Brown, J.C. (1972). *Solar Phys.*, **26**, 441.

Chang, J. S. and Cooper, G. 1970, *J. Comp. Phys.*, **6**, 1.

Cliver, E. W. 1996, in *High Energy Solar Physics*, ed. R. Ramaty *et al.* (New York: AIP), 45.

Davis., L. (1956). *Phys. Rev.*, **101**, 351.

Drake, J. F. (2000). private communication.

Dreicer, H. (1959). *Phys. Rev.*, **117**, 329.

Eichler, D. (1979). *Astrophys. J.*, **229**, 413.

Emslie, A.G. (1978). *Astrophys. J.*, **224**, 241.

Emslie, A.G. and Hénoux, J.-C. (1995). *Ap. J.*, **446**, 371.

Fermi, E. (1949). *Phys. Rev.*, **75**, 1169.

Fisk, L. A. (1978). *Astrophys. J.*, **224**, 1048.

Fuchs, V., Cairns, R.A., and Lashmore-Davies, C.N. (1986). *Phys. Fluids*, **29**, 2931.

Fuchs, V., Shoucri, M., and Teichmann, J. (1988). *Phys. Fluids*, **31**, 2221.

Gisler, G. (1992). in *Particle Acceleration in Cosmic Plasmas*, eds. G. Zank and T. Gaisser (New York: AIP), p. 229.

Gisler, G. and Lemons, D. (1990). *J. Geophys. Res.*, **95**, 14925.

Great Debates in Space Physics (articles by Miller, J. A., Hudson, H. S., and Reames, D. V.) (1995). *Eos Trans. AGU*, **76**, 401.

Hamilton, R. J. and Petrosian, V. (1992). *Astrophys. J.*, **398**, 350.

Holman, G.D. (1985). *Ap. J*, **293**, 584.

Holman, G.D. and Benka, S.G. (1996). *Astrophys. J.*, **400**, L79.

Huba, J. D. (1994). *NRL Plasma Formulary* (NRL/PU/6790–94–265) (Washington, DC: Naval Research Laboratory).

Karimabadi, H. *et al.* (1992). *J. Geophys. Res.*, **97**, 13853.

LaRosa, T. N. and Moore, R. L. (1993). *Astrophys. J.*, **418**, 912.

LaRosa, T. N. Moore, R. L., and Shore, S. N. (1994). *Astrophys. J.*, **425**, 856.

LaRosa, T. N., Shore, S. N., Miller, J. A., and Moore, R. L. (1996). *Astrophys. J.*, **467**, 454.

Litvinenko, Y.E. (1996). *Astrophys. J.*, **462**, 997.

Mandzhavidze, N. and Ramaty, R. (1993). *Nuc. Phys. B.*, **33**, 141.

Martens, P.C.H. and Young, A. (1990). *Astrophys. J. Suppl.*, **73**, 333.

Mason, G. M. *et al.* (2002). *Astrophys. J.*, submitted.

Melrose, D. B. (1974). *Solar Phys.*, **37**, 353.

Miller, J. A. (1997). *Astrophys. J.*, **491**, 939.

Miller, J. A. (2000). in *High Energy Solar Physics: Anticipating HESSI*, eds. R. Ramaty and N. Mandzhavidze (San Francisco: Astro. Soc. Pac), Vol. 206, p. 145.

Miller, J. A. and Ramaty, R. (1987). *Solar Phys.*, **113**, 195.

Miller, J. A. and Roberts, D. A. (1995). *Astrophys. J.*, **452**, 912.

Miller, J. A. and Viñas, A. F. (1993). *Astrophys. J.*, **412**, 386.

Miller, J. A. LaRosa, T. N., and Moore, R. L. (1996). *Astrophys. J.*, **461**, 445.

Miller, J. A. and Reames, D. V. (1996). in *High Energy Solar Physics*, ed. R. Ramaty *et al.* (New York: AIP), p. 450.

Miller, J. A. *et al.* (1997). *J. Geophys. Res.*, **102**, 14631.

Parker, E. N. and Tidman, D. A. (1958). *Phys. Rev.*, **111**, 1206.

Ramaty, R. (1979). in *Particle Acceleration Mechanisms in Astrophysics*, ed. J. Arons, C. Max, and C. McKee, (New York: AIP), p.135.

Ramaty, R. (1986). in *Physics of the Sun* (ed. P.A. Sturrock *et al.*) (Dordrecht: Reidel).

Ramaty, R. and Murphy, R. J. (1987). *Space Sci. Rev.*, **45**, 213.

Reames, D. V. (1990). *Astrophys. J. Suppl.*, **73**, 235.

Share, G. H. and Murphy, R. J. (1997). *Astrophys. J.*, **485**, 409.

Shay, M. A. and Drake, J. F. (1998). *Geophys. Res. Lett.*, **25**, 3759.

Steinacker, J. and Miller, J. A. (1992). *Astrophys. J.*, **393**, 764.

Swanson, D. G. (1989). *Plasma Waves* (San Diego: Academic), Chap. 2.

Tandberg-Hanssen, E.A. and Emslie, A.G. (1986). *The Physics of Solar Flares* (Cambridge University Press. Cambridge).

Temerin, M. and Roth, I. (1992). *Astrophys. J.*, **391**, L105.

Tverskoi, D. A. (1967). *Soviet Phys. JETP*, **25**, 317.

Verma, M. K. *et al.* (1996). *J. Geophys. Res.*, **101**, 21619.

Vestrand, W. T. and Miller, J. A. (1998). in *The Many Faces of the Sun*, ed. K. T. Strong *et al.* (New York: Springer), 231.

Yokoyama, T. (1998). in *Solar Jets and Coronal Plumes* (ESA SP–421), 215.

Yokoyama, T. (1999). private communication.

Zhou, Y. and Mattheaus, W. H. (1990). *J. Geophys. Res.*, **95**, 14881.

15

Radio observations of explosive energy releases on the Sun

M.R. Kundu, S.M. White

Dept. of Astronomy, University of Maryland, College Park MD 20742, USA

15.1 Introduction

For almost 50 years, radio astronomical techniques have made an impressive series of advances in our understanding of solar phenomena. In the early years of solar radio astronomy, most of the efforts were devoted to development of appropriate observational techniques. In recent years, the interest of solar radio research has evolved in two directions. First, solar astronomers have been using large arrays primarily built for cosmic radio astronomy to observe the Sun with arcsec to arcmin resolution at centimeter to metric wavelengths. Observations with such large instruments confirmed many old hypotheses and produced many new results. For example, they confirmed that low harmonic gyroradiation is responsible for radio emission from sunspot-associated active regions. Most importantly, the radio observations could be used to study coronal magnetic fields directly, something optical and X-ray astronomers, who had already achieved arcsecond resolution on the ground and in space, could not easily do. Microwave and millimeter observations are the only way to image electrons of energies greater than a few hundred keV to 1 MeV. Secondly, many highly successful collaborative studies started, based upon space-borne X-ray, UV and white-light data, on the one hand, and high-resolution radio data on the other. For example, the white-light observations of coronal transient events (CME's) with the Skylab, SMM and SOHO Coronagraph experiments, simultaneously with ground-based radio observations by means of radioheliographs and other large interferometers at meter and decameter wavelengths, have significantly changed our view regarding the origin and propagation of coronal mass ejections and the associated phenomena such as flares and shock waves.

The most successful space-borne solar mission in recent years has been the Japanese solar-dedicated satellite Yohkoh. Of particular interest to radio astronomy are two imaging experiments: the Soft X-ray Telescope (SXT), with a pixel size of 2.5″, and the Hard X-ray Telescope (HXT) with a spatial resolution of 5″. Both SXT and HXT have made numerous important discoveries: (1) finding new phenomena on the Sun; and (2) delineating finer details of well-studied problems such as solar flares in order to obtain a better and fuller understanding of their origin. One of the profound discoveries made by SXT is the dynamic nature of the solar corona, revealed by numerous small flare-like brightenings detected in both quiet and active regions. Similarly, HXT has provided good examples of flaring loops including the footpoints where hard X-rays are generated, and along with SXT has shown the time-evolution of flaring loops emitting soft and hard X-rays. Radio astronomers have been carrying out ground-based observations with large telescopes with a view to detect and understand the radio counterparts of many small-scale weak transient phenomena observed by SXT, and to locate the radio emitting material in the hard and soft X-ray emitting flaring loops. Efforts are also underway to find some general pattern in soft X-ray flaring signatures corresponding to various types of radio bursts.

In this chapter, we review primarily radio imaging observations of solar flares and coronal transients and the relationship of radio phenomena with those observed in X-rays – both hard and soft X-rays. This is not a comprehensive review of all the work done at radio wavelengths in coordination with X-ray observations, but rather a summary of some selected results on the relationship between X-ray and radio emissions.

15.2   Flare studies

15.2.1   Millimeter flare emission: comparison with microwave and hard X-rays/gamma-rays

Millimeter-wavelength emission from solar flares is of great interest as a complement to space-based gamma-ray observations because it arises from similar-energy electrons which produce the gamma-rays: electrons with energies above about 1 MeV. This is a consequence of the range of magnetic field strengths in the corona, which requires that any gyrosynchrotron emission at millimeter wavelengths must occur at high harmonics of the gyrofrequency and so can only be produced by electrons with Lorentz factors much greater than unity (White and Kundu, 1992). Figure 15.1 shows what happens to the gyrosynchrotron spectrum of radio emission from nonthermal electrons in a homogeneous magnetic field as one removes the higher energy electrons. When electrons up to 20 MeV are present the spectrum extends well into the millimeter range (the radio spectral index actually increases with frequency due to relativistic effects), but if electrons above 500 keV are removed, the high frequency spectrum (above the spectral peak at 10 GHz in this case) is strongly affected and

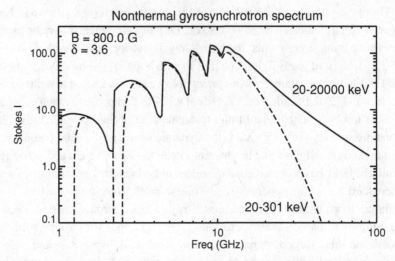

Figure 15.1. A plot showing the effect of the absence of relativistic electrons on the gyrosynchrotron spectrum from a homogeneous source. The solid line shows the spectrum from a source with magnetic field 800 G and nonthermal electron energy spectral index −3.60 when electrons from 20 keV to 20 MeV are included, while the dashed line shows the spectrum when electrons above 300 keV are excluded: the optically thin higher frequencies are severely affected.

increasingly so at higher frequencies. Observations with a millimeter interferometer are very sensitive, and can thus detect MeV electrons even in "typical" solar flares which are barely big enough to produce observable hard X-rays (Kundu *et al.*, 1992, 1994c; Lim *et al.*, 1992). Since high-resolution imaging at gamma-ray energies is difficult, millimeter observations can act as a substitute.

There seem to exist two phases in millimeter burst emission: a nonthermal impulsive phase and a thermal gradual phase. Both phases are often observed in the same flare. Some flares show no nonthermal impulsive phase at millimeter wavelengths, although they seem to show it in 25-50 keV hard X-rays. Even in flares which do show impulsive phase millimeter emission, correlation of the mm emission with electrons of 25-100 keV is often poor. Millimeter emission usually occurs at the steep rise phase of the hard X-ray emitting electrons (50-300 keV). There appears to exist some delay between mm-emission onset and 25-100 keV X-rays (Kundu *et al.* 1994c).

An extensive comparison of observations of solar flares at millimeter wavelengths with hard X-ray data, covering a wide range of flare sizes, shows that energetic electrons of order 1 MeV are produced in flares of all sizes. Further, their behavior is not consistent with their energy distribution being a simple extension of the distribution of the electrons with energies below 200 keV which produce the observed hard X-rays below 100 keV. All the available evidence seems to suggest that the millimeter-emitting electrons form a different population from the HXR-emitting electrons. If they are indeed a different electron population this has obvious implications for the acceleration of electrons in solar flares: it may imply that separate acceleration mechanisms are required for electrons at high- and low-energy ranges.

Using the Berkeley-Illinois-Maryland Array (BIMA), Silva *et al.* (1996) and Raulin *et al.* (1999) produced images of millimeter burst emission (86 GHz) associated with small flares and confirmed the above conclusions which were based upon non-imaging observations. Figure 15.2 shows an example in which both the nonthermal impulsive phase and the thermal gradual phase are observed (Raulin *et al.*, 1999). As is to be expected the thermal gradual phase at 86 GHz coincides quite well in position with the SXR position (as observed with the Yohkoh/SXT telescope). In this particular case, the impulsive mm burst source was also at the same location. The nonthermal emission shows an extended phase which probably corresponds to emission from trapped electrons. Raulin *et al.* also found that the electron energy spectral index derived from hard X-ray observations (5.1) was different from that derived from 86 and 17 GHz observations (2.5-3.2), suggesting two different energetic electron populations – one responsible for millimeter emission and the other responsible for microwave/hard X-ray emission.

15.2.2 Time profiles of millimeter bursts

White (1994) reported a striking morphology in the time profiles of the impulsive phase of millimeter bursts. The first such event reported by White *et al.* (1992) showed a linear rise to a sharp peak in 6 s, followed by an immediate decay of exponential form with a decay constant of 18 s. Since then, a remarkable similarity has been found in the time profiles of emission associated with the impulsive onsets of a good fraction of all flares. In a large fraction of flares studied (50% of the 50 events in which the impulsive phase was well observed by BIMA), the impulsive phase emission at millimeter wavelengths consists of a rapid rise (~ 5 s) linear in time to a sharp peak, followed by an exponential decay with a decay constant of order 45 s. By contrast, the 0.3-1 MeV bremsstrahlung from electrons in the same energy range tends to show impulsive spikes on the same timescale with more symmetric rise and decay profiles. This time profile is not restricted to small flares alone: it has been seen in events ranging from GOES class from less than B3 up to M5. The corresponding range in millimeter flux at the peak is about a factor of over 50, from less than 3 sfu up to 15 sfu.

This linear rise-exponential decay time profile common to impulsive millimeter emission is not generally shared by the microwave emission nor by the hard X-ray emission in the same events. In general the microwave profiles have a more gradual initial rise and a more rounded peak than does the millimeter profile. Hard X-rays in the 25-100 keV range also often show time profiles quite different from the corresponding millimeter emission (Kundu *et al.*, 1994c). Since we interpret the nonthermal impulsive-phase millimeter emission as diagnostic of the MeV- energy electrons, this means that they have a property not shared by the lower energy electrons

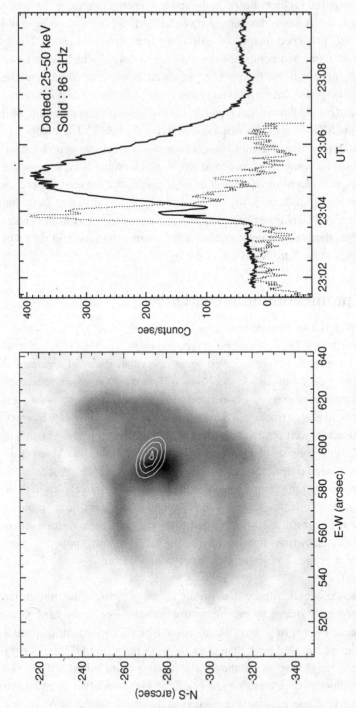

Figure 15.2. Radio observations of a flare observed on 1994 August 16. The left panel shows contours of the 86 GHz radio emission imaged using the Berkeley-Illinois-Maryland Array (BIMA) overlaid on a Yohkoh/SXT soft X-ray image of the flare loop. The right panel shows the time profile of this event in hard X-rays (dotted line, 25–50 keV data from CGRO/BATSE) and at millimeter wavelengths (BIMA 86 GHz data). The millimeter data show extended emission when no hard X-rays are detected, which is attributed to relativistic electrons trapped in the solar corona (Raulin *et al.* 1999).

responsible for the hard X-rays (typically 25-100 keV) and microwaves (probably 100-300 keV).

It seems that for these events the production of MeV-energy electrons follows a very similar pattern. The simplest interpretation of the morphology seems to be that the electrons are accelerated, or somehow injected into a coronal loop, on a timescale of several seconds (the linear rise) but acceleration then ceases abruptly and the number of electrons in the corona decreases exponentially. As wave-particle processes scatter electrons into the loss cone and they precipitate into the corona, the similarity in the rise time and decay time among a large number of events would then impose strong constraints on the mechanism of acceleration and the conditions in which it occurs.

15.2.3 Observations of millimeter and microwave bursts

Millimeter and microwave images of solar flares have been compared for two events by Kundu *et al.* (2000). Both events show very spiky time profiles and a very brief duration. In one of the events the 86 GHz images show two components, resembling the footpoints of a loop: this is the first event we know of in which a compact bipole is seen at 86 GHz. The interest in such an outcome is that the morphology of these sources can be very revealing. Except in the largest flares, millimeter sources will be optically thin and the optical depth will largely determine the fluxes observed. In gyrosynchrotron radiation, the main determinant of optical depth for trapped electrons will be the angle between the line of sight and the magnetic field, and this can be quite different at the two ends of the loop. Gyrosynchrotron calculations suggest that the emissivity depends on, e.g., $\sin^2 \theta$ for an E^{-4} energy distribution (e.g., Dulk 1985). Thus when the millimeter emission is due to electrons trapped in a loop, we expect the two ends of the loop to be bright but not equally so. On the other hand, directly precipitating electrons can experience very different field strengths in the two feet, and since emissivity $\propto B^3$ for an E^{-4} energy distribution, the difference in flux between the two footpoints can be very large.

15.2.4 Simple spiky bursts in microwaves

Kundu *et al.* (2001c) studied the spatial source structures of microwave and millimeter simple spiky bursts with the NoRH at 17 and 34 GHz. These bursts are of short duration, with fast 2-10 s rise times to the peak, followed by a rapid exponential decay. These bursts can be of any intensity, from 1 to tens of sfu; they are often strongly polarized ($\sim$ 50%), and they have similar properties regardless of the nature of the active region in which these bursts originate. The bursts seem to originate in compact sources that have sizes typically no larger than 5″ at 17 GHz and 4″ at 34 GHz. Both the radio images and the photospheric magnetograms provide direct evidence that these compact sources are simple bipolar loops, and in the polarization maps at 17 GHz, the oppositely polarized footpoints are separated by less than 5″. The radio burst source is found to lie over a magnetic neutral line in the photosphere, to within the accuracy of the overlays, and furthermore the orientations of the photospheric magnetic neutral lines are consistent with the orientations inferred from the 17 GHz polarization maps. Nishio *et al.* (1997) studied 14 impulsive flares in order to understand their magnetic field configurations and found that two loops were commonly present in the 17 GHz emission. Those events, while still impulsive, were generally of much longer duration than the simple spiky events. The short-duration, simple spiky time profiles and compact bipolar loop morphology of these bursts are consistent with a scenario in which the radio-emitting non-thermal electrons are produced/accelerated in a compact simple bipolar loop.

15.2.5 Microwave and hard X-ray observations of footpoint emission from flaring loops

Kundu *et al.* (1995) discussed two events observed in May 1993 simultaneously with the Yohkoh/HXT and SXT experiments and by the Nobeyama 17 GHz radioheliograph. Hard X-ray images of these flares were available from the Yohkoh/HXT experiment. Both events at 17 GHz were spiky in their impulsive phase which was of short duration ($\sim$ 1 min). In one case a bipolar structure with two footpoint sources of opposite polarity was observed at 17 GHz. These oppositely polarized sources observed in the May 28, 1993 event coincide in position with the double 'footpoint' sources observed with the HXT experiment in the high energy bands up to M2 (Figure 15.3). For the second event the 17 GHz flaring source is elongated at

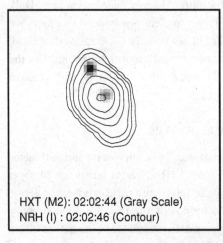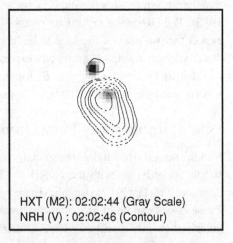

HXT (M2): 02:02:44 (Gray Scale)
NRH (I) : 02:02:46 (Contour)

HXT (M2): 02:02:44 (Gray Scale)
NRH (V) : 02:02:46 (Contour)

Figure 15.3. Hard X-ray images of the 1993 May 28 flare. The panels show overlays of radio images (contours) on X-ray images (grey-scale). In both panels the X-ray image is the Yohkoh/HXT M2-channel (33-53 keV) image. In the left panel the 17 GHz total-intensity flare contours are overlaid, while in the right panel the 17 GHz circularly-polarized flux contours are overlaid. The field of view is 2.6 arcmin in each frame. (From Kundu *et al.* 1995.)

one end, where the weaker HXT source is located: this event is consistent with an asymmetric magnetic structure in the flaring loop such that the stronger HXT source is located above the weaker magnetic field where precipitation of electrons from the loop is higher, whereas no 17 GHz source is seen at this location because of weaker magnetic field (Sakao, 1994). Similar results were obtained by Wang et al. (1995) using the OVRO 1-14 GHz imaging data along with CGRO/BATSE hard X-ray data. These observations are consistent with the long-held belief that the microwaves and hard X-rays are produced by the same group of electrons, although their locations may be separated by large distances.

15.2.6 Double loop configuration of flaring regions

It has been known for a long time that flares occur in regions consisting of many magnetic loops. The multiple magnetic loop configuration of flaring regions has led to the suggestion that at least in some cases a flare is caused by the interaction between loops. Interactions between newly emerging flux and overlying magnetic fields have been known to cause flares. Hanaoka (1996, 1997, 1999a) analyzed a relatively large number of flares using Nobeyama Radioheliograph (NoRH) data along with Yohkoh SXT and Kitt Peak National Observatory magnetogram data. He found that emerging flux appeared near one sunspot-dominated active region, and argued that the interactions between this emerging flux and the overlying loop resulted in the onset of flares and microflares. The main flare site in each case was the location of emerging flux, which also happened to be the site of enhanced soft X-ray brightening. The NoRH images showed that at 17 GHz a compact brightening occurred at the main flare site as well as a brightening near a remote region of opposite polarity. Hanaoka (1999b) further showed that the 17 GHz brightenings in the remote sources (separated from the main flare sites by about 10^5 km) were delayed by some fractions of a second, implying that high energy electrons produced at the main flare site propagated over to the remote site.

Kundu et al. (2001a) discussed a solar flare which exhibits a similar magnetic field topology (Figure 15.4). The flare produces activity in two active regions separated by a large distance (10^5 km). The main flaring region (i.e., the region of the most intense emission at EUV, X-ray and radio wavelengths) contains a compact loop associated with a bipole. The magnetogram data suggested that the flare could be caused by an interaction of magnetic structures resulting from sideward motion of the coronal loops driven by quasi-rotational photospheric motions. The observations are consistent with the remote radio source being produced by electrons which are accelerated in the main flare site and propagate $160''$ to the remote flare site where they produce radio emission in strong magnetic fields, but do not deposit a significant amount of energy by precipitation, either because too few electrons reach the remote site or else they mirror in the strong magnetic field there and do not precipitate at

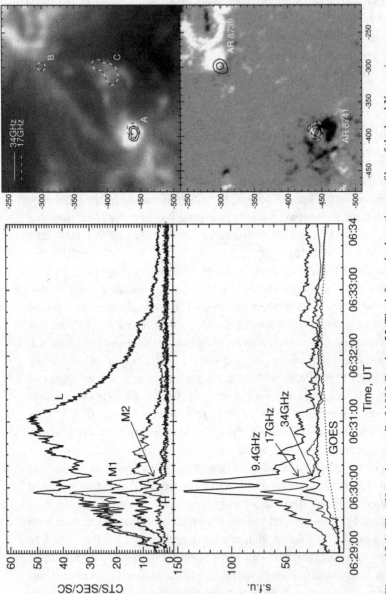

Figure 15.4. The "Hanaoka event" of 1999 October 25. The left panels show time profiles of the hard X-rays (upper left panel, Yohkoh/HXT data: L = 14-23 keV, M1 = 23-33 keV, M2 = 33-53 keV, H = 53-93 keV), radio emission (Nobeyama Radio Observatory data at 9.4, 17 and 34 GHz) and soft X-rays (GOES data converted to predicted thermal bremsstrahlung radio fluxes). The right panels show the radio images of the flare: in the upper panel the total intensity contours (overlaid on a SOHO/EIT 195 Å image) outline the main flare site (source A) where hard and soft X-rays were strong, the remote radio site (B) where only radio emission was detected, and an intermediate location (C) where thermal emission was detected in soft X-rays and in microwaves. The lower panel shows circularly polarized emission at 17 GHz overlaid on a SOHO/MDI magnetogram. (From Kundu et al. 2001a.)

the remote site. An important result of this study which had not previously been demonstrated is that the connection between the main flare site and the remote site is broken at the time of the main impulsive phase. At this time, the radio flux in the remote site begins to drop, rather than rise as it does in the main site; and the main and remote light curves cease to show the same structures with a delay as they do during the rise phase of the flare.

The flaring structure in this event is clearly a double-loop configuration of the type discussed earlier by Hanaoka. In this event the 96-minute cadence magnetogram sequence from SOHO/MDI indicates that the photospheric magnetic fields rotated around the flare site, possibly producing sheared coronal structures which subsequently flared, modulated by the oscillations in the main flare loop.

15.2.7   Modeling of microwave flares

Observations of solar microwave bursts can provide important diagnostics of acceleration processes in the solar corona because the radio emission is produced by energetic electrons accelerated during the flare. It is well known (e.g. see the reviews by Kundu and Vlahos, 1982; Alissandrakis, 1986; Bastian, Benz, and Gary, 1998) that the basic emission mechanism of solar microwave bursts is gyrosynchrotron from mildly relativistic electrons (energies of tens to several hundreds of kilo-electron volts) trapped in flaring loops. Gyrosynchrotron emission offers a powerful diagnostic of physical conditions in flaring regions. Unlike X-ray radiation, it is sensitive to magnetic field strength and orientation and can therefore be used to constrain the coronal magnetic field in the flaring source.

Figure 15.5 shows the appearance of a model flare loop at an optically thick microwave frequency (5 GHz) and an optically thin millimeter frequency (86 GHz). At the optically thick frequency the whole loop is bright, whereas at the higher frequency the strong magnetic field at the footpoints of the loop enhances the millimeter emission there and only the footpoints appear bright. In a loop filled with a population of trapped nonthermal electrons mirroring back and forth between the same magnetic field strength at both ends of the loop, we expect to see two footpoints at very high frequencies. In a loop where beaming produces significant precipitation at one footpoint but not the other, the images may show only one source. Imaging spectroscopy of simple events made with the OVRO solar array (Wang et al., 1994, 1995) has shown that, as a result of the change in magnetic field strength in the loop, low-frequency radio emission is concentrated near the top of the loop, and the centroid moves progressively down toward the footpoint at higher frequencies.

Nindos et al. (2000) studied a simple small flare which was observed at 15 GHz and 5 GHz with the VLA. They tried to reproduce the observed microwave morphology of the VLA maps as well as the OVRO flux spectrum using an inhomogeneous model of gyrosynchrotron emission. At 15 GHz, the flare emission came from two

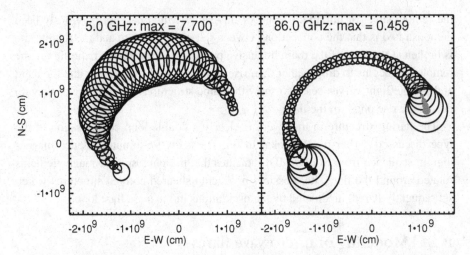

Figure 15.5. A comparison of the appearance of a model flare loop at an optically thick low radio frequency (5 GHz, left) and an optically thin high frequency (86 GHz, right). The size of the circles show the strength of the radio flux at points along the loop. The magnetic field in the loop varies from 800 G at the footpoints to 260 G at the looptop, and the loop is filled with electrons having a power-law energy spectrum up to 20 MeV. (From White 1999.)

compact highly polarized sources which traced the footpoints of the flaring loop. The whole loop itself was seen to emit at 5 GHz. The different morphologies at the two frequencies reflect the properties of gyrosynchrotron emission from mildly relativistic electrons in an inhomogeneous loop.

Nindos *et al.* (2000) tried to reproduce the VLA spatial structure as well as the OVRO radio spectrum using a simple inhomogeneous gyrosynchrotron model. The best fit to the spectral data as well as a qualitatively satisfactory agreement with the morphology of the VLA maps was achieved using a loop in which the magnetic field strength varied from 870 G at the footpoints to 280 G at the looptop. The loop was 25″ high but less than 1″ across. In order to prevent the looptop from being observable at 15 GHz, higher energy electrons (> 210 keV) had to be suppressed. This model reproduced in a satisfactory way most of the features of the VLA observations.

Kundu *et al.* (2001b) studied three additional flares in which loops were observed with the NoRH at 17 GHz and 34 GHz. These events showed relatively little variation in radio intensity along the loops. Classic model loops with magnetic fields that are strong at the footpoints and weaker close to the loop top were unable to reproduce the radio morphologies: it was necessary to keep the magnetic field constant along the loop, but it proved to be very difficult to reproduce the observed radio fluxes given the physical parameters of the loop: they require very small transverse dimensions (≪ 1″) for the loop. These studies indicate that the current status of solar flare modeling is not completely satisfactory when one seeks self-consistent models that reproduce all the observational aspects of a flare in radio, soft X-rays and hard X-rays.

15.3   Small scale energy releases on the Sun

Observations made over the past several years with the soft and hard X-ray telescopes (SXT and HXT) on the Japanese satellite Yohkoh have resulted in many new and important discoveries. For example, they have revealed that the X-ray corona is much more dynamic than had been suspected before. Because of the high spatial resolution and the large dynamic range of SXT, it has been possible to discover varieties of small scale energy releases on the Sun and to study them with extraordinary sensitivity and detail. As a result, we now know of several new phenomena: (a) XBP flares and their extended structures; (b) coronal X-ray jets; (c) Active Region Transient Brightenings (ARTB's); and (d) X-ray plasmoid ejections. Similarly, the SOHO/EIT/SUMER/UVCS experiments have revealed many transients such as eruptive plumes, macro spicules, etc. in EUV images. All these small scale phenomena have their distinctive morphological and physical properties in X-rays, and their interpretations, although related, can take on different variations. Radio observations provide information complementary to that of soft X-ray or EUV images: the latter two are both dominated by thermal emission from coronal plasma, whereas radio observations are also sensitive to nonthermal emission and to much cooler plasma. However, radio observations are not always the most efficient way to search for transients because radio-observed morphological structures of weak transients take rather simple forms. Hence most radio studies of transients use X-ray selected event lists.

In this section we discuss radio observations of small scale energy releases on the Sun. They are essentially weak coronal transients. We concentrate on the transient events identified primarily by Yohkoh/SXT and to a smaller extent by SOHO/EIT. We discuss the observational characteristics of X-ray bright point flares at meter wavelengths and in microwaves, and provide evidence that both thermal and nonthermal processes occur in these small scale flaring events. Similarly, radio observations of X-ray jets in microwaves and at meter wavelengths provide evidence for both thermal and nonthermal processes in these dynamic coronal phenomena. Nonthermal radio emission in the form of metric type III bursts is produced by electron beams propagating along the jet, whereas microwave emission mostly comes from the jet base. We discuss active region transient brightenings (ARTB's) and show that their radio emission can be purely thermal, thermal gyro-resonance or nonthermal gyrosynchrotron radiation. We also discuss quiet Sun transient brightenings located far from active regions and those occurring on network boundaries (also referred to as network flares). We provide evidence that weak plasma ejections following flares is observed at metric wavelengths in the form of some transient continuum emission. Finally, we discuss the time-varying polar brightenings at 17 GHz, and their relationship to polar erupting plumes observed by SOHO-EIT.

15.3.1 XBP flares

Solar X-ray-bright points (XBPs) are compact emitting regions associated with bipolar magnetic fields. At any one time there appear to be dozens of XBPs present on the Sun. Their lifetimes range from a few hours to several days, although only a small number appear to last longer than 2 days. They may be associated with a large fraction of the magnetic flux that emerges to the solar surface. They are known to flare.

From Skylab data (e.g. Golub *et al.* 1974, 1977) it has been known that about 10% of XBPs exhibit a type of sudden, substantial increase in surface brightness which in larger regions would be termed flaring. These flares appear to be impulsive in nature, lasting 2-3 minutes. One of the most important aspects of bright point flares, is whether or not XBP flares produce nonthermal populations of energetic particles, as do ordinary flares. The two methods best suited for detecting the presence of nonthermal electron populations are hard X-ray observations and metric-wavelength radio observations. However, the current sensitivity of hard X-ray detectors limits searches for hard X-rays from flaring XBPs. On the other hand, the production of metric radio emission by nonthermal electron beams is very efficient due to the coherent nature of the emission mechanism, and thus such nonthermal electrons might be more easily detectable at radio wavelengths.

Kundu, Gergely, and Golub (1980) used the Clark Lake Radio Observatory interferometer data to look for short-lived type III bursts at the times of flaring bright points identified in the Skylab images. They found only a 10% association between flaring XBPs and type III bursts. The sensitivity of both radio and soft X-ray detectors has improved since these observations. In particular, Yohkoh-SXT has produced excellent images of XBP's on a regular basis with high time and spatial resolution. Kundu *et al.* (1994a, b) undertook studies of radio emission at meter wavelengths and in microwaves from flaring X-ray-bright points. A search was made for meter-wavelength radio emission from more than a dozen coronal bright points observed in soft X-rays by the Yohkoh/SXT experiment. The radio observations were carried out with the Nancay (France) metric radioheliograph (NRH) at five frequencies in the range 150-450 MHz. Six of the 20 XBP X-ray brightenings showed evidence for an associated radio burst. The metric radio emissions were type III bursts, which are produced by nonthermal beams of electrons: this represents strong evidence that the XBP-flare mechanism is capable of accelerating particles to nonthermal energies, as well as producing the heated material detected in soft X-rays.

15.3.1.1 Metric type III burst emission from an XBP flare in a coronal hole

Kundu *et al.* (1994b) discussed a flare observed in an isolated bright point, well away from an active region. They used SXT soft X-ray images preceding and during the XBP flare. The quiescent soft X-ray emission from this XBP was actually very weak, and it was barely visible prior to the flare. It lies at the eastern edge of a prominent low-latitude "guitar"-shaped coronal hole which dominated the region around disk center on this day. It exhibited an X-ray flare lasting about 10 minutes (see Figure 15.6). Coincident with the XBP X-ray flare, the Nancay radioheliograph observed type III-like radio emission at metric wavelengths. The radio emission lasts only about 10 s at 164 MHz; at 236 MHz the radio emission has a longer duration ($\sim$ 20 sec; Figure 15.6). The brief duration of individual peaks within the profile, their coincidence in time at all frequencies, and the fact that they are seen across a wide frequency range strongly suggest that these bursts are type III's due to beams of nonthermal electrons propagating out through the corona along open magnetic field lines (e.g., Kundu 1965).

An important implication of the association is that electrons accelerated to nonthermal energies in XBP flares have access to open field lines. In addition, the XBP showed a "jet"-like feature extending to the northeast in earlier SXT images, similar to those discussed by Shibata et al. (1992), which also suggests the presence of open field lines connected to the XBP. As in normal flares, this raises the question of whether open field lines are intrinsic to the flare process, or whether accelerated electrons can obtain access to open field lines by drifting across closed field lines.

15.3.1.2 Microwave observations of XBP flares

Coronal X-ray bright points (XBP's) have been studied at other wavelengths. In the microwaves using the VLA, Habbal et al. (1986) and Fu et al. (1987) produced the first maps of coronal bright points at 20 cm and 6 cm wavelengths, respectively, and demonstrated that they were similar in behavior to those seen in soft X-ray and EUV wavelengths (Sheeley and Golub 1979; Habbal and Withbroe 1981). Observed with good spatial resolution, 6 cm BPs appear to have finer structures than at 20 cm (Kundu, Schmahl, Fu 1988). According to Kundu et al. (1988), both 6 and 20 cm emissions from bright points are consistent with optically thin free-free emission from a plasma whose temperature corresponds to EUV and soft X-ray wavelengths. Another similarity of radio- bright points to XBPs is temporal variability in brightness on a timescale of a few minutes to a few hours (Habbal and Harvey 1987).

Using the Yohkoh-SXT, Nitta et al. (1992) made the first direct comparison of X-ray- bright points with bright points in microwaves. Using observations made with the Nobeyama radio heliograph (NoRH) at 17 GHz and the Yohkoh/SXT experiment, Kundu et al. (1994a) reported the first detection of 17 GHz signatures of coronal X-ray bright points (XBP's). This was also the first reported detection of flaring bright points in microwaves. They detected four BP's at 17 GHz out of eight identified in SXT data on 1992 July 31, for which they looked for 17 GHz emission. For one XBP located in a quiet mixed-polarity region, the peak times at 17 GHz and X-rays were very similar, and both were long lasting – about 2 hours in duration. For the quiet region XBP, the gradual, long-lasting and unpolarized emission suggests that the 17 GHz emission is thermal.

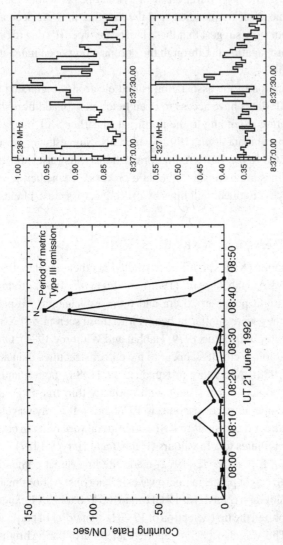

Figure 15.6. Time profiles of the soft X-ray emission from a flaring X-ray bright point (left) together with the time profiles of the associated metric radio emission (right: upper panel, 236 MHz; lower panel, 327 MHz; from the Nançay Radioheliograph). The period of radio emission is marked in the left panel: it is much briefer than the soft X-ray duration. The two light curves on the left panel correspond to two adjacent pixels on SXT image. (From Kundu *et al.* 1994b.)

We should note that soft X-ray observations cannot tell us whether or not XBP flares produce nonthermal as well as thermal emissions. Radio observations are most suitable for distinguishing between thermal and nonthermal emission. Evidence for nonthermal emission can be obtained from observations of type III bursts at meter wavelengths whereas both thermal and nonthermal emissions can be observed in microwaves. As discussed earlier, Kundu *et al.* (1994b) established that type III bursts originated from the same location as the XBP flares in several cases, thus confirming that nonthermal emissions do result from flaring XBPs.

15.3.2   Radio observations of coronal X-ray jets

Among the many discoveries made by Yohkoh-SXT, X-ray jets stand out as one of the most interesting ones. These X-ray jets are transitory X-ray enhancements with well-collimated motion (Shibata *et al.* 1992). In many cases, the jets are associated with small flares at or near their foot points and the motion appears to be a real flow of plasma at temperatures of a few million degrees.

Magnetohydrodynamic simulations of X-ray jets have been carried out by Shibata *et al.* (1992) and Yokoyama and Shibata (1995, 1996). The simulations were based on the magnetic reconnection model in which two separate magnetic field lines of emerging flux and of pre-existing coronal fields come close together by the rising motion of the emerging flux. Due to finite resistivity there is reconnection of the field lines, and by Joule dissipation magnetic energy is partly released as heat to increase the temperature of the plasma such that it is observed as X-ray jets. Yokoyama and Shibata (1995) could reproduce many of the observed characteristics of X-ray jets. In particular they found two types of interaction between emerging flux and coronal fields, which would result in two morphologically different X-ray jets. The most frequent type is the anemone jet type which occurs when emerging flux appears in a coronal hole where magnetic field lines are vertical or oblique. The other is the two-sided-loop type which occurs when emerging flux appears in a quiet region where magnetic field is almost horizontal. In that case, hot plasma is ejected along the coronal loops away from both sides of the emerging flux. Thus, the oblique-field case corresponds to the anemone-type jet, and the horizontal-field case corresponds to the two-sided-loop type jet. Besides the jet, there also results a closed loop structure at the base of the jet due to the interaction between the emerging flux and pre-existing coronal fields.

15.3.2.1 Meterwave observations of jets

Kundu *et al.* (1995) first established that Type III radio bursts are detected in conjunction with jets. Figure 15.7 shows several excellent examples of this phenomenon. The association of type III's with jets establishes that the acceleration of electrons to speeds of $\sim c/3$ (energies of tens of keV) coincides with these plasma flows. The location of type III bursts at the lower frequency (164 MHz) on the invisible or

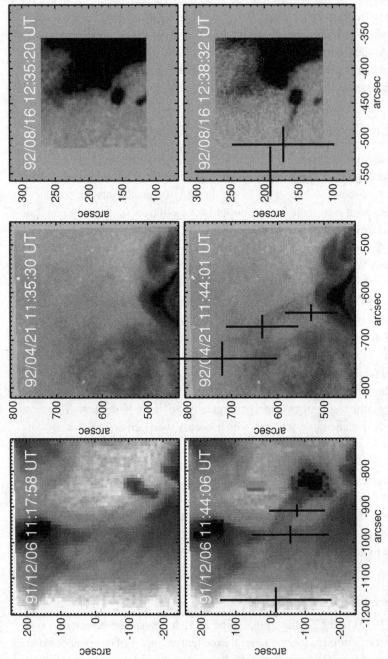

Figure 15.7. Three examples of soft X-ray jets in which metric Type III radio bursts are detected (Kundu *et al.* 1995; Raulin *et al.* 1996). In each case the upper panel shows a Yohkoh/SXT soft X-ray image of the region prior to the jet onset while the lower panel shows the SXT image obtained when jet-like emission is detected. The symbols in the lower panels indicate the positions of the associated Type III emission at 164 (largest symbol), 236 (middle symbol) and 327 MHz (smallest symbol) measured by the Nançay Radioheliograph. Higher frequencies are detected lower in the atmosphere, as expected since the frequency of the radio emission depends on the ambient plasma density.

poorly visible part of the jet suggests that the electron density in that part of the jet is adequate to produce plasma radiation, but not high enough for the jet to be visible in soft X-rays. This is consistent with extrapolation of the decreasing density away from the XBP - the site of jet origin.

While type IIIs were observed in association with anemone-type jets, Kundu *et al.* (1998) detected no nonthermal signature in the form of type III bursts for two-sided-loop type jets. This non-detection of type III bursts is consistent with the results of 2-dimensional MHD simulations of soft X-ray jets. Density gradients, which are necessary conditions in order to produce type III bursts, are unlikely to be present in two-sided-loop type jets where the magnetic field resulting from the reconnection process is almost parallel to the photosphere. On the other hand, density gradients are expected in anemone-type jets. Therefore if the reconnection process provides nonthermal particles, one expects fast drift type III bursts (which are produced by ~ 40 keV electrons) to be associated with "anemone" jets but not with "two-sided-loop" jets, as observed by Kundu *et al.* (1998).

15.3.2.2 A statistical study of jets in microwaves

Nindos *et al.* (1999b) made a statistical study of the 17 GHz properties of 18 X-ray coronal jets as observed by the Yohkoh-SXT. Microwave emission was associated with the majority (75%) of the X-ray jets studied. The radio emission typically came from the base or the base and lower part of the jets. They detected radio emission from almost all jets which showed flare-like activity at their bases. The jets which were not associated with radio emission did not show any significant increase in X-ray emission at their bases. Their data show a general correlation between the X-ray jet fluxes and the associated radio fluxes. The 17 GHz time profiles were gradual and unpolarized, implying that the emission was thermal.

Two of the events showed emission levels that were too high to be explained by the hot material visible in the SXT images. The simulations by Yokoyama and Shibata (1996) predict the existence of cool jets which are adjacent to hot coronal jets for the anemone-type jets as well as for the two-sided loop type jets. The temperature distribution in their simulations is complex and there is always $T_e \sim 10^5$ K plasma between their cool (chromospheric) and hot (coronal) jets. There are no diagnostics for such material on Yohkoh, but it can radiate strongly at microwave frequencies. The observations of Nindos *et al.* (1999) provide the first evidence for the existence of lower temperature material in the two-sided-loop-type jets which cannot be detected by the SXT.

15.3.3 Active region transient brightenings (ARTBs)

Shimizu *et al.* (1992) first reported the frequent occurrence of soft X-ray brightenings in solar active regions. These are easily seen in data from the soft X-ray telescope (SXT; Tsuneta *et al.* 1991) on board Yohkoh, which is particularly sensitive to such brightenings because of its high spatial resolution and high dynamic range. They last from a few minutes to tens of minutes, correspond to tiny enhancements in the GOES 0.5-4 Å channel, have a thermal energy content in the range 10^{25}-10^{29} ergs, and their frequency of occurrence (1-40 events per hour per active region) has a strong correlation with the total soft X-ray flux of the active region (Shimizu *et al.* 1992). Morphologically, they appear about as often in a single loop as in a dual loop configuration, and when they are in a dual loop configuration they often ($\sim 60\%$ of events) show a Y shape, indicative of two loops interacting at their nearer legs (Shimizu *et al.* 1994).

When the rate of occurrence is examined as a function of energy (Shimizu 1994, 1995) – that is, the number of events per energy bin per day, plotted against thermal energy measured with SXT the distribution is in the form of a power law with index 1.5-1.6. This is similar to the power-law index found by Lee, Petrosian and McTiernan (1995) for total hard X-ray counts for flares from the hard X-ray burst spectrometer (HXRBS) and ISEE. There is thus the possibility that soft X-ray transient brightenings could be the low-energy extension of the general flare distribution. If so, the transient brightenings would extend flares to lower energies by at least 2 orders of magnitude below subflares. Furthermore, if transient brightenings are tiny flares, they would presumably be the soft X-ray counterpart to microflares, discovered by Lin *et al.* (1984).

15.3.3.1 Radio (VLA) observations

Using the VLA, Gopalswamy *et al.* (1994) first reported transient brightenings in microwaves (at 2 cm) and their relationship with soft X-ray brightenings observed by Shimizu. These microwave transient brightenings are small-scale energy releases in coronal active regions. They are compact ($\sim 2''$) sources with duration ranging from less than a minute to more than 20 minutes. Their typical microwave flux at 2 cm (0.002-0.025 sfu) is nearly 2 orders of magnitude smaller than that from normal flares. They are also highly polarized, sometimes reaching 100%; they are located close to the spotward footpoints of coronal loops connecting the periphery of the sunspot umbra to nearby regions of opposite magnetic polarity.

Two of the VLA radio brightenings were spatially associated with soft X-ray brightenings, although the 15 GHz time profiles did not match the associated soft X-ray time profiles closely. The majority of the VLA radio brightenings were not associated with soft X-ray increases measured by GOES. The high degree of polarization in the VLA radio data and the association with the penumbra suggest that the radio emission in those cases was thermal gyroresonance emission (radiation at low harmonics of the gyrofrequency) in the strong magnetic fields low in the corona near the sunspot, although Gopalswamy *et al.* (1994) could not rule out the possibility that the emission was nonthermal gyrosynchrotron emission by accelerated electrons (which generally has degrees of polarization lower than 100%). Consistent with the thermal interpretation is the fact that they did not observe radio emission from the whole loop which brightened in soft X-rays, but only from the footpoint anchored in the penumbra, i.e., from the region of strongest magnetic field in the loop. The magnetic field in the microwave source region was found to be 1200-1800 G.

15.3.3.2 Radio (Nobeyama) observations

Using NoRH imaging data at 17 GHz along with Yohkoh/SXT data White *et al.* (1995) made detailed observations of four events in which 17 GHz radio emission was clearly detected. The time profiles of the 17 GHz data were similar to those of the soft X-ray fluxes, and the 17 GHz flux was close to that expected from plasma with the temperature (6×10^6 K) and emission measure derived for the soft X-ray-emitting material from filter ratios. No impulsive nonthermal radio emission was detected from any of the four events, although each was at least GOES class B1 in soft X-rays. Weak hard X-rays may have been detected by GRO/BATSE from the strongest of the events, but not from two others. In all cases the radio emission predicted on the basis of thermal bremsstrahlung emission from the X-ray-emitting material was reasonably close to the peak flux actually observed. Given the uncertainties in both the radio measurements and the derived soft X-ray properties, this agreement suggests that thermal bremsstrahlung is the emission mechanism for the 17 GHz sources.

15.3.3.3 Radio (OVRO) observations

The studies on ARTB's by Shimizu *et al.* (1992), Gopalswamy *et al.* (1994), and White *et al.* (1995), sought a connection with the microflares reported in hard X-rays by Lin *et al.* (1984), by looking for a nonthermal counterpart to the emission. This is an important question, since the soft X-rays are produced by heating of the coronal plasma which could be an entirely thermal process quite different from the nonthermal acceleration needed for microflares. To demonstrate that transient brightenings and microflares are linked, it is important to show that there is an associated release of nonthermal electrons. Attempts to observe hard X-ray emission from SXR transient brightenings, were not initially successful because current hard X-ray detectors are not sensitive enough; subsequently Nitta (1997) showed that they could be detected. The presence of nonthermal electrons can be more easily detected from observations of radio emission, which is generated by nonthermal electrons. Gary *et al.* (1997) used the Owens Valley Radio Observatory (OVRO) Solar Array, operating at 45 frequencies between 1 and 18 GHz, to search for signs of nonthermal electrons associated with ARTB's. The OVRO Solar Array has two advantages for such a study: (1) the frequency range includes the lower frequencies, where significant nonthermal emission should occur; and (2) the broad frequency coverage reduces the possibility of missing events that may have a relatively narrow flux spectrum, and it allows the spectral shape to be used to verify that the emission is nonthermal.

Gary *et al.* found that the transient brightenings were clearly detected in microwaves in 12 of 34 events (35%), possibly detected in another 17 of 34 events

(50%), and only five of 34 events (15%) had no apparent microwave counterpart. As shown in Figure 15.8, the microwave spectra often peak in the range 5-10 GHz (13 of 16 events), and the microwave spectra of some events show narrowband spectra with a steep low-frequency slope. Gary *et al.* (1997) concluded that the emission from at least some events is the result of a nonthermal population of electrons, and that transient brightenings as a whole can therefore be identified as microflares, the low-energy extension of the general flare energy distribution.

15.3.3.4 Transient brightenings in quiet Sun regions

Krucker *et al.* (1997) studied the temporal variations in the soft X-ray (SXR) emission and radio emission above the solar magnetic network of the quiet corona using Yohkoh SXR images with deep exposure and VLA observations in the centimeter radio range. The SXR data show several brightenings, with an extrapolated occurrence probability of one brightening per 3 seconds on the total surface. During the roughly 10 minutes of enhanced flux, total radiative losses of the observed plasma are around 10^{25} ergs per event. These events are more than an order of magnitude smaller than previously reported X-ray bright points or active region transient brightenings. These events thus appear to be flare-like and are called network flares by Krucker *et al.* (1997). Krucker and Benz (1998) extended this work using EIT EUV data and were able to identify even smaller energy releases.

Nindos *et al.* (1999a) studied a set of radio-selected transient brightenings (TB's) as a complement to the more common X-ray selected surveys. Five small impulsive events were identified in a set of VLA observations at 4.5, 1.5 and 0.33 GHz and compared with soft X-ray images from Yohkoh and EUV images from SOHO/EIT. Four of the events were located at the edges of an active region but one was located $100''$ away in a quiet region of the atmosphere. The time profiles of the radio T_B's showed impulsive peaks while the corresponding soft X-ray profiles were gradual. The impulsive radio peaks were up to 35% polarized. Their data favor an interpretation in terms of gyrosynchrotron radiation from mildly relativistic electrons.

Nindos *et al.* concluded that the emission of the TBs was caused by a population of nonthermal, mildly relativistic, electrons. The time profiles of the radio T_B's showed impulsive peaks while the corresponding soft X-ray time profiles were more gradual. Calculations showed that a small population of nonthermal electrons with spectral index $\delta = 3$ reproduced satisfactorily the basic observable characteristics of the radio TBs.

15.3.3.5 Implications of transients for coronal heating

Shimizu (1995) has shown that the collective thermal energy from the events at least down to about 10^{26} erg is not sufficient to heat the active region corona, being from 5 to 10 times too low. Krucker *et al.* (1997) came to a similar conclusion for their smaller events. One possible way around this is that the rate of occurrence of events

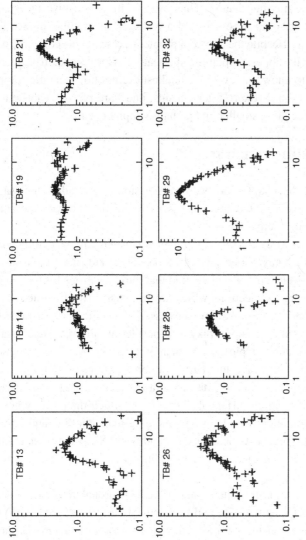

Figure 15.8. OVRO total power spectra for 8 selected active region transient brightenings (Gary *et al.* 1997). The spectral shape is highly variable from event to event, but some of the events (e.g., transient brightening numbers 13, 21, 28, 29, and 32) have a low-frequency slope greater than 2, implying that they must be the result of nonthermal electrons.

per unit energy may increase more steeply at low energies, so that the tiniest events, nanoflares, may supply most of the energy as suggested by Parker (1988). Gary (1999) carried out an analysis of the requirements in this case and found that the distribution of energy releases must increase very sharply below 10^{24} ergs to explain coronal heating (slope > 2.8). Clearly, as one pushes the minimum observed energy to lower energies it becomes more and more difficult for the remaining, unobserved events to supply the missing energy. Thus even as more sensitive measurements push the detection threshold for energy releases lower, the number of small events continues to fall short. Despite the apparent problems, researchers continue to search for smaller events in the hope of explaining the heating of the corona. In the meantime, it is at least useful to know that even the smallest events show a similar proportion of nonthermal emission as their larger cousins, and flare physics can be usefully advanced by studying these smaller and perhaps simpler events.

15.4   Concluding remarks

The main theme of this chapter has been a discussion of the radio observations of explosive energy releases on the Sun. Our discussion focussed on two areas: normal flares, and small-scale energy releases.

We have discussed flare emission at millimeter wavelengths and compared it with microwave and hard X-ray/γ-ray flare emission. Millimeter flare emission is produced by MeV-energy electrons, and we find that these electrons form a population distinct from the energetic electrons which are responsible for the microwave and hard X-ray emission in flares. We have discussed a specific class of radio bursts, namely those which exhibit simple spiky time profiles at millimeter and microwave wavelengths with durations typically less than 30 seconds. We have discussed distinctive properties of microwave and hard X-ray observations of footpoint emission from flaring loops. Finally we have discussed a particular spatial configuration of the microwave flaring region ("Hanaoka" events) in which the primary energy release results from the interaction between emerging magnetic flux and an existing overlying region: such events typically exhibit radio and X-ray emission at the main flare site (the site of the interaction), and in addition radio emission at a remote site up to 2×10^5 km away in another active region.

We have discussed the radio counterparts of weaker coronal transients as observed primarily in X-rays (Yohkoh/SXT) and to some extent in EUV (SOHO/EIT). We find that in all these transients for which there is a radio counterpart, the radio emission consists of both thermal and nonthermal components. This is true for the transients discussed – XBP flares, coronal X-ray jets, ARTB's, quiet Sun transients, and plasmoid ejections. These transients are at least two orders of magnitude weaker than normal flares both in X-ray and radio domains. Thus, one would like to think of them as microflares with different morphological manifestations in the X-ray domain. In the radio domain, one sees distinct signatures of nonthermal and thermal processes in them. In the case of radio thermal emission, one observes more emission than that computed from the plasma parameters of the thermal X-ray plasma; this may simply imply that the cooler material is seen in radio, but not in X-rays. We have found no time-coincident microwave counterparts of eruptive polar plumes or similarly transient EUV phenomena.

References

Alissandrakis, C. E. (1986). *Solar Phys.*, **104**, 207.

Bastian, T. S., Benz, A. O., and Gary, D. E. (1998). *Ann. Rev. Astron. Astrophys.*, **36**, 131.

Dulk, G. A. (1985). *Ann. Rev. Astron. Astrophys.*, **23**, 169.

Fu, Q.-J., Kundu, M. R., and Schmahl, E. J. (1987). *Solar Phys.*, **108**, 99.

Gary, D. E. (1999). in *Nobeyama Symposium on Solar Physics with Radio Observations*, ed. T. S. Bastian, N. Gopalswamy, and K. Shibasaki (Nobeyama Radio Observatory), p. 129.

Gary, D. E., Hartl, M. D., and Shimizu, T. (1997). *Astrophys. J.*, **477**, 958.

Golub, L., Krieger, A. S., Harvey, J. W., and Vaiana, G. S. (1977). *Solar Phys.*, **53**, 111.

Golub, L., Krieger, A. S., Vaiana, G. S., Silk, J. K., and Timothy, A. F. (1974). *Astrophys. J. (Lett.)*, **189**, L93.

Gopalswamy, N., Payne, T. E. W., Schmahl, E. J., Kundu, M. R., Lemen, J. R., Strong, K. T., Canfield, R. C., and de La Beaujardiere, J. (1994). *Astrophys. J.*, **437**, 522.

Habbal, S. R. and Harvey, K. L. (1988). *Astrophys. J.*, **326**, 988.

Habbal, S. R., Ronan, R. S., Withbroe, G. L., Shevgaonkar, R. K., and Kundu, M. R. (1986). *Astrophys. J.*, **306**, 740.

Habbal, S. R. and Withbroe, G. L. (1981). *Solar Phys.*, **69**, 77.

Hanaoka, Y. (1996). *Solar Phys.*, **165**, 275.

Hanaoka, Y. (1997). *Solar Phys.*, **173**, 319.

Hanaoka, Y. (1999a). *Publ. Astron. Soc. Japan*, **51**, 483.

Hanaoka, Y. (1999b). in *Nobeyama Symposium on Solar Physics with Radio Observations*, ed. T. S. Bastian, N. Gopalswamy, and K. Shibasaki (Nobeyama Radio Observatory), p. 229.

Krucker, S., Benz, A. O., Bastian, T. S., and Acton, L. W. (1997). *Astrophys. J.*, **488**, 499.

Krucker, S., and Benz, A. O. (1998). *Astrophys. J.*, **501**, L213.

Kundu, M. R. (1965). *Solar Radio Astronomy* (New York: Interscience Publishers).

Kundu, M. R., Gergely, T. E., and Golub, L. (1980). *Astrophys. J. Lett.*, **236**, L87.

Kundu, M. R., Grechnev, V. V., Garaimov, V. I., and White, S. M. (2001a). *Astrophys. J.*, **563**, 389.

Kundu, M. R., Nindos, A., White, S. M., and Grechnev, V. V. (2001b). *Astrophys. J.*, **557**, 880.

Kundu, M. R., Nitta, N., White, S. M., Shibasaki, K., Enome, S., Sakao, T., Kosugi, T., and Sakurai, T. (1995). *Astrophys. J.*, **454**, 522.

Kundu, M. R., Raulin, J., Nitta, N., Shibata, K., and Shimojo, M. (1998). *Solar Phys.*, **178**, 173.

Kundu, M. R., Schmahl, E. J., and Fu, Q. J. (1988). *Astrophys. J.*, **325**, 905.

Kundu, M. R., Shibasaki, K., Enome, S., and Nitta, N. (1994a). *Astrophys. J. Lett.*, **431**, L155.

Kundu, M. R., Strong, K. T., Pick, M., White, S. M., Hudson, H. S., Harvey, K. L., and Kane, S. R. (1994b). *Astrophys. J. Lett.*, **427**, L59.

Kundu, M. R. and Vlahos, L. (1982). *Space Science Reviews*, **32**, 405.

Kundu, M. R., White, S. M., Gopalswamy, N., and Lim, J. (1992). in *The Compton Observatory Science Workshop*, ed. C. R. Schrader, N. Gehrels, and B. Dennis (Washington, D.C.: NASA CP–3137), p. 502.

Kundu, M. R., White, S. M., Gopalswamy, N., and Lim, J. (1994c). *Astrophys. J. Suppl.*, **90**, 599.

Kundu, M. R., White, S. M., Shibasaki, K., Sakurai, T., and Grechnev, V. V. (2001c). *Astrophys. J.*, **547**, 1090.

Kundu, M. R., White, S. M., Shibasaki, K., and Sakurai, T. (2000). *Astrophys. J.*, **545**, 1084.

Lee, T. T., Petrosian, V., and McTiernan, J. M. (1995). *Astrophys. J.*, **448**, 915.

Lim, J., White, S. M., Kundu, M. R., and Gary, D. E. (1992). *Solar Phys.*, **140**, 343.

Lin, R. P., Schwartz, R. A., Kane, S. R., Pelling, R. M., and Hurley, K. C. (1984). *Astrophys. J.*, **283**, 421.

Nindos, A., Kundu, M. R., and White, S. M. (1999a). *Astrophys. J.*, **513**, 983.

Nindos, A., Kundu, M.R., Raulin, J.P., Shibasaki, K., White, S.M., Nitta, N., Shibata, K., and Shimojo, M. (1999b). in *Nobeyama Symposium on Solar Physics with Radio Observations*, ed. T.S. Bastian, N. Gopalswamy and K. Shibasaki (NRO Report 479), p. 135.

Nindos, A., White, S. M., Kundu, M. R., and Gary, D. E. (2000). *Astrophys. J.*, **533**, 1053.

Nishio, M., Yaji, K., Kosugi, T., Nakajima, H., and Sakurai, T. (1997). *Astrophys. J.*, **489**, 976.

Nitta, N. (1997). *Astrophys. J.*, **491**, 402.

Nitta, N., Bastian, T. S., Aschwanden, M. J., Harvey, K. L., and Strong, K. T. (1992). *Publ. Astron. Soc. Japan*, **161**, L167.

Parker, E. N. (1988). *Astrophys. J.*, **330**, 474.

Raulin, J. P., Kundu, M. R., Hudson, H. S., Nitta, N., and Raoult, A. (1996). *Astron. Astrophys.*, **306**, 299.

Raulin, J., White, S. M., Kundu, M. R., Silva, A. R., and Shibasaki, K. (1999). *Astrophys. J.*, **522**, 547.

Sakao, T. (1994). *Characteristics of Solar Flare Hard X-ray Sources* (Tokyo: The University of Tokyo), Ph. D. Thesis.

Sheeley, Jr., N. R., and Golub, L. (1979). *Solar Phys.*, **63**, 119.

Shibata, K., Nozawa, S., and Motsumoto, R. (1992). *Publ. Astron. Soc. Japan*, **44**, 265.

Shimizu, T. (1994). in *X–ray Solar Physics from Yohkoh*, ed. Y. Uchida *et al.* (Tokyo: Universal Academy Press), p. 33.

Shimizu, T. (1995). *Publ. Astron. Soc. Japan*, **47**, 251.

Shimizu, T., Tsuneta, S., Acton, L. W., Lemen, J. R., Ogawara, Y., and Uchida, Y. (1994). *Astrophys. J.*, **422**, 906.

Shimizu, T., Tsuneta, S., Acton, L., Lemen, J. R., and Uchida, Y. (1992). *Publ. Astron. Soc. Japan*, **44**, L147.

Silva, A. V. L., White, S. M., Lin, R. P., de Pater, I., Shibasaki, K., Hudson, H. S., and Kundu, M. R. (1996). *Astrophys. J. (Lett.)*, **458**, L49.

Tsuneta, S., Acton, L., Bruner, M., Lemen, J., Brown, W., Caravalho, R., Catura, R., Freeland, S., Jurcevich, B., and Morrison, M. (1991). *Solar Phys.*, **136**, 37.

Wang, H., Gary, D. E., Lim, J., and Schwartz, R. A. (1994). *Astrophys. J.*, **433**, 379.

Wang, H., Gary, D. E., Zirin, H., Schwartz, R. A., Sakao, T., Kosugi, T., and Shibata, K. (1995). *Astrophys. J.*, **453**, 505.

White, S. M. (1994). in *High Energy Solar Phenomena: A New Era of Spacecraft Measurements*, ed. J. Ryan and W. T. Vestrand (New York: American Institute of Physics (Conf. Proc. **294**)), p. 199.

White, S. M. (1999). in *Nobeyama Symposium on Solar Physics with Radio Observations*, ed. T. S. Bastian, N. Gopalswamy, and K. Shibasaki (Nobeyama Radio Observatory), p. 223.

White, S. M. and Kundu, M. R. (1992). *Solar Phys.*, **141**, 347.

White, S. M., Kundu, M. R., Shimizu, T., Shibasaki, K., and Enome, S. (1995). *Astrophys. J.*, **450**, 435.

Yokoyama, T. and Shibata, K. (1995). *Nature*, **375**, 42.

Yokoyama, T. and Shibata, K. (1996). *Publ. Astron. Soc. Japan*, **48**, 353.

16

Coronal oscillations

V.M. Nakariakov

Physics Department, University of Warwick, Coventry, CV4 7AL, UK

16.1 Introduction

The magnetically dominated plasma of the solar corona is an elastic medium which can support propagation of various types of waves. For a large class of waves, where wavelengths and periods are large compared to the ion Larmor radius and the gyroperiod, respectively, the waves can be studied using magnetohydrodynamics (MHD). These waves perturb macro-parameters of the coronal plasma, such as density, temperature, bulk velocity and the frozen-in magnetic field. MHD waves in the corona have been intensively investigated for more than two decades in the context of coronal heating and acceleration of the solar wind. It is also believed that the coronal MHD waves play an important role in solar-terrestrial connections. The investigation of the coronal waves is now an essential part of solar physics, space physics, geophysics and astrophysics.

The observational study of coronal MHD oscillations and waves has been undertaken in almost all possible bands for a few decades. In particular, radio pulsations with periods ranging from a fraction of a second to several minutes were reported by many authors (see Aschwanden 1987). Probably, the first EUV observations of MHD waves in the corona were reported by Chapman *et al.* (1972) analyzing data of the GSFC extreme-ultraviolet spectroheliograph on OSO-7 (spatial resolution was $10'' \times 20''$, cadence time was 5.14 s). Mg VII, Mg IX and He II emission intensity periodicities at about 262 s were detected. In the X-ray band, Harrison (1987) with a Hard X-ray Imaging Spectrometer on SMM detected soft X-ray (3.5 to 5.5) keV pulsations with the period of 24 min.

In parallel, significant progress in MHD wave theory has been made. The references to this work are too numerous to mention comprehensively and we refer the interested reader to the reviews of Roberts (1991, 2000).

A new era in observations of the coronal wave activity was opened up by the launching of the SOHO and TRACE spacecraft. For the first time it became possible to combine observational and theoretical efforts in the investigation of MHD waves and oscillations in the corona. The theory provides *quantitative* interpretation of wave phenomena, and the observations can prove the correctness of theoretical constructions. In particular, the implication of the method of MHD coronal seismology became possible.

The aim of this review is to discuss the delicate interplay of MHD wave theory and the observations of coronal waves and oscillations, illustrating it with several examples.

The discussion of other aspects of coronal oscillations and waves can be found elsewhere, in particular, the MHD wave theory (e.g., Goossens, 1991; Roberts, 1991, 2000), observations (e.g., Aschwanden 1987; Aschwanden *et al.* 1999), wave mechanisms of coronal heating (Narain and Ulmschneider 1990, 1996), and prominence oscillations (Oliver, 1999).

16.2   The method of MHD coronal seismology

Despite significant progress in coronal physics over several decades, a number of fundamental questions, for instance, what are the physical mechanisms responsible for the coronal heating, the solar wind acceleration and solar flares, remain to be answered. All these questions, however, require detailed knowledge of physical conditions and parameters in the corona, which cannot yet be measured accurately enough. In particular, the exact value of the coronal magnetic field remains unknown, because of a number of intrinsic difficulties with applications of direct methods (e.g., based upon the Zeeman splitting and gyroresonant emission), as well as indirect (e.g., based upon extrapolation of chromospheric magnetic sources). Also, the coronal transport coefficients, such as volume and shear viscosity, resistivity and thermal conduction, which play a crucial role in coronal heating, are not known even within an order of magnitude.

The detection of coronal waves provides us with a new tool for the determination of the unknown parameters of the corona – *MHD seismology of the corona*. Measurement of the properties of MHD waves and oscillations (periods, wavelengths, amplitudes, temporal and spatial signatures, characteristic scenarios of the wave evolution), combined with a theoretical modelling of the wave phenomena (dispersion relations, evolutionary equations, etc.), can lead to a determination of the mean parameters of the corona, such as the magnetic field strength and transport coefficients. This approach is illustrated in Figure 16.1. Philosophically, the method is similar

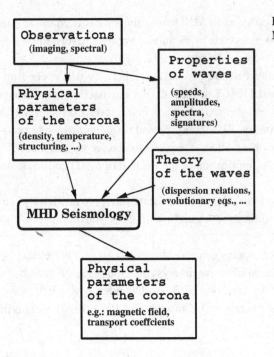

Figure 16.1. The method of MHD coronal seismology.

to the acoustic diagnostics of the solar interior, helioseismology. But, MHD coronal seismology is much richer as it is based upon three different wave modes, namely, Alfvén, slow and fast magnetoacoustic modes. These MHD modes have quite different dispersive, polarization and propagation properties, which makes this approach even more powerful.

Originally, the method of MHD coronal seismology was suggested by Uchida (1970) and Roberts *et al.* (1984), and has recently been applied to obtain estimates of the coronal dissipative coefficients (Nakariakov *et al.* 1999) and the magnetic field (Nakariakov 2000; Nakariakov and Ofman 2001). These implementations are discussed in Section 16.7.3 and Section 16.7.2).

16.3   Detectability of MHD waves in the corona

The very first question is how can MHD waves be detected in the corona of the Sun. This issue is connected with the physical quantities perturbed by the waves. In particular, slow and fast magnetoacoustic waves are compressive and cause perturbations of plasma density. As emission is proportional to the density, the waves can be detected as emission variations by imaging telescopes. An important characteristic of the phenomenon is the angle between the direction of the wave propagation and the line of sight (LOS). Consequently, imaging telescopes allow one to observe magnetoacoustic waves propagating with a sufficiently high angle to the LOS. In certain circumstances, practically incompressible Alfvén waves can also be detected with an imaging telescope, if perturbations of the magnetic field have a component perpendicular to the LOS. Indeed, as the magnetic field is frozen-in the coronal plasma, the perpendicular displacement of the field can be highlighted by variation of emission intensity.

The LOS component of the plasma velocity perturbed by a wave causes positive and negative Doppler shifts in the emission spectrum. Thus, as both compressive magnetoacoustic and incompressive Alfvén waves perturb the plasma velocity, they can be detected by telescopes with spectral resolution.

Another constraint is imposed on observations of coronal waves by the telescope resolution. To observe waves *directly*, the spatial (e.g., the pixel size) and temporal (e.g., the exposure and the cadence time) resolution of the telescope should be several times less than the expected wavelengths and periods. The parameters of modern spaceborne coronal telescopes are summarized in Table 16.1. Much better time resolution can be reached in white light ground-based observations of the corona during eclipses.

Therefore, the currently available resolution does not allow one to detect *directly* waves with periods less than a few minutes and lengths shorter than a few megameters. It is also important to note that the spatial and temporal resolution must be reached *simultaneously*. For example, consider a global standing fast magnetoacoustic mode of a coronal loop and take the phase speed of this wave to be about the Alfvén speed (i.e., 1 Mm/s), then we see that for a short loop (i.e., 30 Mm), the wave period is about 60 s. Consequently, the global standing fast magnetoacoustic modes of short and medium length loops cannot be detected.

Table 16.1. *Spatial and temporal resolutions of spaceborne coronal telescopes*

Spacecraft Instrument	Spatial Res. (pixel size)/″	Temporal res. (max cadence)/s	Bandpass
SOHO			
EIT	2.6	30	EUV
CDS	1 × 2	30	EUV
UVCS	12–300	sec – hr	EUV/FUV/WL
LASCO-C1	5.6	60	WL
LASCO-C2	11.4	75	WL
LASCO-C3	56.0	75	WL
SUMER	1	≈ 10	EUV/FUV
YOHKOH/SXT	4	few seconds	SX
YOHKOH/HXT	≈ 60	0.2	HX
TRACE	0.5	10	EUV/FUV/WL

Note that all the coronal waves are directly observed very near the threshold of the currently achievable resolution, for example 2.5″ in space and 30 s in time for SOHO/EIT and 0.5″ in space and 20 s in time for TRACE.

Thus, short period coronal waves (less than a few minutes) can be observed only *indirectly*, in particular, by non-thermal broadening of coronal emission lines. Broadening of spectral lines emitted by optically thin plasmas, measured as effective temperature, T_{eff}, is formed by two effects: thermal broadening and non-thermal broadening associated with the Doppler shift due to unresolved LOS motions

$$T_{\text{eff}} = T_i + C \frac{m_i}{2k} \langle V_{\text{LOS}}^2 \rangle, \tag{16.1}$$

where T_i is the temperature of the line forming ion, k is the Boltzmann constant, V_{LOS} is the LOS velocity, $2/3 < C < 1$ is a constant defined by polarization and other effects.

In the long wavelength/long period limit, there is one additional observational constraint. The large wave periods and long wavelengths may not be observable because of the finite duration of the observational sequence or the restricted field of view of the instrument used.

16.4   Compressive waves in polar plumes

16.4.1   Observations

Polar plumes are ray-like bright features observed in coronal holes. They originate from near-surface layers arising from predominantly unipolar magnetic regions, where small bipoles reconnect with the unipolar flux concentrations, and spread out up to at least several solar radii. The magnetic field topology is believed to be close to a magnetic monopole and the density of the plasma to be enhanced in the plumes (e.g., DeForest and Gurman 1998). The global structure of polar plumes remains quiescent for up to a few weeks. However, short-term variations of the plume brightness with periods of about a few minutes have been known for the last two decades (e.g., Withbroe 1983).

A possible way to identify coherent motions in imaging telescope data is to use a stroboscopic method. The emission intensities along a chosen path, taken in different instants of time are laid side-by-side to form a time–distance map. DeForest and Gurman (1998) applied this method to SOHO/EIT 171 Å data. They extracted a vertical strip of the middle few pixels of several plumes and constructed time-distance maps. These maps contained diagonal stripes. Such a diagonal stripe exhibits an EUV brightness disturbance which changes its position in time and, consequently, propagates along the path. The variation of the intensity in the disturbances suggests that they are perturbations of density and, consequently, the disturbances are compressive. Thus, compressive disturbances propagating along polar plumes have been identified. Summarizing the findings: outwardly propagating perturbations of the intensity (plasma density) at distances of 1.01-1.2 $R_\odot$ from the Sun are detected, they are gathered in quasiperiodic groups of 3-10 periods, with periods of about 10-15 min, the duty cycle is roughly symmetric, the projected speeds are about 75-150 km/s, the amplitude (in density) is about 2-4% of the background and grows with height.

These waves are probably of the same nature as the polarized brightness (density) fluctuations with periods of about 9 min, detected in coronal holes at the height about 1.9 $R_\odot$ by Ofman et al. (1997, 1998) using the white light channel of SOHO/UVCS. The signal was integrated over a $14'' \times 14''$ area with an exposure time of 60-180 s and a cadence time of 90-210 s. Developing this study and analyzing several more UVSC data sequences with a 75-125 s cadence time, Ofman et al. (2000) determined the fluctuation periods to be 7-10 min. The measurement of the propagation speed of the fluctuations indicated values in the range of 160-260 km/s at 2 $R_\odot$. This range is consistent with the phase speed of slow magnetoacoustic waves (possibly, accelerated by nonlinearity). Perhaps, these waves are also observed in the transition region with SOHO/CDS (Banerjee et al. 2000) as periodic (10-25 min) intensity fluctuations of the O v 629 Å line, corresponding to a formation temperature of 0.25 MK.

16.4.2   Interpretation as slow magnetoacoustic waves

The observations demonstrate that the compressive perturbations propagate with the speed, which is slightly less than the expected sound speed, 152 km/s. Also, the observed periodicities are much shorter then the acoustic cutoff periods in the corona, $P_c = 4\pi C_s / \gamma g$, where C_s is the sound speed, $\gamma = 5/3$ the adiabatic index and $g = 274$ m/s^2 the gravitational acceleration. This suggests that the compressive perturbations can be interpreted as slow magnetoacoustic waves. The theory of slow magnetoacoustic waves in polar plumes has been developed in Ofman et al. (1999) and Ofman et al. (2000). The model incorporates several physical mechanisms of the wave evolution, namely the dissipation due to viscosity, effects of gravitational stratification and magnetic field divergence, and weak nonlinearity. The last mechanism is important because slow magnetoacoustic waves propagating upwards in a stratified atmosphere are amplified and consequently becoming more and more nonlinear.

In this model, the magnetic field is taken to be radially divergent

$$B_0(r) = \frac{B_{00} R_\odot^2}{r^2} \qquad (16.2)$$

where B_{00} is the magnetic field strength at the base of the corona ($r = R_\odot$) and r is the radius vector (see Figure 16.2). Such a configuration corresponds to a magnetic monopole and is not correct for modeling of a global structure of the solar magnetic field. However, the configuration models very well the *local* magnetic structure of a coronal hole. The temperature, T, and, consequently, the sound speed, C_s, are taken

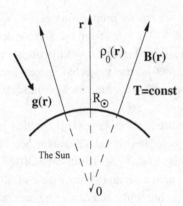

Figure 16.2. The model for a coronal hole as a gravitationally stratified atmosphere with vertical magnetic field.

to be constant. The hydrostatic equilibrium profile of the gravitationally stratified density is

$$\rho_0(r) = \rho_{00} \exp\left\{ -\frac{R_\odot}{H} \left[1 - \frac{R_\odot}{r} \right] \right\}, \tag{16.3}$$

where H is the *scale height*, $H/\text{Mm} \approx 50T/\text{MK}$, and ρ_{00} is density at $r = R_\odot$. The vertical dependence of the gravitational acceleration is $g = GM_\odot/r^2$, where $M_\odot$ the mass of the Sun and G the gravitational constant.

Spherically symmetric slow magnetoacoustic waves, which perturb the density ρ and the vertical component V_r of the plasma velocity are described by the wave equation

$$\frac{\partial^2 \rho}{\partial t^2} - \frac{C_s^2}{r^2} \frac{\partial}{\partial r} \left(r^2 \frac{\partial \rho}{\partial r} \right) - g \frac{\partial \rho}{\partial r} = RHS \tag{16.4}$$

where nonlinear and dissipative terms are gathered on the right hand side and represented by RHS.

When the considered wavelengths, λ, are much less than both the scale height, H, and $R_\odot$, equation (16.4) can be solved in the WKB (or the single wave) approximation, introducing the small parameter $\epsilon = \lambda/H \ll 1$. The nonlinear, quadratic in the weakly nonlinear limit, and dissipative terms can be considered to be of the same order in ϵ.

Following an upwardly propagating wave and passing to the moving frame of reference

$$\xi = r - C_s t, \qquad R = \epsilon r, \tag{16.5}$$

we obtain

$$\frac{\partial \rho}{\partial R} + \left(\frac{1}{R} + \frac{g(R)}{2C_s^2} \right) \rho + \frac{1}{\rho_0(R)} \rho \frac{\partial \rho}{\partial \xi} - \frac{2\eta_0}{3C_s\rho_0(R)} \frac{\partial^2 \rho}{\partial \xi^2} = 0, \tag{16.6}$$

where η_0 is the compressive viscosity coefficient. When C_s is a function of r, it is more convenient to use a temporal variable $\tau = t - \int C_s^{-1}(r)dr$ instead of the spatial variable ξ.

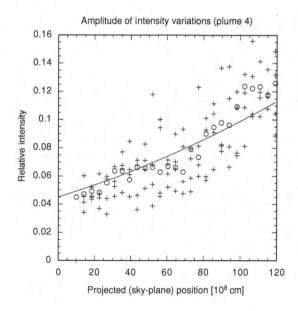

Figure 16.3. Dependence of the compressive wave amplitude relative to the background intensity on the height in a polar plume (Ofman *et al.* 1999).

Equation (16.6) is the *spherical Burgers equation*. Neglecting the nonlinear and dissipative terms, we obtain the ideal linear solution

$$\rho = \rho(R_\odot)\frac{1}{R}\ \exp\left[-\frac{R_\odot}{2H}\left(1-\frac{R_\odot}{R}\right)\right].$$ (16.7)

This expression shows that the absolute amplitude of the density perturbation decreases with height, whilst *the ratio of the density perturbation amplitude and the local equilibrium density value grows*

$$\frac{\rho}{\rho_0} \propto \frac{1}{R}\ \exp\left[\frac{R_\odot}{2H}\left(1-\frac{R_\odot}{R}\right)\right].$$ (16.8)

Such a behaviour has been confirmed observationally, see Figure 16.3. In the presence of dissipative processes, the dissipation rate should be subtracted from the growth rate, consequently the difference between the observational and theoretical curve $\rho(R)/\rho_0(R)$ can be explained by dissipation.

The growth of the relative amplitude of the slow wave is accompanied by the nonlinear generation of higher harmonics and wave steepening. In the regions with high gradients, the dissipative term, the last term, in equation (16.6) becomes significant and it leads to enhanced dissipation of the waves.

The nonlinear regime of the slow magnetoacoustic waves in plumes was investigated by Ofman *et al.* (1999, 2000) and Cuntz and Suess (2001). It was found that weakly nonlinear slow waves quickly steepen and deposit energy and momentum in the lower parts of the plumes, below 1.3-1.6 $R_\odot$, depending upon the amplitude and the period.

16.5   Search for Alfvén waves

16.5.1   Theoretical aspects

In the linear regime, Alfvén waves are incompressive and very weakly dissipative. This means they can propagate very long distances and deposit energy and momentum far from their source. Concerning the generation of Alfvén waves, they can easily be excited by various dynamical perturbations of magnetic field lines. This makes Alfvén waves to be a promising tool for heating and diagnostics of open magnetic structures.

Consider purely vertical, linearly polarized Alfvén waves in a vertical open magnetic structure in a spherically stratified isothermal static atmosphere (see Figure 16.2) described by the hydrostatic equilibrium (16.2) and (16.3). From (16.2) and (16.3), the vertical profile of the Alfvén speed C_A is given by

$$C_A(r) = \frac{B_0(R_\odot)R_\odot^2}{r^2[4\pi\rho_0(R_\odot)]^{1/2}} \exp\left(\frac{R_\odot(r - R_\odot)}{2Hr}\right). \tag{16.9}$$

In this geometry, linear Alfvén waves perturb the ϕ-components of the magnetic field and velocity and are described by the wave equation,

$$\frac{\partial^2 V_\phi}{\partial t^2} - \frac{B_0(r)}{4\pi\rho_0(r)r}\frac{\partial^2}{\partial r^2}\left[rB_0(r)V_\phi\right] = 0. \tag{16.10}$$

As in Sec. 16.4.2, we apply the WKB approximation, neglecting coupling of upwards and downwards propagating waves and introducing the small parameter $\epsilon = \lambda/H \ll 1$. Following an upwards propagating wave and passing to the moving frame of reference

$$\tau = t - \int \frac{dr}{C_A(r)}, \qquad R = \epsilon r, \tag{16.11}$$

equation (16.10) is reduced to

$$\frac{\partial V_\phi}{\partial R} - \frac{R_\odot^2}{4H}\frac{1}{R^2}V_\phi = 0, \tag{16.12}$$

with the solution

$$V_\phi = V_\phi(R_\odot)\exp\left(\frac{R_\odot(R - R_\odot)}{4HR}\right), \tag{16.13}$$

which is the linear Alfvén wave evolutionary equation.

The same result can be obtained by the following geometrical reasonings. The radial component of the Poynting flux carried out by the wave is

$$\mathcal{P}_r \propto (\mathbf{B} \times \mathbf{V} \times \mathbf{B})_r = B_0 V_\phi B_\phi = V_\phi^2 B_0^2/C_A. \tag{16.14}$$

On the other hand, the Poynting flux of a linear spherical wave in a non-dissipative medium is proportional to $1/r^2$. Consequently, V_ϕ^2/C_A is constant, which gives us

$$V_\phi \propto \rho_0^{-1/4}(r), \tag{16.15}$$

which coincides with the dependence given by (16.13).

Thus, in the spherically stratified atmosphere, the relative Alfvén wave amplitude grows with height and nonlinear effects come into play. Also, it is worth taking into account dissipative effects. Assuming that both the nonlinearity and dissipation are weak, we can add these effects to the evolutionary equation

$$\frac{\partial V_\phi}{\partial R} - \frac{R_\odot^2}{4H} \frac{1}{R^2} V_\phi - \frac{1}{4C_A(C_A^2 - C_s^2)} \frac{\partial V_\phi^3}{\partial \tau} - \frac{\nu}{2C_A^3} \frac{\partial^2 V_\phi}{\partial \tau^2} = 0 \tag{16.16}$$

(see Nakariakov *et al.* 2000 for the rigorous derivation). Equation (16.16) is a spherical analog of the *scalar Cohen–Kulsrud–Burgers* equation. The main difference of this equation and the spherical Burgers equation (16.6) is the order of the nonlinearity. The Cohen–Kulsrud–Burgers equation is of cubic nonlinearity.

The scenario of the wave evolution with height is as follows: the amplitude grows and the nonlinearity generates higher harmonics, which leads to the wave overturning and consequent nonlinear dissipation. The waves of higher amplitudes and smaller periods reach the nonlinear dissipation regime closer to the base of the corona. Various regimes of the Alfvén wave amplitude evolution are investigated by Nakariakov *et al.* (2000). The main conclusion is that Alfvén waves with initial amplitudes of 25 km/s and periods less than 1 min are subject to efficient nonlinear dissipation within 5-7 $R_\odot$ from the coronal base. The essential element of this scenario is the generation of secondary perturbations of density and field-aligned plasma motions, which should not be confused with slow magnetoacoustic waves, as they propagate with the Alfvén, not the sound speed. These results are consistent with full-MHD numerical simulations of the Alfvén wave dynamics in one-dimensional (Torkelsson and Boynton 1998) and two-dimensional (Ofman and Davila 1997, 1998) open coronal structures.

16.5.2 Observational aspects

There is no direct evidence of Alfvén wave activity in the corona, mainly due to genuine difficulties with their detection. However, the waves are observed *in situ* further out in the polar solar wind as magnetic fluctuations (e.g., Balogh *et al.* 1995). Development and implementation of indirect methods for the wave detection in the corona are based upon the theory discussed in Sec. 16.5.1.

Non-thermal broadening of the UV and EUV coronal lines, which can be associated with unresolved LOS wave motions with the amplitude of about 20-30 km/s, has been known for 25 years, since Skylab (e.g., Feldman *et al.* 1976) and rocket experiments (see, e.g., Hassler *et al.* 1990). There is some discrepancy in results found using different instruments (see, e.g., Doschek *et al.* 2001 for discussion). However, the results clearly show the presence of the unresolved LOS motions caused by waves and/or turbulence in the corona.

Banerjee *et al.* (1998) using SOHO/SUMER, measured the evolution of non-thermal broadening of the emission line Si VIII in interplume regions of polar coronal holes and found that the amplitude of the LOS component of the plasma velocity grows with height from 27 km/s near the surface to 46 km/s at 180 Mm from the surface. In the polar regions the magnetic field is believed to be almost perpendicular to the LOS. Therefore, the broadening can be interpreted as the transverse wave motion associated with either fast or Alfvén waves. The fact that the wave amplitude increases with height makes the Alfvén wave interpretation more relevant. Indeed, the observed growth rate of the waves, compared with the observationally determined density profile, is in excellent agreement with the theoretical prediction (16.15).

Similar amplitudes of the LOS velocities (20-30 km/s) but on the disc were measured by Chae *et al.* (1998) using SOHO/SUMER and Esser *et al.* (1999) using SOHO/UVCS (LOS velocities of 20-23 km/s at 1.35-2.1 $R_\odot$).

Developing this study and combining the Si VIII nonthermal broadening with SOHO/UVCS O VI line measurements, Doyle *et al.* (1999) found that the amplitude of unresolved thermal broadening (possibly, the Alfvén wave amplitude) is growing up to 1.2 $R_\odot$, then has a plateau to 1.5 $R_\odot$, and then grows sharply again. The authors suggested that this phenomenon could be associated with nonlinear overturning of the waves, but a rigorous theoretical modelling is still required.

16.6   Compressive waves in long loops

16.6.1   Observations and interpretation

Applying the stroboscopic method to the analysis of temporal variation of EUV emission from coronal loops, Berghmans and Clette (1999), with SOHO/EIT, have observed compressive propagating disturbances. This discovery has been confirmed by Berghmans *et al.* (1999) and De Moortel *et al.* (2000) with TRACE. Probably, the same phenomenon was observed by Nightingale *et al.* (1999) in TRACE 171Å. The results turned out to be very similar to the polar plume waves discussed above (see Figure 16.4).

Summarizing the observational findings: (a) Perturbations of the intensity (plasma density), propagating *upwards* along long coronal loops have been detected at the EIT 195 Å (Berghmans and Clette 1999) and TRACE 171 Å (Berghmans *et al.* 1999, De Moortel *et al.* 2000) bandpasses. (b) The projection, perpendicular to the line of sight, of the propagation speed is about 65-165 km/s (Berghmans and Clette 1999), or > (70 ± 16) km/s (De Moortel *et al.* 2000). (c) The amplitudes are less than 10% in intensity (less than 5% in density) in both the 171 Å and 195 Å bandpasses. (d) The

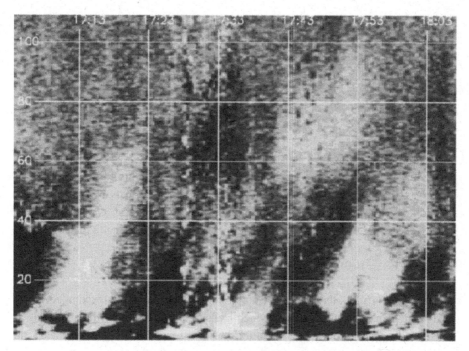

Figure 16.4. A typical time–distance diagram showing evolution of EUV emission intensity along a coronal loop with time. The diagonal stripes are the propagating disturbances interpreted as slow magnetoacoustic waves. (Courtesy E. Verwichte.)

disturbances are quasi-periodic with the periods about 180-420 s (De Moortel *et al.* 2000). The periods are well below the acoustic cut-off period, which is about 87 min. (e) In most cases, only upwards propagating disturbances have been detected (from the footpoints to the apex of the loop).

The interpretation of this phenomenon in terms of slow magnetoacoustic waves was provided by Nakariakov *et al.* (2000). Taking into account effects of weak non-linearity, weak dissipation connected with viscosity and thermal conductivity, density stratification, and neglecting coupling of the slow magnetoacoustic mode with other wave modes (in other words, applying the single wave approximation), neglecting 2D effects (and, consequently, wave dispersion) and loop curvature, and assuming that the wavelength is much less than the density scale height H $(= C_s^2(\gamma g)^{-1})$, an evolutionary equation has been derived

$$\frac{\partial V}{\partial s} - \frac{1}{2H(s)}V + \frac{\gamma + 1}{2C_s}V\frac{\partial V}{\partial \xi} - \frac{R_\odot \rho_0(0)\bar{\eta}}{2\rho_0(s)}\frac{\partial^2 V}{\partial \xi^2} = 0, \tag{16.17}$$

where V is the longitudinal velocity, s is the distance along the loop, measured from a footpoint, $\xi = s - C_s t$ is the running coordinate, and and

$$\bar{\eta} = \frac{1}{\rho_0(0)C_s R_\odot}\left[\frac{4\eta_0}{3} + \frac{\kappa_\parallel(\gamma - 1)^2}{\mathcal{R}\gamma}\right] \tag{16.18}$$

is the normalized dissipation coefficient connected with the thermal conduction $\kappa_{\|}$ and the volume viscosity η_0, other notations are standard. Perturbations of other physical values in the wave are expressed through V,

$$\rho = \frac{\rho_0}{C_s}V, \quad p = C_s\rho_0 V, \quad T = \frac{C_s(\gamma - 1)}{\gamma\mathcal{R}}V. \tag{16.19}$$

Equation (16.17) is the modified Burgers equation and, in general, does not have an exact analytical solution. However, it is not difficult to solve the initial value problem for this equation numerically. It was found that for the observed wave amplitudes, less than 10% in the perturbations of intensity, corresponding to less than 5% in density, and longitudinal velocity perturbations, nonlinear effects are insignificant.

From the linearized evolutionary equation, the amplitude of a linear harmonic wave $V \propto \cos k\xi$, where k is the wave number, evolves according to the expression

$$V(s) = V(0)\exp\left[\int_0^s \left(\frac{1}{2H(x)} - \frac{k^2\bar{\eta}\rho_0(0)R_\odot}{2\rho_0(x)}\right)dx\right]. \tag{16.20}$$

According to (16.20), the slow wave amplitude is determined by the competition of two phenomena, the stratification, which leads to amplification of the amplitude at the ascending part of the wave trajectory, and the dissipation always decreasing the wave amplitude.

Figure 16.5 shows the evolution of the amplitude of a harmonic slow magnetoacoustic wave with distance along the loop for two different values of the viscosity and three different values of the wave period. The main result is that for expected values of the viscosity ($\bar{\eta} \approx 5 \times 10^{-4}$), the wave amplitude begins decreasing well before the wave approaches the loop apex. That explains why no downward propagating waves are ever detected.

Similar results were obtained during Joint Observing Programme JOP-80, when observations of the phenomenon in two wavelength bands, 171 Å and 195 Å, were undertaken with TRACE and SOHO/EIT (Robbrecht et al. 2001). The propagating disturbances along the same path measured in different bandpasses, corresponding to different plasma temperatures were found to have different propagation speeds. The

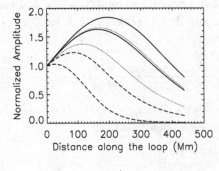

Figure 16.5. The initial amplitude $V(0) = 0.02\,C_s$, wave periods: 900 s (the solid curves), 600 s (the dotted curves) and 300 s (the dashed curves). The upper curve of each kind corresponds to the normalized dissipation coefficient $\bar{\eta} = 4 \times 10^{-4}$, and the lower curve to 1×10^{-3}. (Nakariakov et al. 2000.)

waves measured in the cooler bandpass, 171 Å (Fe ix/x), were systematically slower than the waves measured in hotter bandpass (dominated by Fe xii). The projected speeds found are typically 95 km/s for the 171 Å band (the corresponding sound speed is 152 km/s) and 110 km/s for the 195 Å band (the corresponding sound speed is 192 km/s). The projection angle required to explain the difference between the observed and the predicted speeds of the waves is about 60° for both bandpasses. This angle is consistent with the observed angle determined a few days later when the active region structure crossed the limb.

16.6.2 Seismologic implications

The observations of the slow magnetoacoustic waves in coronal loops allow us to use the measured properties of the waves for seismologic purposes.

The difference in the wave speeds, measured along the same path, found for bandpasses corresponding to different temperatures suggests that the corona may have even a finer structure than can be detected by TRACE. Within a TRACE pixel there may be unresolved hot and cool structures, in form of concentric shells or closely packed bundles of magnetic threads with the characteristic scale significantly less than 360 km (the TRACE pixel size) across the field.

More elaborate seismologic methods for coronal plasma diagnostic with slow magnetoacoustic waves can be developed. For example, the actual dependence of a loop cross-section area on height remains unknown. Does the loop expand with height, or is it a bundle of magnetic threads and only the distance between separated threads increases with the height? Direct coronal observations do not give the answer. However, a divergence of the magnetic field with height and the resulting increase of the loop cross-section would affect the propagation of the waves. Mathematically, this would lead to the appearance of a third term in expression (16.20). The analysis of the evolution of the wave amplitude with height can provide information on the loop structure.

If there is a possibility of measuring both the density and the temperature perturbations of the wave, it would be possible, by applying formulae (16.19), to determine the adiabatic index γ in the plasma filling the loop.

16.7 Flare-generated oscillations of coronal loops

16.7.1 Observations

On 14th of July 1998, the TRACE imaging telescope registered, in both the 171 Å and 195 Å bandpasses, decaying oscillating displacements of hot coronal loops in the active region AR8270 (Aschwanden et al. 1999, Nakariakov et al. 1999). These oscillations happened shortly after a solar flare and, most probably, were generated by the flare. The mechanism of the excitation remains hidden, but it can be connected

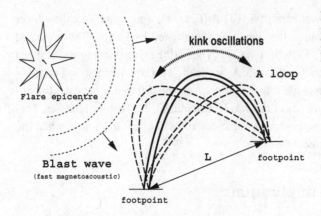

Figure 16.6. A possible mechanism for excitation of kink oscillations of coronal loops by a flare.

with a blast wave generated in the flare epicentre. Another possible option is that the loop footpoints are perturbed by a sunquake. Oscillations of different loops were not in phase. The highest amplitude was seen near loop apices. A possible sketch of the phenomenon is shown in Figure 16.6. Later on, a similar event was reported by Schrijver and Brown (2000) for the flare on 4th of July 1999. During 2001, several new examples have been found in TRACE EUV data (e.g., 15th of April and 15th of May 2001), which have not yet been analyzed. Also, similar phenomena were found by O'Shea *et al.* (2001) in SOHO/CDS data.

For the example on 14th of July, the analysis of the loop displacement shows that the oscillations are almost harmonic with the period of about 256 s (in frequency about 4 mHz). About three strongly damped oscillations were observed. The period remains almost constant during these oscillations. The amplitude of these displacements is several megameters, whereas the distance between the loop footpoints is estimated to be about 83 Mm. The displacement amplitude is several times larger than the loop cross-section radius, which is about 1 Mm. Similar quasi-periodic oscillations have been found for several loops at the distance of several megameter to 60-70 Mm from the flare epicentre (Aschwanden *et al.* 1999). Taking into account these observational findings, the oscillations have been interpreted as a *kink global standing mode* of a loop.

The second example, on 4th of July 1999, shows a behaviour similar to that of the event on 14th of July 1998. The estimated period of the decaying oscillations was ≈360 s (see Nakariakov and Ofman 2001 for details). The distance between the footpoints was estimated to be about 120 Mm.

Taking the observed periods, P, and loop lengths, L, and noting that the wavelength of a global standing mode is double the length of the loop, we estimate the phase speed required as

$$\frac{\omega}{k} = \frac{2L}{P} \approx \begin{cases} (1020 \pm 132) \text{ km/s (14th July 1998)} \\ (1030 \pm 410) \text{ km/s (4th July 1999)} \end{cases} \tag{16.21}$$

The kink modes are the eigenmodes of a straight magnetic cylinder and have been investigated in great detail (see, e.g. Roberts *et al.* 1983, Roberts 1991, 2000). Neglecting the effect of the loop curvature, and taking the wavelength to be much larger than the loop cross-section diameter, we find that the phase speed of the waves observed is the kink speed,

$$
C_k \equiv \left(\frac{2}{1 + \rho_e/\rho_0} \right)^{1/2} C_{A0}, \tag{16.22}
$$

where ρ_e and ρ_0 are the densities of the plasma outside and inside the loop, $C_{A0} = B_0/(4\pi\rho_0)^{1/2}$ is the Alfvén speed in the loop and B_0 is the magnetic field. Here we have assumed that the magnetic field is the same inside and outside the loop, because the parameter β, the ratio of the kinetic and magnetis pressures, is believed to be very small in the corona.

16.7.2 Determination of the magnetic field

The expression for the kink speed (16.22) contains two unknown parameters, the Alfvén speed C_{A0} and the density ratio ρ_e/ρ_0. From observational measurements of C_k and by considering the density ratio as a parameter, we can determine the Alfvén speed in the loop. Fortunately, for small ρ_e/ρ_0, the dependence of the kink speed on the density ratio is weak. Assuming $\rho_e/\rho_0 = 0.1$, we obtain $C_A = (756 \pm 100)$ km/s for a kink speed of (1020 ± 132) km/s estimated for the event on 14th of July, 1998.

The Alfvén speed is defined by the magnetic field strength and the density of the medium. Consequently, we can estimate the value of the magnetic field in the loop

$$
B_0 = (4\pi\rho_0)^{1/2}C_{A0} = \frac{\sqrt{8\pi}L}{P}\sqrt{\rho_0(1 + \rho_e/\rho_0)}. \tag{16.23}
$$

The determination of the magnetic field is weakly sensitive to uncertainties in the determination of the density, because the magnetic field is proportional to the square root of the density. Figure 16.7 shows the dependence of the magnetic field on the density for different values of the kink speed and shows that for a quite wide range of plasma densities, $(1\text{-}6) \times 10^9$ cm^{-3}, the magnetic field is in the range of $(4\text{-}30)$ G. For the oscillating loop of 4th of July 1999 we obtain a value of (13 ± 9) G (Nakariakov and Ofman 2001).

This magnetic field strength, obtained by the method of MHD coronal seismology, corresponds to the expected value of the field in coronal loops. Moreover, the absence of a direct method for a determination of weak magnetic fields makes this estimation very important for coronal physics. Indeed, the direct methods based upon the observation of gyroemission in the radio band allow measurement of only very strong fields (greater than several hundred gauss) and have relatively poor spatial resolution, not better than $5''$. Radio methods based upon the Faraday rotation fail closer than $5\ R_\odot$ to the limb. Modern spectroscopic methods measuring the field by

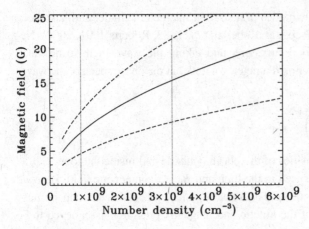

Figure 16.7. Determination of the magnetic field for the event on 4th of July 1999. The solid curve corresponds to the kink speed of the oscillations of 1030 km/s, the dashed curves give the maximal uncertainties and the vertical dotted lines give the interval of the density inside the loop estimated via the difference of the 171 Å and 195 Å emission measures. (Nakariakov and Ofman 2001.)

Zeeman and Handle effects have very poor spatial resolution ($> 1'$) and require a very long time of signal integration. Currently used indirect methods for the determination of the coronal magnetic field include extrapolation of photospheric magnetic sources which are measured with the Zeeman effect in photospheric lines. However, this approach has a number of theoretical difficulties and ambiguities.

The seismologic method has several important advantages: (a) the precise relation of the measurement to a specific coronal structure, and as a consequence high spatial resolution (less than $1''$), (b) both on-the-limb and off-limb measurements are possible, (c) determination of the absolute value of the coronal magnetic field strength that includes all three components.

16.7.3   Determination of transport coefficients

The observations provide us with another important parameter of the oscillations, the decay time. Determination of the damping of the modes allows us to obtain information about the nature of the dissipation in the coronal plasma. In particular, the viscous and magnetic Reynolds numbers, the parameters which are vital in developing coronal heating theories, can be estimated.

The decay time of the oscillations can be determined by approximating the amplitude vs time data points by a best-fitted curve, e.g., an exponentially decaying harmonic curve. The observationally determined decay time can then be combined with theoretically or numerically determined scaling laws connecting the decay time with parameters of the coronal plasma.

In particular, Nakariakov *et al.* (1999) estimated the shear viscosity Reynolds number to be 2×10^5-1.3×10^6. This value of the Reynolds number is about eight orders of magnitude smaller than previous theoretical estimates suggest and is, most probably, connected with coronal micro-turbulence. Such a situation is not unusual in astrophysics. For example, the implementation of turbulent viscosity is necessary for the understanding of physical properties of accretion disks.

16.8   EIT or coronal Moreton waves

Another challenge for theorists is connected with the interpretation of EIT or coronal
Moreton waves. They are compressive disturbances of EUV brightness propagating
quasi-radially from the initiation region (Thompson *et al.* 1998, 1999), and are believed to be associated with chromospheric eruptions. A typical example is shown in
Figure 16.8.

These waves are promising diagnostic tools for seismology of the corona, in particular for the mapping of Alfvén speeds. However, this phenomenon provides us
with a puzzle: the registered speed of these waves is 200-400 km/s, which is very
much below the expected fast magnetoacoustic speed. A qualitative explanation of

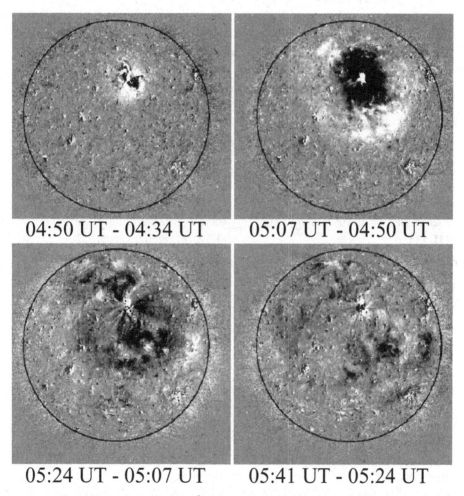

Figure 16.8. SOHO/EIT 195Å running difference images of the large scale wave
transient associated with a coronal Moreton waves (Thompson *et al.* 1998, Copyright
1998 American Geophysical Union).

this discrepancy, based upon interaction of the waves with random structuring, was suggested by Murawski *et al.* (2001). Phase and group speeds of fast magnetoacoustic pulses are indeed decreased by randomly distributed coronal structures, e.g., the loops. However, the decrease in the speeds was found to be only 15-25%, which is insufficient to explain the observations. Also, the model constructed is one-dimensional and linear, and more elaborated two- and three-dimensional and nonlinear models are required.

Another mechanism responsible for the decrease of the propagation speed, connected with wave refraction in higher layers of the corona (Uchida 1970), was recently modelled by Wang (2000).

16.9   Conclusions

Recently achieved spatial and temporal resolution performance of coronal telescopes, allows *systematic* observational investigations of various wave and oscillitory phenomena in the corona. The combination of the observationally gained knowledge with MHD wave theory and direct numerical simulations of the phenomena gives rise to a new method of investigation of coronal plasmas, the MHD seismology of the corona. This makes the investigation of coronal MHD waves and oscillations to be an interesting, rapidly developing and promising branch of solar physics.

References

Aschwanden, M. J. (1987). *Solar Phys.*, **111**, 113.

Aschwanden, M. J., Fletcher, L., Schrijver, C. J., and Alexander, D. (1999). *Astrophys. J.*, **520**, 880.

Balogh, A., Smith, E.J., Tsurutani, B.T., Southwood, D. J., Forsyth, R. J., and Horbury, T. S. (1995). *Science*, **268**, 1007.

Banerjee, D., Teriaca, L., Doyle, J. G., and Wilhelm, K. (1998). *Astron. Astrophys.*, **339**, 208.

Banerjee, D., O'Shea, E., and Doyle, J. G. (2000). *Solar Phys.*, **196**, 63.

Berghmans, D. and Clette, F. (1999). *Solar Phys.*, **186**, 207.

Berghmans, D., Clette, F., Robbrecht, E., and McKenzie, D. (1999). in *ESA SP-448* ed. A. Wilson.

Chae, J., Schühle, U., and Lemaire, P. (1998). *Astrophys. J.*, **505**, 957.

Chapman, R. D., Jordan, S. D., Neupert, W. M., and Thomas, R. J. (1972). *Astrophys. J.*, **174**, L97.

Cuntz, M. and Suess, S.T. (2001). *Astrophys. J.*, **549**, L143.

DeForest, C. E. and Gurman, J. B. (1998). *Astrophys. J.*, **501**, L217.

De Moortel, I., Ireland, J., and Walsh, R. W. (2000). *Astron. Astrophys.*, **355**, L23.

Doschek, G. A., Feldman, U., Laming, J. M., Schühle, U., and Wilhelm, K. (2001). *Astrophys. J.*, **546**, 559.

Doyle, J. G., Teriaca, L., and Banerjee, D. (1999). *Astron. Astrophys.*, **349**, 956.

Esser R., Fineschi, S., Dobrzycka, D. *et al.* (1999). *Astrophys. J.*, **510**, L63.

Feldman, U., Doschek, G. A., VanHoosier, M. E., and Purcell, J. D. (1976). *Astrophys. J. Suppl.*, **31**, 445.

Goossens, M. (1991). in *Advances in Solar System Magnetohydrodynamics*, ed. E. R. Priest and A. W. Hood (CUP, Cambridge).

Harrison, R. A. (1987). *Astron. Astrophys.*, **182**, 337.

Hassler, D. M., Rottman, G. J., Shoub, E. C., and Holzer, T. E. (1990). *Astrophys. J.*, **348**, L77.

Murawski, K., Nakariakov, V.M., and Pelinovsky, E.N. (2001). *Astron. Astrophys.*, **366**, 306.

Nakariakov, V. M. (2000). in *Waves in Dusty, Solar and Space Plasmas*, ed. F. Verheest *et al.* (AIP CP537, New York).

Nakariakov V.M. and Ofman L., (2001). *Astron. Astrophys.*, **372**, L53.

Nakariakov V.M., Ofman, L., and Arber, T.D. (2000). *Astron. Astrophys.*, **353**, 741.

Nakariakov, V.M., Ofman, L., DeLuca, E.E., Roberts, B., and Davila, J.M. (1999). *Science*, **285**, 862.

Nakariakov, V.M., Verwichte, E., Berghmans, D., and Robbrecht, E. (2000). *Astron. Astrophys.*, **362**, 1151.

Narain, U. and Ulmschneider, P. (1990). *Space Science Rev.*, **54**, 377.

Narain, U. and Ulmschneider, P. (1996). *Space Science Rev.*, **75**, 453.

Nightingale R.W., Aschwanden M.J., and Hurlburt N.E. (1999). *Solar Phys.*, **190**, 249.

Ofman, L. and Davila, J.M. (1997). *Astrophys. J.*, **476**, 357.

Ofman, L. and Davila, J.M. (1998). *J. Geophys. Res.*, **103**, 23677.

Ofman, L., Nakariakov, V.M., and DeForest, C.E. (1999). *Astrophys. J.*, **514**, 441.

Ofman, L., Nakariakov, V.M., and Sehgal, N. (2000). *Astrophys. J.*, **533**, 1071.

Ofman, L., Romoli, M., Poletto, G., Noci, G., and Kohl, J. L. (1997). *Astrophys. J.*, **491**, L111.

Ofman, L., Romoli, M., Poletto, G., Noci, G., and Kohl, J. L. (1998). *Astrophys. J.*, **507**, L189.

Ofman, L., Romoli, M., Poletto, G., Noci, G., and Kohl, J. L. (2000). *Astrophys. J.*, **529**, 592.

Oliver, R. (1999), in *ESA SP-448*, ed. A. Wilson, p. 870.

O'Shea, E., Banerjee, D., Doyle, J. G., Fleck, B., and Murtagh, F. (2001). *Astron. Astrophys.*, **368**, 1095.

Robbrecht, E., Verwichte, E., Berghmans, D., Hochedez, J.F., Poedts, S., and Nakariakov, V.M. (2001). *Astron. Astrophys.*, **370**, 591.

Roberts, B. (1991). in *Advances in Solar System Magnetohydrodynamics*, ed. E. R. Priest and A. W. Hood (CUP, Cambridge).

Roberts, B. (2000). *Solar Phys.*, **193**, 139.

Roberts, B., Edwin, P. M., and Benz, A. O. (1983). *Nature*, **305**, 688.

Roberts, B., Edwin, P. M., and Benz, A. O. (1984). *Astrophys. J.*, **279**, 857.

Schrijver, C.J. and Brown, D. S. (2000). *Astrophys. J.*, **537**, L69.

Thompson B.J., Plunkett S.P., Gurman J.B., Newmark J.S., St Cyr O.C., and Michels

D.J. (1998). *Geophys. Res. Lett.*, **25**, 2465.

Thompson B.J. *et al.* (1999). *Astrophys. J.*, **517**, L151.

Torkelsson, U. and Boynton, G.C. (1998). *Mon. Not. R. Astron. Soc.*, **295**, 55.

Uchida, Y. (1970). *Pub. Astron. Soc. Japan*, **22**, 341.

Wang, Y. M. (2000). *Astrophys. J.*, **543**, L89.

Withbroe, G. L. (1983). *Solar Phys.*, **89**, 77.

17

Probing the Sun's hot corona

K.J.H. Phillips

NRC Senior Research Associate, NASA Goddard Space Flight Center, Greenbelt, MD 20771, USA

B.N. Dwivedi

Department of Applied Physics, Institute of Technology, Banaras Hindu University, Varanasi-221005, India

17.1   The solar corona

The tenuous atmosphere of the Sun extending above the visible surface or photosphere, from which much of its radiation is emitted, is much hotter than the photosphere, against physical expectation: one does not expect the temperature to rise with
increasing distance from an energy source (Dwivedi and Phillips, 2001). The photosphere has a temperature of 6000 K and, immediately above it, the chromosphere,
an irregularly shaped layer up to 9000 km above the photosphere, has a temperature
which rises from a value of 4400 K, the temperature minimum (500 km above the
photosphere), to temperatures of up to 20 000 K, where the Lyman lines of neutral
hydrogen dominate the total radiation. Extending above the chromosphere and into
the solar system is the solar corona and its dynamic extension the solar wind which
flows outwards throughout the solar system and beyond, forming the 'heliosphere'.
In this chapter we deal with the inner corona, where the temperatures are typically $(1\text{-}2) \times 10^6$ K but much more in localized 'active' regions (the photospheric
counterparts of which are the sunspot groups). Both the chromosphere and corona
are highly structured, with clear evidence for an association of magnetic fields which
are revealed at the photospheric layers by the Zeeman splitting of magnetically sensitive Fraunhofer lines. The coronal structures are generally loops or large arches,
with footpoints in the photosphere, but there are also nearly radial structures called
streamers extending out to very large distances. The degree of regularity of the
corona, at least as defined by its large-scale structures, is a function of solar activity: at sunspot minimum there are long stable streamers extending out in equatorial
regions while over the polar regions shorter radial 'plumes' occur, but at sunspot

Figure 17.1. The white-light corona seen during the June 21, 2001 total eclipse from Angola. Image taken with a neutral density radial filter by Jean Mouette, member of the team led by Dr Serge Koutchmy. There was a high level of sunspot activity at the time so the corona does not display the regular appearance and polar plumes that are present at solar minimum. (Courtesy J. Mouette, S. Koutchmy, Institute of Astronomy, Paris.)

maximum there are many loops and so-called helmet streamers which are associated with active regions occurring at a range of latitudes. The solar minimum or quiet-Sun appearance of the white-light corona during total eclipses is strongly reminiscent of a bar magnet's pattern, showing the intimate link between the corona and magnetic fields.

The corona's high temperature means that it is better visible at ultraviolet and X-ray wavelengths, and thus many spacecraft and rocket instruments over the years have studied its character. However, it is also visible to the naked eye during the rare circumstances of a total eclipse, when the Moon covers the bright photosphere. The corona then appears as a pearly-white structure all around the Moon's limb (Figure 17.1). The white-light emission is due to Thomson-scattered photospheric light off fast-moving free electrons in the corona. The spectrum of this radiation – the so-called K (Kontinuierlich) corona – is like that of the photosphere but without the Fraunhofer lines, since they have line profiles which are highly Doppler-broadened (up to about 30 nm) by the free electrons so that they cannot be distinguished against the continuous spectrum. There is a faint extra component, the F (Fraunhofer) corona,

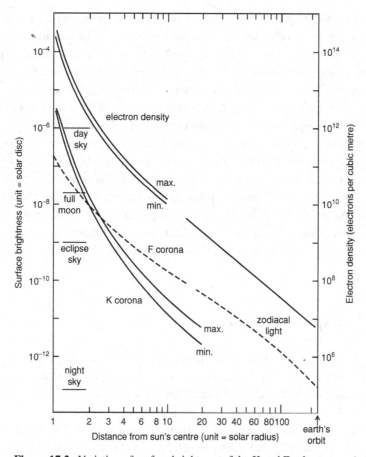

Figure 17.2. Variation of surface brightness of the K and F solar corona (solar maximum and minimum levels for K corona) with distance from the Sun's centre (unit = 1 solar radius, or 696 000 km). Levels of sky brightness for a clear atmosphere are also indicated along the left vertical axis. The top curves show the variation of electron density (m^{-3}) on the same distance scale, with values indicated on the right vertical axis. Note that the F corona continues into the zodiacal light at large radial distances.

due to dust particles in the interplanetary space which scatter photospheric radiation also (Figure 17.2). The low temperature of the dust particles means that the spectrum of this component is photospheric, including the Fraunhofer lines.

The first clues that the corona might be an unusually hot environment were obtained during total eclipses in the nineteenth century, which many astronomers at the time made expeditions to observe (Phillips 1995). C.A. Young and W. Harkness studied the corona during the 1869 total eclipse and found a bright emission line at 530.3 nm (known as the 'green' line) which had not previously been seen in laboratory spectra. Several more unidentified lines became evident in spectra obtained in subsequent eclipses, and a new element named 'coronium' was suspected to be the reason for these spectral lines. As the years passed, however, it was clear that coronium (as well as 'nebulium', discovered from spectral lines emitted by certain gaseous nebulae) could not be easily admitted into the periodic table of elements. Many years later, in the 1930s and 1940s, the lines of coronium and nebulium were finally reproduced in the spectra of very hot spark sources in the laboratory. It was then realized that the corona must have a temperature of about a million kelvins, as judged from the temperature of the spark environments. The clinching argument was the discovery by Edlén that the green line was due, not to an unknown element, but to 13-times-ionized iron (Fe XIV in spectroscopic notation). The high state of ionization was proof of the corona's extreme temperature. The line is normally 'forbidden' by the usual rules of atomic transitions but the low particle densities in the corona in fact allows the upper state in the transition to decay to the ground state by the emission of a single photon. Edlén's work also led to the identification of other forbidden coronal lines as being due to multiply ionized atoms of familiar elements such as Fe and Ni.

17.2   The spacecraft era

The temperature of the corona is in fact so high that it emits copious amounts of X-ray and extreme-ultraviolet radiation which can only be observed from above the Earth's atmosphere with rockets and satellites, as was eventually found in pioneering studies soon after World War Two. R. Tousey and colleagues at the US Naval Research Laboratory in the late 1940s obtained the first ultraviolet spectra of the solar atmosphere, finding that the hydrogen Lyman-α line emitted by the chromosphere was a strong feature at 121.6 nm. X-ray emission was first detected by T.R. Burnight in 1949 using a pinhole camera pointed at the Sun on board a rocket. Spacecraft built by the US and Soviet space agencies in the 1960s and 1970s dedicated to solar observations added much to our knowledge of the Sun's atmosphere, particularly the manned NASA *Skylab* mission of 1973-74. Ultraviolet and X-ray telescopes on board gave the first high-resolution images of the chromosphere and corona and the intermediate transition region (temperatures between 10^4 to 10^6 K). Images of active regions revealed a complex of loops which evolved greatly over their several-day lifetimes, being especially rapidly changing (periods of hours) while the region was young and developing. Ultraviolet images of the quiet Sun showed that the transition region and chromosphere followed the 'network' character previously known from visible-wavelength chromospheric Ca II K-line images, with concentrations of emission generally situated at the boundaries of photospheric supergranules, which are revealed by Doppler-shifted Fraunhofer lines to be the tops of large-scale (diameters about 30 000 km) convection currents beneath the solar surface. The X-ray images showed that the quiet-Sun corona consisted, in the main, of diffuse large-scale loops but with huge regions over the poles, sometimes extending down to

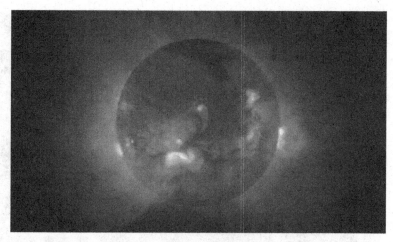

Figure 17.3. *Yohkoh* soft X-ray telescope (SXT) image of the Sun's corona on May 8, 1992 (when the Sun was relatively active and close to the peak of Cycle 22). (*Courtesy Yohkoh SXT Team.*)

equatorial latitudes, where there was a depletion of emission, the so-called coronal holes. Comparison with photospheric magnetograms (maps showing magnetic field strengths derived from Zeeman measurements) showed clearly the association of the X-ray structures with the field, loops for example connecting regions of opposite magnetic polarity and coronal holes being regions of a single polarity where the field lines extend far out into interplanetary space and, as was established with other spacecraft measurements, giving rise to fast (i.e. speeds up to 800 km/s) solar wind streams.

In more recent times, the spatial resolution of spacecraft instruments has steadily improved to levels that are nearly equal to what can be achieved with ground-based solar telescopes. The Japanese *Yohkoh* spacecraft, launched in 1991, has on board the US/Japan Soft X-ray Telescope (SXT), which images X-rays from active and quiet-Sun regions and flares (i.e. sudden releases of energy in active regions) with a resolution of about 2 arcsec (1 arcsec = 725 km at mean solar distance; the mean solar angular diameter is 32 arcmin). X-rays with wavelengths in the range 0.2–2 nm are sensed by the SXT. *Yohkoh*, which is in a low-Earth orbit, continues to operate at the present time, and the SXT instrument has obtained many thousands of images (see the example in Figure 17.3) as well as considerable amounts of data from the other instruments on board, which are mainly for detecting X-ray emission during flares.

The ESA/NASA *Solar and Heliospheric Observatory* (*SOHO*) was launched in 1995 into an orbit about the inner Lagrangian (L1) point situated some 1.5×10^6 km from the Earth on the sunward side. Its twelve instruments therefore get an uninterrupted view of the Sun, unlike the instruments on *Yohkoh*. Apart from a period in 1998 when the spacecraft was temporarily lost to ground-station contact, there has been

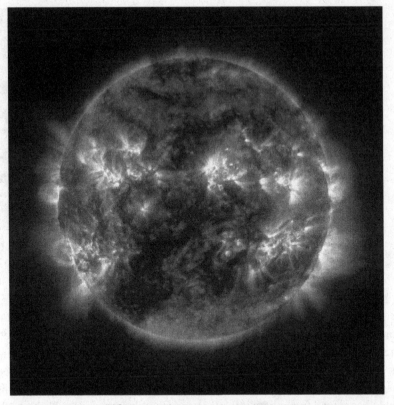

Figure 17.4. Image taken at a wavelength of 17.1 nm (includes emission lines of Fe IX and Fe X, temperature 650 000 to 1 000 000 K) by the *SOHO* Extreme-ultraviolet Imaging Spectrometer (EIT) instrument on June 7, 1999. (*Courtesy SOHO EIT Team.*)

continuous operation since launch. There are several imaging instruments, sensitive to light from visible-light wavelengths to the extreme-ultraviolet (EUV). The Extreme-ultraviolet Imaging Telescope (EIT), for instance, uses normal-incidence optics to get full-Sun images several times a day in the wavelengths of lines emitted by the coronal ions Fe IX/Fe X, Fe XII, Fe XV (emitted in the temperature range 600 000 – 2 500 000 K) as well as the chromospheric He II 30.4 nm line (see Figure 17.4). The spatial resolution is about 2 arcsec. The Coronal Diagnostic Spectrometer (CDS) and the Solar Ultraviolet Measurements of Emitted Radiation (SUMER) are two spectrometers operating in the EUV region, capable of obtaining temperatures, densities and other information about the emitting structures from spectral line ratios. The Ultraviolet Coronagraph Spectrometer (UVCS) has been making spectroscopic observations of the extended corona from 1.25 to 10 solar radii from the Sun's centre, which has allowed the measurement of densities, velocity distributions, and outflow velocities of hydrogen, electrons, and several minor ions. The striking difference in the width of line profiles seen on the disk and in a polar coronal hole from UVCS and SUMER instruments has led to the discovery of the large velocity anisotropy observed in coronal holes and its possible interpretation as solar wind acceleration by ion-cyclotron resonance. The extremely broad O VI line implies velocities up to 500 km s^{-1}, which corresponds to a kinetic temperature of 200 million K. The three coronagraphs making up the Large Angle and Spectroscopic Coronagraph (LASCO) view the white-light corona with high resolution out to distances of 30 solar radii (1 solar radius = 696 000 km). Movies of the corona from LASCO show the large-scale structures in the corona as they rotate with the rest of the Sun (the 'synodic' solar rotation period, i.e. as viewed from the Earth, is 27 days or so, with some latitude dependence), but more particularly they show the large ejections of coronal mass in the form of huge bubbles, moving out with velocities of several hundred km s^{-1} that, on colliding with the Earth in particular, give the well known magnetic storms and associated phenomena that have become a matter of widespread concern for telecommunications in recent years and the burgeoning of 'space weather' studies.

The *Transition Region and Coronal Explorer* (*TRACE*), another highly successful spacecraft observatory, is operated by the Stanford–Lockheed Institute for Space Research, and was launched in 1998. The spatial resolution is of order 1 arcsec (i.e. double the pixel size of 0.5 arcsec), and there are wavelength bands covering the Fe IX, Fe XII and Fe XV lines as well as the hydrogen Lyman-α line at 121.6 nm. Movies made by the concatenation of particularly active region images reveal a vast wealth of detail, with the coronal loops showing continual brightenings and motions. A remarkable feature is the very small diameters of the loops, often no more than the spatial resolution, so are equal to or smaller than 1 arcsec. Figure 17.5 shows one image of a newly developed active region as it rotated on the Sun's north-east limb.

17.3   Heating of the corona: theory

There is a clear evidence that the corona is heated by the dissipation of magnetic fields, although the evidence is not direct. One piece of information that we are still unable to measure is the corona's magnetic field strength. As indicated earlier, we can measure with considerable accuracy the photospheric magnetic field and produce magnetograms as a result. These are routinely available in, for example, the NOAA *Solar-Geophysical Data Bulletin* based on ground-based observatories and the Michelson Doppler Interferometer (MDI) instrument on *SOHO*. Most magnetograms are based on line-of-sight field measurements, but for vector magnetographs, all three components of the photospheric magnetic field can be deduced (B_x, B_y, B_z) and therefore current density $\mathbf{J}$ ($= \nabla \times B$). Although, eventually, Zeeman measurements of infrared emission lines may give important information, in practice almost the only way at present in which the coronal field can be deduced is through

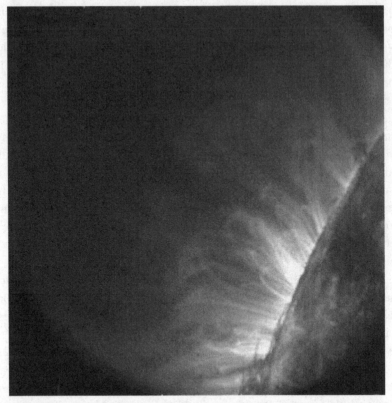

Figure 17.5. Coronal loops of active region AR9511 rotating on to the Sun's north-east limb, June 21, 2001, seen in ultraviolet light by the *TRACE* spacecraft. Dark loop-like structures near the limb are the cool prominences in absorption in this image taken in the light of emission lines due to Fe IX and Fe X (wavelength approximately 17.1 nm). *(Courtesy K. Reeves and TRACE Team.)*

extrapolations of the photospheric field through the assumption of a potential ($\nabla \times \mathbf{B} = 0$, i.e. current-free) or more generally a force-free ($\mathbf{J} \times \mathbf{B} = 0$) field. (A recent work by Nakariakov and Ofman 2001 has shown that the field strength can also be found from flare-generated oscillations of coronal loops which has been observed in the *TRACE* data on a number of occasions.) It is clear from force-free field calculations or merely the appearance of photospheric magnetograms that the field in active regions is more complex than in quiet regions. It is also known that the active region corona is appreciably hotter (typically 4×10^6 K, depending on the nature of the active region) than in quiet regions (2×10^6 K, less in coronal holes). There does seem, then, to be a qualitative relation between field strength and heating.

Unfortunately, there is a significant problem with magnetic field heating, that is, to release the magnetic energy $B^2/2\mu$ (μ is the permeability) into heat energy of particles $(3/2)NkT$ (N is the particle number density, k Boltzmann's constant and T temperature). Such energy transfer requires the diffusion of the magnetic field, which implies a resistive plasma. However, the coronal plasma is, on the contrary, highly conducting. Using Spitzer's classical expression for plasma resistivity $(1/\sigma)$ at temperature T (K),

$$1/\sigma = 10^3 \times T^{-1.5} \quad \text{ohm m} \tag{17.1}$$

we find for $T = 2 \times 10^6$ K,

$$1/\sigma = 4 \times 10^{-7} \quad \text{ohm m} \tag{17.2}$$

only a factor 20 or so higher than a highly conducting solid like copper at room temperature. To illustrate the effect of this high conductivity, we use the induction equation of magnetohydrodynamics (MHD),

$$\frac{\partial \mathbf{B}}{\partial t} = \nabla \times (\mathbf{v} \times \mathbf{B}) + \eta \nabla^2 \mathbf{B} \tag{17.3}$$

where $\eta = 1/(\mu\sigma)$. If the first term can be neglected, (i.e. if v, the plasma velocity, is very small), the diffusion time τ for a magnetic field is given by :

$$\tau = \frac{L^2}{\eta} \tag{17.4}$$

With $\mu \simeq \mu_0 = 1.26 \times 10^{-6}$ H m^{-1} (the permeability of free space) and L of the order of the dimensions of the visible structures in the corona, we find that the time τ is extremely long (many years). Only if the characteristic distance L over which diffusion occurs is as short as a few meters is the diffusion time is a realistic value (tens of s). Expressed another way, the first (advection) term in the induction equation (17.3) is generally much larger than the second (diffusive) term. Defining the magnetic Reynolds (or Lundquist) number R_m, by

$$R_m = \frac{|\nabla \times (\mathbf{v} \times \mathbf{B})|}{|\eta \nabla^2 \mathbf{B}|}, \tag{17.5}$$

which measures how tied the magnetic field is to the plasma, R_m is normally about 10^{12} for the corona unless the length scales are very small. Only then can magnetic reconnection be achieved.

It is known that very small length scales do occur in the region of neutral points or current sheets, where there are steep magnetic field gradients which give rise to large currents. It is thought, then, that such geometries are important for coronal heating if this is by very small energy releases known as nanoflares (Parker 1988). Some 10^{16} J would be released in a nanoflare, i.e. 10^{-9} of a large solar flare, and many energy releases such as this occurring all over the corona, quiet regions as well as active regions, could account for heating of the corona. Most likely this mechanism would not apply to coronal hole regions where the field lines are open to interplanetary space.

The above reasoning applies equally to the competing theory that the corona is heated by magnetohydrodynamic waves, which are generated by photospheric motions and are hypothesized to be damped in the corona. In this case, we need conditions such that the magnetic field changes occur in a shorter time than, say, the Alfvén wave transit time across a closed structure like an active region or quiet Sun loop. The literature for wave heating of the corona is very considerable, but we may briefly summarize it by stating that the waves, generated by turbulent motions in the solar convection zone or photosphere or perhaps by other means, may be surface waves in a loop geometry, or body waves which are guided along the loops and are trapped. The work of Porter *et al.* (1994) shows that only short-period fast-mode and slow-mode waves (periods less than 10 s) could be responsible for heating since only for them are the damping rates high enough. It has recently been suggested that small-scale reconnection occurring in the chromospheric network creates high-frequency Alfvén waves (Axford and McKenzie, 1997), but if such waves exist, they will be absorbed preferentially by the minor heavy ions, so it is unclear whether there is actually enough wave energy left over for the heating and the acceleration of protons and alpha particles (Cranmer, 2000).

A landmark observation made with the *TRACE* spacecraft (Nakariakov *et al.* 1999) has perhaps alleviated the constraints on the theoretically extremely low value of the corona's resistivity or sheer viscosity. A large flare occurred in an active region on July 14, 1998, and nearby fine loops emitting in the extreme ultraviolet were observed to undergo a series of damped oscillations, period just over 4 minutes. The decay time of the oscillations leads to an estimate of the magnetic Reynolds number, about 10^5 to 10^6, i.e. several orders of magnitude smaller than the estimates above based on classical theory. Equivalently, if the oscillations decay by viscous dissipation, then the estimate for the dimensionless Reynolds number is correspondingly much less than the classical value. The implication is that if the resistivity in the corona is at an enhanced level (or the resistivity and viscosity) then the dissipation of MHD waves or magnetic reconnection could proceed much more readily and the coronal heating problem is thereby less formidable than was hitherto supposed.

17.4   Observational evidence: transient brightenings

Early observations with the High Resolution Telescope and Spectrograph (HRTS) by Brueckner and Bartoe (1983) showed that the profiles of the strong C IV (transition region) ultraviolet (154.8, 155.1 nm) emission line pair, emitted at 100 000 K, showed much dynamic activity that could be broadly classified into turbulent events (velocities up to 250 km s^{-1} in small areas) and jets (velocities up to 400 km s^{-1}). The energy contained in the jets amounts to a 'microflare' (i.e. up to 10^{19} J), and it was considered that shock waves generated by a jet could heat up the corona. Enough energy and mass are contained in the jets, assumed to occur over the whole Sun, to satisfy the requirements of the corona and solar wind.

Such phenomena are just an example of the many transient events that occur in the solar atmosphere. Shimizu (1995), in studies of the *Yohkoh* SXT data for active

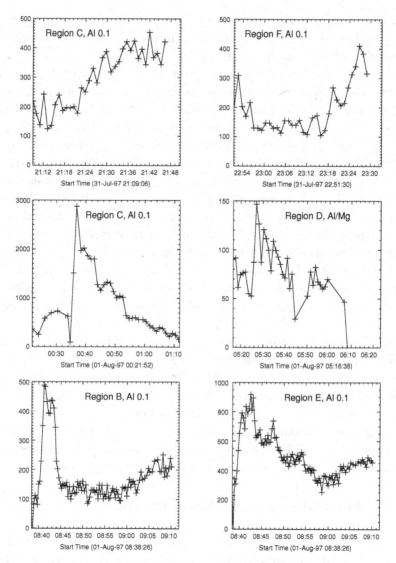

Figure 17.6. Light curves of soft X-ray quiet Sun 'network flares', observed with the Yohkoh SXT instrument outside of active regions during a very quiet period in July/August 1998. (*Courtesy P. Preś.*)

regions, found numerous small brightenings in active region loop structures having energies of the order 10^{20} J, i.e. comparable to microflares. Similar microflares have been noted at higher X-ray energies by Lin *et al.* (1984). Even smaller events – 'network flares' – have been noted outside of active regions by Krucker *et al.* (1997) and later by Preś and Phillips (1999) (see Figure 17.6) in studies of *Yohkoh* SXT and EIT/SOHO data. Here, the energies of the events are much less (down to only 100 times a nanoflare), but then the energy requirements of the quiet solar corona is correspondingly less. At solar minimum and within coronal holes, where the soft X-ray background is very small indeed, Koutchmy *et al.* (1997) have seen tiny coronal 'flashes' which have energies of order 10 times a nanoflare.

There is a possibility that the energy supplied by these small events in active regions and the quiet Sun is adequate to heat the entire corona, though there is still conflicting evidence on this. The energy spectrum of these brightenings is of significance for studying how important they are for coronal heating. If the number distribution is expressed as a power law in energy (apparently a good approximation for the observed distribution of thermal energies), $N(E) = N_0 E^{-\gamma}$, and has $\gamma > 2$, then the implication is that numerous very small flares might contribute to coronal heating, but if $\gamma < 2$, then the implication is that a few very large flares would heat the corona – not a feature that is observed (one does not see the general corona heat up in response to a large flare). The latest indications (Parnell and Jupp 1999, Harra *et al.* 1999) based on SOHO data are equivocal on this point.

If we sum the energies of in the various types of quiet Sun brightenings, both in soft X-rays and in the extreme ultraviolet (EUV), we obtain values listed in Table 17.1 based on data given by Krucker *et al.* (1997), Preś and Phillips (1999) and Harra *et al.* (2000).

According to Acton (cited by Haisch and Schmitt 1996) the total radiative power in soft X-rays (1–300Å) of the entire quiet Sun corona was several 10^{19} W during the solar minimum of Cycle 22 (1994/5). Thus, it appears that the numerous EUV quiet-Sun transients that have been observed by *SOHO* are energetically sufficient for accounting for the X-ray luminosity of the solar corona on the basis of the Parker nanoflare concept. However, much more energy (approximately 10^{22} W) is needed to account for energy losses from the corona by conduction and also acceleration of the solar wind, so coronal heating by nanoflares is still an open question.

17.5   Physical characteristics of the corona

It is clear from eclipse or spacecraft images of the corona (see Figures 17.1, 17.3 and 17.4) that the corona is highly structured, and that the hot plasma making up the corona is confined to intricate magnetic field patterns. Particle densities and temperatures can be derived for different coronal regions using a variety of methods.

Table 17.1. *Energetics for EUV and X-ray transient events in the quiet Sun corona (approximate estimates)*

No. of events hr^{-1}	Thermal energy rate (W)	Event type
18000	3×10^{19}	EUV transient brightenings
1200	2×10^{18}	Small X-ray network flares
100	2×10^{18}	Large X-ray network flares

For example, measurements of the surface brightness (i.e. flux per unit solid angle) of the white-light corona yield electron densities, since the emission is by Thomson scattering of photospheric light from free electrons. Near the base of the corona the measured electron densities are a few times 10^{14} m^{-3}, although this is strongly dependent on whether the feature is a quiet region (smaller densities) or within a complex of active region loops (larger densities). Plasma densities are smaller in coronal holes than in quiet regions. The density falls off rapidly with height (see Figure 17.2): at five solar radii from Sun centre the density is 10^{11} m^{-3}, while at the distance of the Earth's orbit (where the corona is in the form of the free-flowing solar wind) the density is less than 10^7 m^{-3}. As indicated earlier, temperatures vary in the corona from place to place, with maximum values in highly complex active regions (up to 4×10^6 K) and minimum values in the polar coronal holes (slightly less than or equal to 10^6 K).

To illustrate the fact that the coronal gas is strongly dominated by the magnetic field, we find that the plasma beta, i.e. ratio of magnetic pressure $B^2/2\mu$ to gas pressure NkT (N is the particle number density and k is the Boltzmann constant), is much smaller than unity. For the inner part of a typical coronal active region where the magnetic field is 0.01 T and particle density 10^{17} cm^{-3}, the magnetic pressure is about 40 P, but the gas pressure only 5 P (beta about 0.12). Further out into the corona the gas pressure is generally still less than the magnetic pressure but Gary (2001) has recently shown that at very large distances from the photosphere (>100,000 km) the plasma beta may under certain circumstances become rather larger than one.

The measurement of densities and temperatures using X-ray or ultraviolet data is possible, particularly using line intensities, as indicated in section 17.2. Most lines in these regions are optically thin, although there are notable exceptions such as the hydrogen Lyman-α line and other strong resonance lines. The intensity of an optically thin emission line is a function of temperature, through the ionization fraction of the emitting ion (a strongly peaked curve) and the excitation rate from the ground level of the ion to the upper level giving rise to the line emission. Excitation of most lines in the X-ray and ultraviolet ranges is due to collisions of free electrons with the emitting ions. Very roughly we can assign a single temperature to an ion–for example three-times-ionized carbon, which gives rise to the C IV 154.8/155.1 nm line doublet, has a temperature of about 100 000 K, approximately the maximum fractional abundance of the C^{+3} ion in 'ionization equilibrium' (a balance between the collisional ionization processes and the radiative and dielectronic recombination processes which occur in the solar corona). These fractional abundances can be calculated using atomic data, and there are many publications which give values as a function of temperature (in general there is practically no density dependence) (e.g., Arnaud and Rothenflug, 1985; Arnaud and Raymond, 1992; Mazzotta et al., 1998).

The excitation of the lines can be calculated to much higher accuracy than was possible about 20 years ago as there are now sophisticated atomic codes which take into account auto-ionizing resonances in the collisional cross sections. Among these is the close-coupling R-matrix code developed at Queen's University, Belfast. As a result, there are a number of line pairs in especially the EUV part of the spectrum which are sensitive to electron densities. This fact is very useful as nearly all of the other methods of obtaining densities are indirect. To illustrate the latter point, a measured line intensity of a feature leads to a value for the 'emission measure' $N_e^2 V$ (V is the volume and N_e the electron number density). This combined with a measured value for V (e.g. from image data) gives N_e. However, this technique is often quite imprecise since the presence of fine structure within the feature, if unresolved by the instrument taking the observations, renders the value of N_e to be merely a lower limit. Account then has to be taken of a 'filling factor', often much less than one. EUV spectral diagnostics is discussed in the next chapter by Dwivedi *et al.*

The CDS instrument on *SOHO* has been widely used to obtain densities, and it is now possible to construct maps of regions of the Sun showing electron densities (Gallagher *et al.*, 1999). Two examples of suitable density-sensitive lines in the wavelength range of the CDS are those due to Si IX and Si X, both emitted around 1.3×10^6 K. This value of temperature makes them ideal for studying the quiet corona. Several scans using a pair of lines emitted by each ion were made around the Sun's limb with the CDS instrument in February 1996, and Figure 17.7 shows electron densities plotted against position angle around the Sun's limb. It clearly

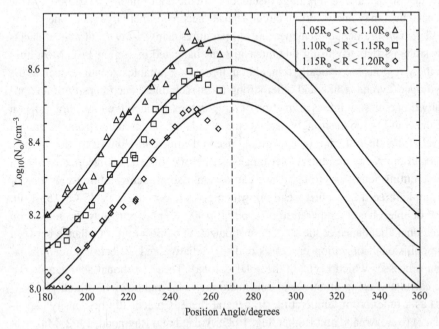

Figure 17.7. Electron density plot showing variation of electron density with position angle round the Sun, from west limb (PA 270 degrees) to south pole (PA 180 degrees) and a range of radial distances (different symbols). The smooth curves are the models of Lima *et al.* (1997).

shows the presence of higher densities in low-latitude regions (N_e around 4×10^{14} m^{-3}), while near the south pole (position angle 180^0) the electron density is at least a factor four lower. The density profile with position angle at three radial distances out to 1.2 solar radii agrees remarkably well with analytical model that has been developed for Sun-like stars by Lima *et al.* (1997).

17.6   Observational evidence: wave motions

Despite the fact that the nanoflare hypothesis seems promising from energetic considerations, MHD waves may well contribute to the heating of the corona, and it is important to look for signatures of them. It is, for example, unlikely that nanoflares could heat the corona in the regions of open field lines such as occur in polar coronal holes, yet it appears that the corona is still hot in these open-field regions. A magnetoacoustic wave would most likely produce a compression and rarefaction of the gas it passes through and so give rise to a periodic modulation. This could be seen in white light (enhancement of Thomson scattering) or green-line emission (enhancement of collisional excitation) or in X-rays and ultraviolet line or continuum emission. With white-light modulations the emission enhancement would be proportional to the density enhancement but with line or X-ray or ultraviolet continuum emission the enhancement is proportional to the square of the density enhancement because of the two-body nature of the emission process in each case.

A basic observational difficulty seems to be that theoretically only very short-period waves (of order 1 minute or less) are likely to be effective in heating the corona. However, spacecraft imaging, limited as it is by the rate at which data are telemetered to the Earth, is necessarily rather slow. It takes about 2 min for any instrument on *SOHO* to produce an image of even relatively small portions of the Sun. Some investigations have found indications of changes occurring over tens of minutes – Wood *et al.* (1998) found 30-minute-timescale variations in compact loop structures with the *SOHO* LASCO instrument.

If the theoretical predictions are correct about short-period waves, then it would seem that ground-based instrumentation during total eclipses can at present offer the only hope of observing modulations of the appropriate period. There are now very high speed CCD cameras that are capable of imaging in this way, and telemetry is of course not an issue, only computer capabilities. A pioneer in searching for short-period oscillations in the green-line corona has been Pasachoff, who has used photomultiplier tubes and fibre optics during several eclipses around the world since the 1980s (e.g. Pasachoff and Ladd 1987). Analysis of his best results indicate the presence of of a slight peak in Fourier spectra at frequencies 0.5–1 Hz. This has been seen in more recent eclipses, including the 1998 eclipse in the Caribbean. Other measurements using ground-based white-light coronagraphs have been taken, notably by Koutchmy *et al.* (1983) at the US National Solar Observatory, Sacramento

Figure 17.8. The Solar Eclipse Coronal Imaging System (SECIS). A telescope (top left) collects light from the eclipsed Sun, which passes through lenses, a beam splitter and a green-line interference filter to one of two CCD cameras (centre of picture), while the beam reflected by the beam splitter passes to a second CCD camera without filter (above camera 1). Images at up to 70 frames/second can be obtained by each camera which are grabbed and stored by a specially adapted PC (top right).

Peak. Here searches were made for periodic modulations in both the intensity and velocity of the green line, with evidence of periods equal to 43, 80 and 300 s (the 300 s period may be related to a photospheric oscillation frequency).

While Pasachoff continues to develop his instrument with colleagues at Williams College, Massachusetts, a group led by one of us (KJHP) and which includes the Rutherford Appleton Laboratory, Queen's University, Belfast, and the Astronomical Institute, University of Wroclaw in Poland (Phillips *et al.*, 2000) have been developing a fast-imaging system with charge-coupled device (CCD) cameras that can image up to 70 frames a second with a specially adapted computer that 'grabs' images, placing the data on to large-capacity hard discs for later analysis (see Figure 17.8). The cameras were developed by EEV, a UK company specializing in CCD cameras for scientific and technical use, and the computer hardware and software were developed by Carr Crouch Computer Company, UK. The system, called the Solar Eclipse Coronal Imaging System (SECIS), has been tested and has been very successful in obtaining high-speed imaging of the solar corona with a coronagraph and during the recent total eclipses. During the eclipse of August 11, 1999, some 12 700 images, both in white-light and in the green line, were obtained in the 2m. 21s of totality, while during the eclipse of June 21, 2001 some 16 000 images were obtained in 3m. 13s. Recent results (Williams *et al.*, 2002) indicate the presence of travelling

MHD waves down the leg of a loop seen in the August 1999 eclipse images with wave periods of 6s. This is a period short enough to be of interest for coronal heating, according to Porter *et al.*'s (1994) theoretical work.

17.7 Conclusions

A large amount of information concerning the physics of the solar atmosphere has been made available in recent years from highly successful spacecraft and from ground-based instruments. The exact reason for the heating of the solar corona is still not known for certain, but there is much evidence from spacecraft observations that, in regions where there are complex magnetic field geometries or at least the presence of closed loops, heating by numerous small flare-like releases of energy is adequate to explain the X-ray luminosity if not the total energy requirements of the corona. The damping of MHD waves may also be an important contributor, but the short periods can at present only be examined with ground-based instruments.

References

Arnaud, M. and Rothenflug, R. (1985). *Astron. Astrophys. Suppl.*, **60**, 425.

Arnaud, M. and Raymond, J. (1992). *Astrophys. J.* **398**, 394.

Aschwanden, M.J., Fletcher, L., Schrijver, C.J., and Alexander, D. (1999). *Astrophys. J.*, **520**, 880.

Axford, W.I. and McKenzie, J. F. (1997). In *Cosmic Winds, and the Heliosphere*, ed. M.S. Matthews, A.S. Ruskin, and M.L. Guerrieri, p. 31 Tucson, AZ: University of Arizona Press.

Brueckner, G.E. and Bartoe, J.-D.F. (1983). *Astrophys. J.*, **272**, 329.

Cranmer, S.R. (2000). *Astrophys. J.*, **532**, 1197.

Dwivedi, B.N. and Phillips, K.J.H. (2001). *Sci. Am.* **284**, 40.

Gallagher, P.T., Mathioudakis, M., Keenan, F.P., Phillips, K.J.H., and Tsinganos, K. (1999). *Astrophys. J.*, **524**, L133.

Gary, G.A. (2001). *Solar Phys.*, **203**, 71.

Harra, L.K., Gallagher, P.T., and Phillips, K.J.H. (1999). *Astron. Astrophys.*, **362**, 371.

Haisch, B., and Schmitt, J.H.M.M. (1996). *Publ. Astron. Soc. Pac.*, **108**, 113.

Koutchmy, S., Zugzda, Y.D., and Locans, V. (1983). *Astron. Astrophys.* **120**, 185.

Koutchmy, S., Hara, H., Suematsu, Y., and Reardon, K. (1997). *Astron. Astrophys.*, **320**, L33.

Krucker, S., Benz, A.O., Bastion, T.S., and Acton, L.W. (1997). *Astrophys. J.*, **488**, 499.

Lima, J.J.G., Priest, E.R., and Tsinganos, K. (1997). in *The Corona and Solar Wind near Minimum Activity* SP-404, p. 521, Noordwijk: European Space Agency.

Lin, R.P., Schwartz, R.A., Kane, S.R., Pelling, R.M., and Hurley, K.C. (1984). *Astrophys. J.*, **283**, 421.

Mazzotta, P., Mazzitelli, G., Colafrancesco, S., and Vittorio, N. (1998). *Astron. Astrophys. Suppl.*, **133**, 403.

Nakariakov, V.M., Ofman, L., DeLuca, E.E., Roberts, B., and Davila, J.M. (1999). *Science*, **285**, 862.

Nakariakov, V.M. and Ofman, L. (2001). *Astron. Astrophys.*, **372**, L53.

Parker, E.N. (1988). *Astrophys. J.*, **330**, 474.

Parnell, C.E. and Jupp, P.E. (2000). *Astrophys. J.*, **529**, 554.

Pasachoff, J.M. and Ladd, E.F. (1987). *Solar Phys.*, **109**, 365.

Phillips, K.J.H. (1995). *Guide to the Sun*, Chapter 1, Cambridge University Press.

Phillips, K.J.H., Read, P.D., Gallagher, P.T., Keenan, F.P., Rudawy, P., Rompolt, B., Smartt, R.N., Pasachoff, J.M., and Babcock, B.A. (2000). *Solar Phys.*, **193**, 259.

Phillips, K.J.H. and Dwivedi, B.N. (1999). *Curr. Sci.*, **77**, 1511.

Porter, L.J., Klimchuk, J.A., and Sturrock, P.A. (1994). *Astrophys. J.*, **435**, 482.

Preś, P. and Phillips, K.J.H. (1999). *Astrophys. J.*, **510**, L73.

Shimizu, T. (1995). *Astron. Soc. Japan*, **47**, 251.

Williams, D.R., Phillips, K.J.H., Rudawy, P., Mathioudakis, M., Gallagher, P.T., O'Shea, E., Keenan, F.P., Read, P., and Rompolt, B. (2001). *Mon. Not. R. Astron. Soc.*, **326**, 428.

Williams, D.R., Mathioudakis, M., Gallagher, P.T., Phillips, K.J.H., McAteer, R.T.J., Keenan, F.P., and Katsiyannis, A.C. (2002). *Mon. Not. R. Astron. Soc.*, **336**, 747.

Wood, B.E. and Karovska, M. (1998). *Astrophys. J.*, **503**, 432.

18

Vacuum-ultraviolet emission line diagnostics for solar plasmas

B.N. Dwivedi, A. Mohan
Department of Applied Physics, Institute of Technology, Banaras Hindu University, Varanasi-221005, India
K. Wilhelm
Max-Planck-Institut für Aeronomie, 37191 Katlenburg-Lindau, Germany

18.1 The Sun in the ultraviolet emission lines

The Sun presents us with a thousand-fold face. Depending on the ways we observe it, whether it be through a ground-based telescope, during an eclipse, or from a space observatory, we see a different Sun. One can distinguish the wavelengths in which it is observed: in X-ray, ultraviolet, visible, infrared, or radio emissions and even by non-photon instruments (e.g., neutrino detectors). One can distinguish the time when it is observed: near its maximum or minimum activity. Still this multi-faceted Sun of ours is a single celestial object, and it is this idea of the Sun as whole that has been described in the previous articles of this book. In this chapter, we present the diagnostics of solar plasmas in the EUV emission lines.

Generally it is assumed that the EUV (extreme-ultraviolet) spectral region covers the range 100 Å to 1200 Å, while the FUV (far-ultraviolet) region covers 1200 Å to 2000 Å. Here we concentrate on the overlapping wavelength region 150 Å to 1600 Å, called the vacuum ultraviolet (VUV), which is dominated by emission lines from the transition region ($T_e > 2 \times 10^4$ K), corona ($T_e > 10^6$ K) and flares ($T_e > 3 \times 10^6$ K). The UV range at longer wavelengths from 1600 Å to 3000 Å is characterized by spectral emissions and absorption lines from the photosphere and chromosphere. The UV and VUV radiation has been measured during the last 50 years by a large number of space programmes, both solar (e.g., OSO, Skylab, SMM, SOHO, and TRACE; recently reviewed by Wilhelm, 2002) and astrophysical (e.g., IUE, EXOSAT, ROSAT, HST, EUVE). These missions together with rocket-borne instrumentation have produced a wealth of spectral data. The SOHO spacecraft, which was launched in December 1995, includes four VUV instruments for studying the solar atmosphere. The Coronal Diagnostic Spectrometer (CDS) (160 Å to 800 Å) is designed to determine the physical parameters of the solar plasma (Harrison *et al.*, 1995). The Solar Ultraviolet Measurements of Emitted Radiation (SUMER) instrument (465 Å to 1610 Å) can resolve fine structures in the chromosphere and transition region, as well as flows and wave motions (Wilhelm *et al.*, 1995). The EUV Imaging Telescope (EIT) has provided superb images and movies of the Sun in four spectral bands centred on the lines He II 304 Å, Fe IX/X 171 and 174 Å, Fe XII 195 Å, and Fe XV 284 Å, covering the temperature range 7×10^4 K to 3×10^6 K (Delaboudinière *et al.*, 1995). The Ultraviolet Coronagraph Spectrometer (UVCS) can observe the solar corona from its base out to 10 solar radii ($R_\odot$) (Kohl *et al.*, 1995). The UVCS spectroscopic diagnostics are based on the measurement of the intensities and spectral profiles of the resonantly and Thomson-scattered H I Lyman-α line, as well as the collisionally excited and resonantly-scattered Li-like resonance lines of O VI at 1032 Å and 1037 Å; it can also detect other coronal lines, such as Si XII 499 Å and 521 Å and Fe XII 1242 Å. The Large Angle Spectroscopic Coronagraph (LASCO) provides images of the solar corona from 1.1 $R_\odot$ to 30 $R_\odot$ in several wavelength bands in the visible 5300 Å to 6000 Å (Brueckner *et al.*, 1995). UVCS and LASCO have provided fascinating observations of streamers, coronal mass ejections (CMEs) and many other large-scale dynamic phenomena. While presenting VUV diagnostics of solar plasmas in this chapter, we take illustrations from the SUMER spectrograph.

18.1.1 SUMER spectrograph

A full description of the SUMER spectrograph and its performance are available (Wilhelm *et al.*, 1995, 1997; Lemaire *et al.*, 1997). Briefly, the SUMER instrument probes the Sun in the VUV light from 465 Å to 1610 Å with high spatial and spectral resolution. This wavelength range contains emission lines and continua from the chromosphere through the transition region to the corona, thereby providing a unique opportunity to study plasmas of the solar atmosphere. An example is shown in Figure 18.1. The spatial resolution is close to 1″ (about 715 km at the Sun) while the spectral resolution element (one pixel) is about 44 mÅ in first order of diffraction, which corresponds to 0.11 mm on the scale of Figure 18.1. The spectral resolution, which is twice as high in second order, can further be improved for relative (and, under certain conditions, for absolute) measurements to fractions of a pixel, allowing the measurement of Doppler shifts corresponding to plasma bulk velocities of about 1 km s^{-1} along the line of sight. Turbulent speeds are obtained from determining line broadenings. There now exists a vast literature over the last five years on the results from SUMER observations. We do not attempt to review this literature in this article. Instead, we present some new results from SUMER on solar plasmas that have added a new dimension to a better understanding of the solar mysteries, especially coronal heating and the solar wind acceleration.

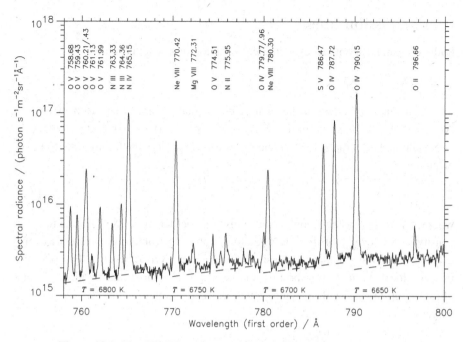

Figure 18.1. The SUMER spectrum of the quiet Sun in the wavelength range from 758 Å to 800 Å in first order. It was obtained with detector B near the centre of the solar disk on 11 November 2001 and shows emission lines with formation temperatures from 30 000 K to 810 000 K above the H I Lyman continuum. Under the assumption of black body-radiation, the continuum is emitted by plasmas at temperatures between 6650 K and 6800 K as the Planck curves (dashed lines) indicate. The wavelengths of the lines, taken from Curdt *et al.* (2001), are given in ångström.

18.2   Atomic processes

The atomic processes in the high-temperature (electron temperature: $T_e > 2 \times 10^4$ K) and low-density (electron density: $N_e < 10^{12}$ cm^{-3}) plasmas of the solar atmosphere have been studied in detail in recent decades. Comprehensive reviews are available (e.g., Mariska, 1992; Dwivedi, 1994; Mason and Monsignori-Fossi, 1994; Mason *et al.*, 1997) that can be referred to in this context. We apply the set of equations here specifically to spatially-resolved observations of the Sun. We assume that the spectral lines are emitted from optically-thin plasma regions, which is valid for many lines in the outer atmospheres of the Sun and stars. The notation used is such that the Fe XIV spectrum is emitted by Fe^{+13}, the element iron with thirteen electrons stripped off. Taking the Fe^{+13} ion as an example, the ground configuration $3s^2 3p$ has two levels, $^2P_{1/2}$ and $^2P_{3/2}$. The transition between these two levels gives rise to the coronal green line at 5303 Å, which is observed by LASCO. The transitions between the excited configurations $3s3p^2$ and $3s^2 3p$ are at around 300 Å to 400 Å (seen by CDS) and those between $3s^2 3d$ and $3s^2 3p$ are at shorter wavelengths around 200 Å.

18.2.1 Emission lines

Radiation can be spontaneously emitted by an ion in an excited state

$$X_j^{+m} \rightarrow X_i^{+m} + h\nu_{ji} \tag{18.1}$$

where an ion X of charge state m in a bound state j emits a photon of energy $\Delta\varepsilon_{ij} = h\nu_{ji} = hc_0/\lambda_{ji}$ (with c_0 the speed of light) to arrive at a lower energy state i. The radiant power density of a spectral line with wavelength λ_{ji} is given by

$$\varphi(\lambda_{ji}) = n_j(X^{+m})A_{ji}\Delta\varepsilon_{ij} \tag{18.2}$$

where A_{ji} is the Einstein spontaneous emission coefficient, $n_j(X^{+m})$ is the number density of the level j of the ion. The atomic physics problem thus reduces to the calculation of the population density of the upper excited level j. For plasmas in LTE (local thermal equilibrium) this is relatively simple and requires the application of the Saha-Boltzmann equation. The number density n_j is usually expressed as

$$n_j(X^{+m}) = \frac{n_j(X^{+m})}{n(X^{+m})} \frac{n(X^{+m})}{n(X)} \frac{n(X)}{n(H)} \frac{n(H)}{n_e} n_e \tag{18.3}$$

where $n_j(X^{+m})/n(X^{+m})$ is the population of level j relative to the total number density, $n(X^{+m})$, of ion X^{+m} and is a function of electron temperature and density; $n(X^{+m})/n(X)$ is the ionization ratio of the ion X^{+m}, which is predominantly a function of temperature (Arnaud and Rothenflug, 1985; Arnaud and Raymond, 1992; Mazzotta et al., 1998); $n(X)/n(H)$ is the element abundance relative to hydrogen which varies in different astrophysical plasmas and also in different solar features; $n(H)/n_e$ is the hydrogen abundance relative to electron density (≈ 0.8); n_e is the electron number density.

In low-density plasmas the collisional excitation processes are generally faster than ionization and recombination time scales, therefore the collisional excitation is dominant over ionization and recombination in producing excited states. The number density population of level j must be calculated by solving the statistical equilibrium equations for a number of low-lying levels and including all the important collisional and radiative excitation and de-excitation mechanisms.

The irradiance of a spectral line at a certain distance, r, is given by

$$E(\lambda_{ji}) = \frac{1}{4\pi r^2} \int_V \varphi(\lambda_{ji}) dV \tag{18.4}$$

where V is the total volume of emission, and if the radiation is assumed to be isotropic. For spatially-resolved observations, the relevant volume element, dV, is seen under a certain solid angle element $d\Omega = dS/r^2$, where dS is the projected area of dV.

Consequently, we can conclude that the radiance of a line in this simple case defined through

$$L(\lambda_{ji}) = \frac{dE(\lambda_{ji})}{d\Omega} = \varphi(\lambda_{ji})\frac{dV}{4\pi dS} \tag{18.5}$$

does not depend on the distance, r, from which the observation is made (cf., Wilhelm et al., 1998a). For optically-thin plasmas, $s = dV/dS$ can be very large in the corona. This is the well-known line-of-sight problem, which cannot be resolved without additional knowledge of or assumptions about $\varphi(\lambda_{ji})$ along the viewing direction. In this case the averaged radiant power density is

$$\langle\varphi(\lambda_{ji})\rangle = \frac{1}{s}\int_s \varphi(\lambda_{ji})ds \tag{18.6}$$

and

$$L(\lambda_{ji}) = \langle\varphi(\lambda_{ji})\rangle\frac{s}{4\pi}. \tag{18.7}$$

18.2.2 Coronal model approximation

In the coronal model approximation the assumption is made that the population of the upper level of transition j occurs mainly via collisional excitation from the ground state g and that the radiative decay overwhelms any other depopulation process. The statistical equilibrium equations can be solved as a two-level system for each transition

$$n_g(X^{+m})n_e C^e_{gj} = n_j A_{jg} \tag{18.8}$$

where the electron collisional excitation rate coefficient is C^e_{gj} . If $A_{jg} \gg n_e C^e_{gj}$, then the population of the upper level j is negligible in comparison with that of the ground level g, so $n_g(X^{+m})/n(X^{+m}) \approx 1$. For a typical EUV transition in Fe^{13+}, at coronal densities and temperatures, we find that A_{jg} is approximately 10^{10} s^{-1}, whereas $N_e C^e_{gj}$ is around 1 s^{-1}.

The radiant flux density in the line for the coronal model is given by

$$\varphi(\lambda_{jg}) = \frac{n(X^{+m})}{n(X)}\frac{n(X)}{n(X)}\frac{n(H)}{n_e}C^e_{gj}\Delta\varepsilon_{gj}n_e^2 \tag{18.9}$$

and so the line radiance in eq. (18.5) is proportional to n_e^2.

18.2.3 Electron collisional excitation and de-excitation

The collisional excitation rate coefficient for a Maxwellian electron velocity distribution with a temperature, T_e, is given by

$$C_{ij}^e = \frac{8.63 \times 10^{-6}}{T_e^{1/2}} \frac{\gamma_{ij}(T_e)}{\omega_i} \exp\left(\frac{-\Delta\varepsilon_{ij}}{kT_e}\right) \tag{18.10}$$

where ω_i is the statistical weight of level i, k is the Boltzmann constant and γ_{ij} is the thermally-averaged collision strength

$$\gamma_{ij}(T_e) = \int\limits_0^\infty \Omega_{ij} \exp\left(\frac{-\varepsilon_j}{kT_e}\right) d\left(\frac{\varepsilon_j}{kT_e}\right). \tag{18.11}$$

Here the collision strength, $\Omega_{ij} = \Omega_{ji}$, is a symmetric, dimensionless quantity, which is related to the electron excitation cross-section, and ε_j is the energy of the scattered electron relative to the final energy state of the ion.

The solution of the electron-ion scattering problem is complex and the reader is referred to the reviews mentioned above and the references therein. New calculations have been carried out as a part of the Iron Project (http://www.am.qub.ac.uk/projects/iron) for many coronal ions, including Fe^{13+} and Fe^{11+}. Fe^{11+} is a particularly useful ion for SOHO, since different lines from Fe XII spectrum are observed by CDS, SUMER, UVCS and EIT. The CHIANTI atomic database and analysis software (Dere et al., 1997) is the result of an international collaboration (USA/Italy/UK) to provide a comprehensive dataset for ions of astrophysical interest seen in the wavelength region $\lambda > 50$ Å. There are three basic sets of files – the observed and theoretical energy levels, the wavelengths and radiative data for each transition in an ion and electron collisional data for each transition. CHIANTI can be accessed via the web sites at NRL, Cambridge or Arcetri (e.g., http://wwwsolar.nrl.navy.mil/chianti.html).

18.2.4 Proton collisional excitation and de-excitation

The proton collisional excitation and de-excitation rates should also be considered. They become comparable with electron collisional processes only for transitions where $\Delta\varepsilon_{ij} \ll kT_e$. This happens for transitions between fine structure levels at high temperatures, for example, the Fe XIV transition in the ground configuration: $3s^2 3p$ ($^2P_{1/2} - {}^2P_{3/2}$). An excellent review of proton-ion collisions has been published by Reid (1988).

18.2.5   Ionization balance

The degree of ionization of an element is obtained by equating the ionization and recombination rates

$$n^{+m}(q_{col} + q_{au} + q_{ct}) = n^{+m+1}(\alpha_r + \alpha_d + \alpha_{ct}). \tag{18.12}$$

The dominant processes in optically-thin plasmas are collisional ionization, (q_{col}), excitation followed by autoionization (q_{au}), radiative recombination (α_r), di-electronic recombination (α_d), and charge transfer (α_{ct}, q_{ct}).

The process of radiative recombination is

$$X_n^{+m+1} + e \rightarrow X_{n'}^{+m} + h\nu \tag{18.13}$$

with n and n' the quantum states of the ions. The inverse process is photoionization, which is, however, not a dominant process for the transition regions and coronae of stars.

The process of di-electronic recombination

$$X_n^{+m+1} + e \rightarrow \left(X_{n'}^{+m}\right)^{**} \rightarrow X_{n''}^{+m} + h\nu \tag{18.14}$$

leads to a capture of an electron by an ion with charge $+m+1$ to form a doubly-excited state ()** of the ion with charge $+m$. It can then either autoionize again or undergo a spontaneous radiative transition of the inner excited electron to a state below the first ionization limit. Di-electronic recombination is the dominant recombination mechanism at high temperatures.

18.3   Plasma diagnostics

Without knowledge of the densities, temperatures, and elemental abundances of space plasmas, almost nothing can be said regarding the generation and transport of mass, momentum and energy. Thus, since early in the era of space-borne spectroscopy, we have faced the task of inferring plasma temperatures, densities, and elemental abundances for hot solar and other astrophysical plasmas. A fundamental property of hot solar plasmas is their inhomogeneity. The emergent radiances of spectral lines from optically-thin plasmas are determined by integrals along the line of sight through the plasma. Spectroscopic diagnostics of the temperature and density structures of such plasmas using emission-line intensities is usually described in two ways. The simplest approach, the line-ratio diagnostics, uses an observed line intensity ratio to determine density or temperature from theoretical density or temperature sensitive line-ratio curves, based on an atomic model and taking account of physical processes for the formation of lines. The line-ratio method is stable, leading to well-defined values of T_e or n_e, but in realistic cases of inhomogeneous plasmas these are hard to interpret, since each line pair yields a different value of density or temperature. The more general Emission Measure (EM) method recognizes that observed plasmas are better described by distributions of temperature or density along the line of sight, and poses the problem in inverse form. The EM function is the solution to the inverse problem and is a function of T_e, n_e, or both. Derivation of EM functions, while more generally acceptable, is unstable to noise and uncertainties in spectral and atomic data. The exact relationship between the two approaches has never been explored in depth, although particular situations were discussed (Brown et al., 1991), and the mathematical relationship between these two approaches has recently been reported (McIntosh et al., 1998).

Line shifts give information about the bulk motions of the plasma in the solar atmosphere. Systematic red shifts in transition-region lines have been observed in both solar and stellar spectra of late-type stars. On the Sun, outflows of coronal material have been correlated with coronal holes. In the upper panel of Figure 18.2, a raster scan is shown obtained by SUMER in the Ne VIII 770 Å line on 21 September 1996. The emission line is formed at 630 000 K and observed in second order. The polar coronal hole at this time can clearly be seen in the central EIT Fe IX/X image as well as some bright points and polar plumes. In the lower panel, a Doppler velocity map, derived from the Ne VIII observations is shown, where outflow regions with speeds higher than 5 km s^{-1} are encircled by contours. The same contours are used as overplot in the upper panel in order to demonstrate that the outflow occurs in the dark regions of the coronal hole (Wilhelm et al., 2000). The zero point is adjusted to give no Doppler shift above the limb at about 20″.

18.3.1 Emission measure analysis

The radiant power density for an allowed transition in the simple coronal model, can be re-written in the form

$$\varphi(\lambda_{jg}) = \frac{n(X)}{n(H)} \frac{n(H)}{n_e} G(T, \lambda_{jg}) \Delta \varepsilon_{gj} n_e^2 \qquad (18.15)$$

where

$$G(T, \lambda_{jg}) = \frac{n(X^{+m})}{n(X)} C_{gj}^e \qquad (18.16)$$

is called the contribution function. In general, this function is strongly peaked in temperature, because the ionic fraction also has a peak there.

Following Pottasch (1963), one can assume that each spectral line is emitted from a uniform volume V, over a temperature range ΔT (usually about 0.3 on a lg T_e/K scale) around the temperature T_{max} corresponding to the peak value of its contribution function, $G(T_{max})$. Taking a constant value of $\langle G(T) \rangle$ for the contribution function over ΔT, the radiance can be expressed as

$$L(\lambda_{jg}) = \frac{1}{4\pi} \frac{n(X)}{n(H)} \frac{n(H)}{n_e} \Delta \varepsilon_{gj} \langle EM \rangle \langle G(T) \rangle \qquad (18.17)$$

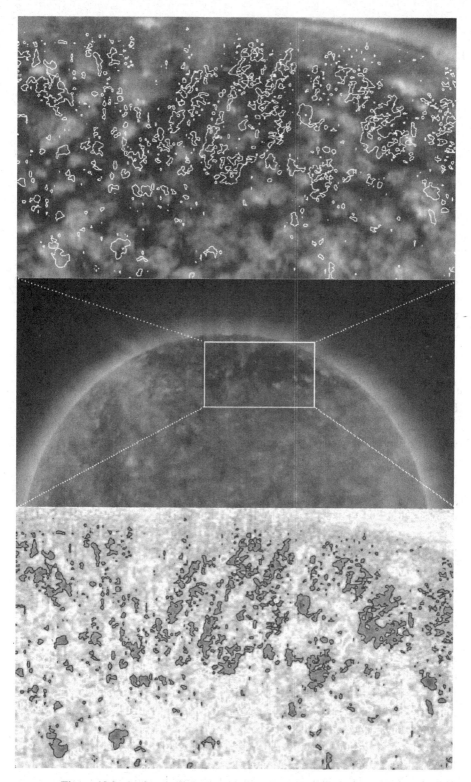

Figure 18.2. Outflow in dark regions of a polar coronal hole on 21 September 1996. The upper panel displays a Ne VIII 770 Å line radiance map with ouflow contours taken from the bottom panel, where the Ne VIII Dopplergram is given with outflow contours at 5 km s^{-1}. An Fe IX/x image of EIT is shown in the centre panel.

where

$$\langle G(T) \rangle = \frac{1}{\Delta T} \int_{\Delta T} G(T, \lambda_{jg}) dT \qquad (18.18)$$

and the emission measure $\langle EM \rangle$ is defined for spatially-resolved observations as

$$\langle EM \rangle = \int_s n_e^2 ds. \qquad (18.19)$$

The average electron density can be crudely deduced from

$$\langle n_e^2 \rangle = \langle EM \rangle / s \qquad (18.20)$$

assuming that the spectral line is emitted over a homogeneous, isothermal volume estimated from images at that temperature. This method can also be used to derive the isothermal emission measure as a function of temperature using lines emitted at different values of T_{max}.

18.3.2 Electron-density diagnostics

Experience from solar observations is that the plasma often exists in the form of distinct, but in most cases unresolved filamentary structures, even down to the best spatial resolution which has yet been obtained. At one extreme is the solar transition region, where only a very small fraction of the observed emitting volume is actually filled with plasma. The situation is a little better in the corona, but even here the filamentary nature of the emission is evident. The determination of electron density from spectral line radiance ratios from the same ion, makes no assumption about the size of the emitting volume or the element abundance. It therefore provides a powerful and important diagnostic for the solar plasma. Spectral lines may be grouped into different categories according to the behaviour of the upper-level population: allowed lines collisionally excited from the ground level (coronal model discussed in Section 18.2.2); forbidden or intersystem lines originating from a metastable level, m; allowed lines excited from a metastable level.

For forbidden and intersystem transitions the radiative decay rate is generally very small ($A_{mg} \approx 1 \text{ s}^{-1}$ to 100 s^{-1}), collision de-excitation then becomes an important depopulating mechanism ($A_{mg} \approx n_e C_{mg}^e$) and may even be the dominant mechanism; moreover the population of the metastable level becomes comparable with the population of the ground level. For small electron densities the radiance consequently has the same dependence on the density as an allowed line ($L \sim n_e^2$). For very large values of electron densities the collisional de-population dominates, and the metastable level is in Boltzmann equilibrium with the ground level. The line radiance then is proportional to n_e. For intermediate values of electron density the population of the metastable level is significant and the radiance varies as n_e^β, $1 < \beta < 2$.

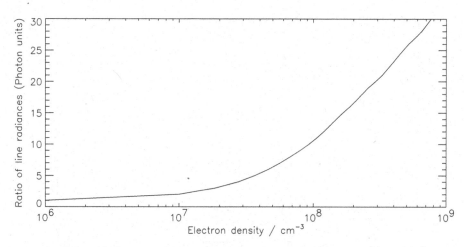

Figure 18.3. Electron density diagnostics with the Si VIII $L(1445 \text{ Å})/L(1440 \text{ Å})$ pair has been performed in many SUMER studies. The dependence of the ratio on the density is shown here after Doschek *et al.* (1997).

More information can again be obtained from the referenced review papers. Here we demonstrate the method in Figure 18.3 with the Si VIII ratio $L(1445 \text{ Å})/L(1440 \text{ Å})$, which has been widely used on SOHO.

18.3.3 Electron-temperature diagnostics

The simplest method of deducing plasma temperatures is to assume ionization equilibrium. Since many ions are formed over the same range of temperature, line ratios can be plotted as a function of temperature of the emitting isothermal plasma.

A more accurate determination of electron temperature can be obtained from the radiance ratio of two allowed lines excited from the ground level g but with significantly different excitation energy. The ratio is given by

$$\frac{L_{jg}}{L_{kg}} = \frac{\Delta \varepsilon_{gj}\, \gamma_{gj}}{\Delta \varepsilon_{gk}\, \gamma_{gk}} \exp\left(\frac{\Delta \varepsilon_{gk} - \Delta \varepsilon_{gj}}{kT_{e}} \right). \tag{18.21}$$

The ratio is sensitive to the change in electron temperature if $(\Delta \varepsilon_{gk} - \Delta \varepsilon_{gj}) > kT_{e}$ assuming that the lines are emitted by the same isothermal volume with the same electron density. Such spectral lines are far apart in wavelength and it may be necessary to use lines from different instruments. This gives rise to major uncertainties in the derived temperature due to the relative calibration responsivities. The Mg X ratio $L(706 \text{ Å})/L(750 \text{ Å})$, shown in Figure 18.4, provides some advantages in this respect. Good temperature diagnostics in the VUV spectral range are the ratios of O V $L(629 \text{ Å})/L(172 \text{ Å})$, O V $L(1218 \text{ Å})/L(629 \text{ Å})$, O VI $L(1032 \text{ Å})/L(173 \text{ Å})$.

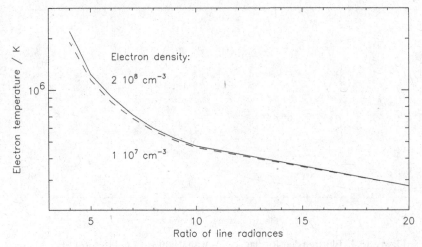

Figure 18.4. Temperature diagnostics within the SUMER wavelength range can be performed with the help of the Mg IX $L(706\ \text{Å})/L(750\ \text{Å})$ line ratio. The ratio as shown has been taken from the Atomic Data and Analysis Structure (ADAS, cf., Summers and O'Mullane, 2000).

18.3.4 Abundance determination

Extensive research regarding the abundances of elements in the solar atmosphere has recently been reviewed by Feldman and Laming (2000). The status of coronal abundances relative to hydrogen is not entirely settled. There is a significant body of evidence that abundances are correlated with the first ionization potential (FIP), giving rise to the so-called FIP effect. This effect consists of systematic differences between photospheric and coronal element abundances. The latter show enhancements of a factor of about four in the so-called low-FIP elements (FIP $\leq$ 10 eV), while abundances for the high-FIP elements (FIP $\geq$ 10 eV) remain constant between the photosphere and the corona. It is believed that this could reflect the ionization and acceleration processes for the solar corona, low down in the solar atmosphere. So far, no theoretical model has been able to satisfactorily explain this effect. One approach to determining element abundances is to use the detailed shape of the DEM distribution for ions from the same element and apply an iterative procedure to normalizing the curves for different elements. Another procedure is to use the radiance ratios for individual spectral lines which have very similar $G(T)$ functions, but different FIPs. For example, the neon atom, with a FIP of 21.6 eV, has a closed shell structure $(1s^2 2s^2 2p^6)$ which is difficult to ionize, whereas it is much easier to ionize magnesium $(1s^2 2s^2 2p^6 3s^2)$, with a FIP of 7.6 eV. Dwivedi *et al.* (1999a,b) presented results from a study of VUV off-limb spectra obtained on 20 June 1996 with the SUMER spectrograph. They recorded Ne VI and Mg VI intercombination/forbidden lines, which provided a good possibility to study the relative element abundance of Ne and Mg in transition region emission in the corona. Mohan *et al.* (2000) extended this investigation taking account of other low-FIP/high-FIP pairs such as K/Ar, Si/Ar and S/Ar present in the spectra. They found that the Mg/Ne relative abundance is highly variable in the complex, cool core of an active region, and seems to be strongly correlated with line radiance patterns; this might suggest that each (spatially unresolved) plasma structure has its own peculiar composition. Mg abundance enhancements relative to Ne reach up to a factor of 8.8.

18.3.5 Spectral line profiles

Line broadenings give information about the dynamic nature of the solar and stellar atmospheres. The transition region spectra are characterized by broadened line profiles. The nature of this excess broadening puts constraints on possible heating processes, and provides information on wave propagation in the solar wind. If the lines from optically-thin plasmas can be fitted with Gaussian profiles, the spectral radiance, L_λ, can be expressed by

$$L_\lambda = \frac{1}{\Delta\lambda\sqrt{\pi}}\exp(-(\lambda - \lambda_0)^2/(\Delta\lambda)^2). \qquad (18.22)$$

Then $L(\lambda_o) = \int L_\lambda d\lambda$ is the line radiance of a line at λ_o and $\Delta\lambda$ is its Doppler width given by

$$(\Delta\lambda)^2 = \frac{\lambda_o^2}{c_o^2}\left(\frac{2kT_i}{M} + \xi^2\right) + \Delta_I^2 \qquad (18.23)$$

for a Maxwellian velocity distribution of the ion temperature, T_i, usually assumed to be the electron temperature corresponding to peak abundance of the ion. Here M is the ion mass, Δ_I is the Gaussian instrumental width, and ξ is the most probable non-thermal velocity.

18.4   Some new results from SUMER

18.4.1 Coronal holes and the solar wind

It has been established by Krieger et al. (1973) that the high-speed solar wind originates from coronal holes, which are well-defined regions of strongly-reduced extreme-ultraviolet (VUV) and soft X-ray emissions (Zirker, 1977). More recent data from Ulysses shows the importance of the polar coronal holes, particularly at times near the solar minimum, when a dipole field dominates the magnetic configuration of the Sun. In Figure 18.2 a polar coronal hole can be seen with bright points and polar plumes. The mechanism for accelerating the fast wind to the high values observed, of the order of 800 km s^{-1}, is not understood quantitatively. The Parker model (Parker, 1958) is based upon a thermally-driven wind. To reach such high velocities, temperatures of the order (3 to 4) MK would be required near the base of the corona. However, other processes are available for accelerating of the wind, for example, the direct transfer of momentum from MHD waves, with or without dissipation. This process results from the decrease of momentum of the waves as they enter less dense regions, coupled with the need to conserve momentum of the total system. If this transfer predominates, it may not be necessary to invoke very high coronal temperatures at the base of the corona.

In reality very little information was available on the density and temperature structure in coronal holes prior to the SOHO mission. Data from Skylab was limited, due to the very low intensities in holes and the poor spectral resolution, leading to many line blends. Skylab was able to follow temperatures up to nearly 1 MK and no further and the interpretation of the data is quite uncertain. High-resolution VUV observations from instruments on SOHO provided the opportunity to infer the density and temperature profile in coronal holes. Wilhelm *et al.* (1998b) observed polar coronal holes with the SUMER spectrograph in the Si VIII forbidden lines at 1445.75 Å and 1440.49 Å and the Mg IX inter-system lines at 706 Å and 750 Å (cf., Figures 18.3 and 18.4) to determine density and temperature structure via line-ratio spectroscopic diagnostics. Figure 18.5 shows the electron densities deduced. Comparing the electron temperatures with the ion temperatures, the authors concluded that ions are extremely hot and the electrons are relatively cool. This result is in agreement with the result from UVCS (Kohl *et al.*, 1997).

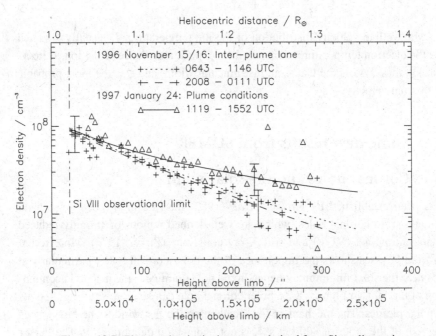

Figure 18.5. Electron density in the corona derived from Si VIII line-ratio observations (Wilhelm *et al.*, 1998b). Polar plumes and inter-plumes conditions are shown.

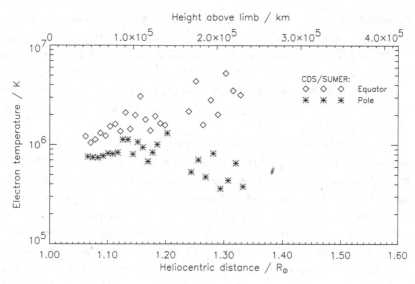

Figure 18.6. Summary plot of the coronal temperature measurements using the O VI $L(173\ \text{Å})/L(1032\ \text{Å})$ line ratio as a function of height above the pole (stars) and the equator (diamonds) (cf., David *et al.*, 1998).

Using the two SOHO instruments CDS and SUMER, David *et al.* (1998) have measured electron temperatures as a function of height above the limb in a polar coronal hole. Observations of two lines from the same ion, O VI 1032 Å from SUMER and O VI 173 Å from CDS, were made to determine the electron temperature gradient in a coronal hole. They deduced temperatures of around 0.8 MK close to the limb, rising to a maximum of less than 1 MK at 1.15 $R_\odot$, then falling to around 0.4 MK at 1.3 $R_\odot$ (Figure 18.6). It seems that present observations preclude the existence of temperatures over 1 MK at any height near the centre of a coronal hole. Wind acceleration by temperature effects is therefore inadequate as an explanation of the high-speed wind, and it becomes essential to look towards other effects, probably involving the momentum and the energy of Alfvén waves.

18.4.2   The "red/blue" Sun

In the so-called transition region the temperature rises sharply from about 30 000 K to coronal values near 1 MK and above. It has been known since the Skylab era that there is a net red shift in transition region lines (Doschek *et al.*, 1976). It has recently been reported that the red shift peaks at about 12 km s^{-1} near 150 000 K and extends into the hotter regions where Ne VIII emission line is formed (Brekke *et al.*, 1997; Chae *et al.*, 1998). Ne VIII belongs to the Li-sequence and has a strong 2s–2p resonance line at 770 Å which can be observed by SUMER in both first and second order. A laboratory wavelength of 770.409 Å $\pm$ 0.005 Å was used for this line. Dammasch *et al.* (1999) and Peter and Judge (1999) made accurate wavelength measurements of this line recorded in second order together with several S I and C I lines in first order, which have well-known wavelengths. Assuming that there is no net Doppler flow along the line of sight at and above the limb and eliminating other residual errors, a rest wavelength of 770.428 Å ± 0.003 Å was derived. With 1 mÅ corresponding to 390 m s^{-1}, this new result moves the Doppler shift of Ne VIII towards the blue by 6.6 km s^{-1}. This immediately implies that there is no net downflow at this temperature in quiet-Sun regions.

As mentioned above, the fact that the fast-speed solar wind originates from coronal holes (open magnetic field regions in the corona) had been widely accepted, even before Ulysses. Little direct observational evidence from low altitudes had been obtained to support this view, before Hassler et al. (1998) found the Ne VIII emission blue shifted in the north polar coronal hole along the magnetic network boundary and at network boundary interfaces compared to the average quiet-Sun flow. These Ne VIII observations reveal the first two-dimensional coronal images showing velocity structure in a coronal hole, and provide strong evidence that coronal holes are indeed the source of the fast solar wind. The apparent relationship to the chromospheric magnetic network, as well as the relatively large outflow velocity signatures at the intersections of network boundaries at midlatitudes, is a first step in better understanding the complex structure and dynamics at the base of the corona and the source region of the solar wind.

18.4.3 Explosive events

The universe abounds with explosive energy releases that can heat plasma to millions of degrees, and accelerate particles to relativistic velocities. Such occurrences are not uncommon on our own star, too. Examples include solar flares, coronal mass ejections, chromospheric and coronal microflares, etc. In many cases, the magnetic field seems to be the only source of energy available to power these cosmic explosions. While it is well established that the Sun has a large reservoir of magnetic energy, the reason for its release is still debated. The widely accepted explanation for explosive energy release is a process known as magnetic reconnection. This effectively involves the cutting and reattachment of magnetic lines of force. Many place the reconnection on a sound observational and theoretical footing. Sceptics, however, argue that there is no definite proof for the reconnection taking place on the Sun. It is for this reason that new results from SUMER for the magnetic reconnection on the Sun is so important. They provide the best evidence to date for the existence of bi-directional outflow jets, a fundamental part of the standard reconnection model. Explosive events were first seen by Dere et al. (1991) in the ultraviolet spectra obtained with the NRL's High Resolution Telescope and Spectrograph flown on several rocket flights and Spacelab 2. They were found to be short-lived (60 s), small-scale (1 500 km), high-speed (±150 km s^{-1}) flows that occur very frequently at all locations on the entire surface of the Sun. Their importance lies in the fact that they probably represent the high-energy tail of a spectrum of network events that occur on scales unobservable with current techniques. It is also noted that explosive events are associated with freshly emerged magnetic field and their Doppler velocities are roughly equal to the Alfvén speed in the chromosphere. The suggestion is that the events result from magnetic reconnection. Possible evidence for bi-directional nature of the flows had been noted in earlier spectroscopic data but the examples were not as clear. This was so because the structure of the flow could not be resolved due to limited time and space coverage by previous space experiments.

The SUMER spectrograph has made it possible to observe the chromospheric network continuously over an extended period and to discern the spatial structure of the flows associated with these explosive events. The observation that explosive events are bi-directional jets provides new evidence that they result from magnetic reconnection above the solar surface. From simultaneous magnetic field and ultra-violet measurements, it has already been suggested that explosive events are often found on the chromospheric network boundary and seem to be associated with the cancellation of photospheric magnetic fields. The network consists of curtains of very strong magnetic flux tubes. All flux tubes are anchored by their footpoints to the photosphere. The continuous motions in the photosphere mean that field lines of opposite polarity are naturally drawn together. If such flux tubes are pushed together, a current sheet forms. In a finite resistivity plasma, a small region near the neutral region may collapse and create a thin reconnection region. From this region, plasma is ejected in both directions along the field lines with velocity of the order of the Alfvén speed (the Alfvén speed depends on magnetic field strength and plasma density). Electrical resistance to this current flow liberates energy from the system to heat the plasma, much as the filament of a light bulb is heated. Interpreting the evolution of jets seen in the Si IV 1393 Å line profiles, Innes *et al.* (1997) have shown that explosive events involve bi-directional jets ejected from small sites above the solar surface. The structure of these plasma jets evolves in the manner predicted by theoretical models of magnetic reconnection. This lends support to the view that magnetic reconnection is the fundamental process for accelerating plasma on the Sun. There exists a vast collection of data obtained from the SUMER spectrograph and similar bi-directional jets are expected to be seen in the solar atmosphere wherever reconnection takes place. The present and future observations of this kind are likely to provide new clues to a better understanding of how the Sun's magnetic energy feeds its million-degree hot corona and the solar wind.

18.4.4   Sunspot transition region oscillations

The sunspot transition region between the chromosphere and the corona oscillates. This may reveal crucial information about the physics of sunspots. The first detailed study of the oscillations in the sunspot transition region was presented by Gurman *et al.* (1982) mainly based on observations of eight sunspots in the C IV 1548 Å line with the UVSP instrument on the Solar Maximum Mission. They observed oscillations with periods of 129 s to 173 s with no sign of shocks and suggested that the oscillations are caused by upward-propagating acoustic waves. Observations with instruments on SOHO have revived the interest for the transition region oscillations. Intensity oscillations in the two minute range were recently reported (Rendtel *et al.*, 1998). Based on observations with a 15 s time resolution Fludra (1999) has observed three minute intensity oscillations and suggested that oscillations occur in sunspot plumes. Combining observations of intensity and line-of-sight velocity oscillations in three transition region lines, Brynildsen *et al.* (1999) found that their observations of NOAA 8156 were compatible with the hypothesis that the three minute oscillations in sunspot umbra transition region are caused by linear, upward-propagating, progressive acoustical waves. This appears to be in conflict with the upward propagating shock waves observed in the sunspot chromosphere (Bard and Carlsson, 1997). Hence, either the wave amplitude decreases abruptly between the chromosphere and the transition region or considerable differences in the three minute umbra oscillations exist between different sunspots. Typical sunspot oscillations observed with SUMER in the O v 629 Å line by Brynildsen and co-workers are shown in Figure 8 of Dwivedi *et al.* (1999c).

18.4.5 Solar flare observed by SUMER

Imaging and spectroscopy together provide us with very powerful tools to enrich our understanding of the Sun's outer atmosphere. SUMER spectra with emission lines formed in a wide temperature range are displayed in Figure 18.7 for a quiet-Sun region and the corona above a flare on 9 May 1999. Note in particular the Fe XVIII

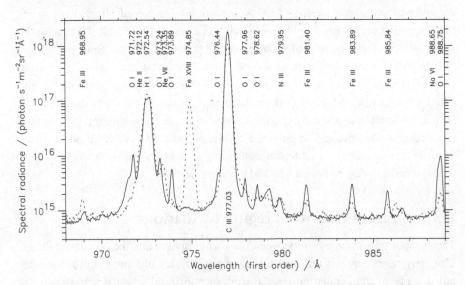

Figure 18.7. Comparison of a quiet-Sun radiance spectrum (solid line) with a coronal flare spectrum (dotted line). The H I Lyman-β and C III lines are of the same brightness in the flaring corona, whereas the chromospheric O I lines cannot be seen.

line at 975 Å with a formation temperature of 6×10^6 K in the flare spectrum and its complete absence in the quiet-Sun emission. The corresponding EIT image of the million-degree corona is shown as Figure 1 in Feldman *et al.* (2000).

18.5   Conclusions

In conclusion, after a review of general plasma diagnostic methods, we present, as examples mostly SUMER results on plasma density and temperature structures.

Indications are found for hot ions and cool electrons in coronal holes. The "blue" Sun, observational evidence for magnetic reconnection on the Sun, the FIP effect in the corona, and sunspot transition-region oscillations are also presented in this article. These phenomena are crucial to a better understanding of how the Sun's magnetic energy heats its million degree corona and feeds the solar wind. It will take some time to digest all these data to decipher what tricks the Sun is performing.

Acknowledgements

A. Mohan thanks the CSIR, New Delhi, for the award of its Senior Research Associateship. The SUMER project is financially supported by the Deutsches Zentrum für Luft- und Raumfahrt (DLR), the Centre National d'Etudes Spatiales (CNES), the National Aeronautics and Space Administration (NASA), and the European Space Agency's (ESA) PRODEX programme (Swiss contribution). SOHO is a project of international cooperation between ESA and NASA. We thank K.J.H. Phillips for many comments and suggestions.

References

Arnaud, M. and Rothenflug, R. (1985). *Astron. Astrophys. Suppl. Ser.*, **60**, 425.

Arnaud, M., and Raymond, J.C. (1992). *Astrophys. J.*, **398**, 394.

Bard, S., and Carlsson, M. (1997). in *Proceedings of Fifth SOHO Workshop*, ed. A.Wilson, ESA SP-404, p. 189.

Brekke, P., Hassler, D.M., and Wilhelm, K. (1997). *Solar Phys.*, **175**, 349.

Brueckner, G.E., Howard, R.A., Koomen, M.J. *et al.* (1995). *Solar Phys.*, **162**, 357.

Brown, J.C., Dwivedi, B.N., Almleaky, Y.M., and Sweet, P.A. (1991). *Astron. Astrophys.*, **249**, 277.

Brynildsen, N., Leifsen, T., Kjeldseth-Moe, O., Maltby, P., and Wilhelm, K. (1999). *Astrophys. J.*, **511**, L121.

Chae, J., Yun, H.S., and Poland, A.I. (1998). *Astrophys. J. Suppl. Ser.*, **114**, 151.

Curdt, W., Brekke, P., Feldman, U. *et al.* (2001). *Astron. Astrophys.*, **375**, 591.

Dammasch, I.E., Wilhelm, K., Curdt, W., and Hassler, D.M. (1999). *Astron. Astrophys.*, **346**, 285.

David, C., Gabriel, A.H., Bely-Dubau, F., Fludra, A., Lemaire, P., and Wilhelm, K. (1998). *Astron. Astrophys.*, **336**, L90.

Delaboudinière, J.-P., Artzner, G.E., Brunaud, J. *et al.* (1995). *Solar Phys.*, **162**, 291.

Dere, K.P. *et al.* (1991). *J. Geophys. Res.*, **96**, 9399.

Dere, K.P., Landi, E., Mason, H.E., Monsignori-Fossi, B.C., and Young, P.R. (1997). *Astron. Astrophys. Suppl. Ser.*, **125**, 149.

Doschek, G.A., Feldman, U., and Bohlin, J.D. (1976). *Astrophys. J.*, **205**, L177.

Doschek, G.A., Warren, H.P., Laming, J.M. *et al.* (1997). *Astrophys. J.*, **482**, L109.

Dwivedi, B.N. (1994). *Space Sci. Rev.*, **65**, 289.

Dwivedi, B.N., Curdt, W., and Wilhelm, K. (1999a). *Astrophys. J.*, **517**, 516.

Dwivedi, B.N., Curdt, W., and Wilhelm, K. (1999b). in *Proceedings of 8th SOHO Workshop*, ed. J.-C. Vial and B. Kaldeich-Schürmann, ESA SP-446, p. 293.

Dwivedi, B.N., Mohan, A., and Wilhelm, K. (1999c). *Curr. Sci.*, **77**, 1521.

Feldman, U. and Laming, J.M. (2000). *Phys. Scripta*, **61**, 222.

Feldman, U., Curdt, W., Landi, E., and Wilhelm, K. (2000). *Astrophys. J.*, **544**, 508.

Fludra, A. (1999). *Astron. Astrophys.*, **344**, L75.

Gurman, J.B., Leibacher, J.W., Shine, R.A., Woodgate, B.E., and Henze, W. (1982). *Astrophys. J.*, **253**, 939.

Harrison, R.A., Sawyer, E.C., Carter, M.K. *et al.* (1995). *Solar Phys.*, **162**, 233.

Hassler, D.M., Dammasch, I.E., Lemaire, P. *et al.* (1998). *Science*, **283**, 810.

Innes, D.E., Inhester, B., Axford, W.I., and Wilhelm, K. (1997). *Nature*, **386**, 811.

Kohl, J.L., Esser, R., Gardner, L.D. *et al.* (1995). *Solar Phys.*, **162**, 313.

Kohl, J.L., Noci, G., Antonucci, E. *et al.* (1997). *Solar Phys.*, **175**, 613.

Krieger, A.S., Timothy, A.F., Roelof, E.C. (1973). *Solar Phys.*, **29**, 505.

Lemaire, P., Wilhelm, K., Curdt, W. *et al.* (1997). *Solar Phys.*, **170**, 105.

Mariska, J.T. (1992). *The solar transition region*, Cambridge University Press.

Mason, H.E. and Monsignori-Fossi, B.C. (1994). *Astron. Astrophys. Rev.*, **6**, 123.

Mason, H.E., Young, P.R., Pike, C.D. *et al.* (1997). *Solar Phys.*, **170**, 143.

Mazzotta, P., Mazzitelli, G., Colafrancesco, S., and Vittorio, N. (1998). *Astron. Astrophys. Suppl. Ser.*, **133**, 403.

McIntosh, S.W., Brown, J.C., and Judge, P.G. (1998). *Astron. Astrophys.*, **333**, 333.

Mohan, A., Landi, E., and Dwivedi, B.N. (2000). *Astron. Astrophys.*, **364**, 835.

Parker, E.N. (1958). *Astrophys. J.*, **128**, 664.

Peter, H., and Judge, P.G. (1999). *Astrophys. J.*, **522**, 1148.

Pottasch, S.R. (1963). *Astrophys. J.*, **137**, 945.

Reid, R.H.G. (1988). *Adv. in Atomic and Mol. Phys.* **25**, 251.

Rendtel, J., Staude, J., Innes, D.E., Wilhelm, K., and Gurman, J.B. (1998). in *A Crossroads for European Solar and Heliospheric Physics*, ed. R.A. Harris, ESA SP-417, p. 277.

Summers, H.P. and O'Mullane, M.G. (2000). in *Atomic and Molecular Data and their Applications*, ICAMDATA, ed. K.A. Berrington and K.L. Bell, AIP Proceedings **543**, p. 304.

Wilhelm, K., Curdt, W., Marsch, E. *et al.* (1995). *Solar Phys.*, **162**, 189.

Wilhelm, K., Lemaire, P., Curdt, W. *et al.* (1997). *Solar Phys.*, **170**, 75.

Wilhelm, K., Lemaire, P., Dammasch I.E. *et al.* (1998a). *Astron. Astrophys.*, **334**, 685.

Wilhelm, K., Marsch, E., Dwivedi, B.N. *et al.* (1998b). *Astrophys. J.*, **500**, 1023.

Wilhelm, K., Dammasch, I.E., Marsch, E., and Hassler, D.M. (2000). *Astron. Astrophys.*, **353**, 749.

Wilhelm, K. (2002). *J. Atm. Sol. Terr. Phys.*, in press.

Zirker, J.B. (1977). (ed.) *Coronal Holes and High Speed Wind Streams* (Colorado Associated Univ. Press, Boulder).

19

Solar wind

E. Marsch, W.I. Axford, J.F. McKenzie

Max-Planck-Institut für Aeronomie, 37191 Katlenburg-Lindau, Germany

19.1 The solar wind

The solar wind is the continuous outflow of completely-ionized gas from the solar corona. It consists of protons and electrons, with an admixture of a few percent alpha particles and much less abundant heavy ions in different ionization stages. The hot corona typically has (base) electron and proton temperatures of 1-2 MK and expands radially outward into interplanetary space, with the flow becoming supersonic within a few solar radii. Because the solar wind plasma is highly electrically conductive, the solar magnetic field lines are dragged away by the flow, and due to solar rotation are wound into spirals. The wind attains a constant terminal speed, and its density then decreases radially in proportion to the square of the radial distance. The interaction between the solar wind and the interstellar medium involves a termination shock located near 100 AU, where the declining ram pressure of the wind matches the pressure of the interstellar plasma and magnetic field.

The heating of the solar corona and the acceleration of the solar wind are not yet fully understood. Here we briefly review the history of the subject; discuss important new observations; describe some of the recent attempts to model the solar wind; identify problem areas and address debated issues; and assess the state of the field and future research perspectives.

19.2   Basic energy considerations

19.2.1   Historic retrospective: Parker's model

Solar wind theory essentially began with the work of (Parker, 1958), who was the first to show that in the presence of electron heat conduction the solar corona cannot be hydrostatic. He developed polytropic, spherically-symmetric fluid models of the solar wind which involved a subsonic-to-supersonic transition in the flow. This critical point is located at a few solar radii. He also showed that the interplanetary magnetic field lines are Archimedean spirals, being wound up ever more tightly the farther away from the Sun they are. He applied this model of the flow and field to describe the propagation of solar-flare particles and the modulation of the cosmic rays. This early work is summarized in his book *Interplanetary Dynamical Processes* (Parker, 1963).

19.2.2   Problems with a polytropic single-fluid model

The essential physics of the heating and acceleration is missing in a polytropic model. In particular, if $\gamma = 5/3$, as it has to be, there is no critical point and the fast flow cannot be adequately described. The two-fluid model of (Hartle and Sturrock, 1968) produced a wind which was much too slow, and yielded protons adiabatically cooled to as low as 40000 K at 1 AU, and electrons kept too hot (350000 K) by classical heat conduction. Therefore it was suggested by (Holzer and Axford, 1970) that extended coronal heating should be allowed for explicitly by adding an exponentially-varying heat source to the proton energy equation. In this case the temperature of the plasma in the outer corona $(3-5)$ R_s becomes very high (~ 10 MK).

How to get the heavy species up to the solar wind speed was addressed by (Ryan and Axford, 1975). It was shown that (1) the heavy ions must be heated selectively (at least proportional to mass), (2) their temperatures become very high, (3) Coulomb collisions are damaging in that they cause heat transfer from the very hot heavies to the relatively cool protons in the inner corona, (4) Coulomb friction is important in the same region and possibly at larger distances, and (5) it was argued that some sort of resonant acceleration is involved which is difficult to analyse because of the differential speeds of the ions. This paper made over-simplified assumptions about the electrons but the arguments concerning the ions are still valid.

The anisotropic two-fluid model of (Leer and Axford, 1972) included an exponentially-varying heat source and aimed at producing hot protons by extended coronal heating. The resulting interplanetary proton anisotropy was found to be in fair agreement with the early *in-situ* observations, which showed that the perpendicular was greater than the parallel temperature, thus indicating that perpendicular heating must be taking place continuously.

To study the effects of transverse magnetic fluctuations (Belcher, 1971), Alfvén waves (evolving according to the WKB theory with no dissipation) were included in the models. The wave pressure acts on the flow and in this way increases the acceleration of the solar wind. This effect gives rise to a faster wind, but the wind tends to be rather dilute (because the energy flux in the waves is limited). The acceleration obtained is somewhat gentle and mainly occurs beyond the Alfvén point (at about $10\ R_s$). General studies of energy addition and re-distribution in the corona were carried out by (Leer and Holzer, 1980) and showed that energy deposition in subsonic flow increased the mass flux with little effect on the asymptotic speed, whereas the same in the supersonic region had the opposite effect - an increase of the wind speed but none in the particle flux. However, in these studies the base density was fixed, and the mass flux forced to be consistent with it. But the mass flux is largely determined below the coronal base, outside the realm of the solution considered. Heating beyond the critical point affects nothing because the flow has lost contact with its source.

Such studies show that electrons may remain hot because of their high heat conduction and that, although protons (and other ions) can be accelerated by magneto-hydrodynamic wave pressure, it is necessary that they are heated preferentially in the corona. This can be concluded from a simple consideration of the energetics of a polytropic model of coronal expansion, in which the sum of the specific enthalpy, binding gravitational energy and kinetic energy is conserved. This conservation law takes the simple form:

$$\frac{\gamma}{\gamma - 1} 2k_B T_c = m_p \left(\frac{GM_s}{R_s} + \frac{1}{2}V^2 \right) , \tag{19.1}$$

with Boltzmann's constant, k_B, the Sun's mass, M_s, its radius, R_s, the terminal wind speed, V, and coronal temperature, T_c. The escape speed from the solar surface $V_\infty = 618$ km/s.

The above constraint requires a coronal energy per proton-electron pair of about 5 keV, to release the fast wind from the Sun's gravitational potential well and attain its high asymptotic speed. If $\gamma = 5/3$, there is no critical point, so that if $T_c = 1$ MK the corona appears gravitationally bound (in the absence of electron heat conduction - Parker's point contra Chapman). A temperature of about $T_c = 11$ MK, is required according to (19.1), if $V = 750$ km and $\gamma = 5/3$. Isothermal models, for which $\gamma \to 1$, are unphysical because the internal energy is not conserved (formally the enthalpy diverges), and the asymptotic flow speed becomes logarithmically infinite. In any case, the key issue of the cause of coronal heating is not addressed in simple polytropic models.

19.2.3   Energy requirements on heavy ions

Given that Coulomb friction (Geiss, 1982) in the dilute, hot corona is usually too weak to couple the ions together tightly, it appears rather difficult to drag out such heavy ions as He^+, He^{2+}, or multiply-charged ions of any other heavier element, against the Sun's gravitational attraction. To achieve equal proton and heavy-ion bulk speeds in the distant wind (Ryan and Axford, 1975), the coronal velocity distributions should overlap sufficiently, which roughly requires about equal effective thermal speeds of a proton-electron pair and a heavy ion dressed by its electron cloud, that is

$$T_i + ZT_e = A(T_p + T_e) \,, \tag{19.2}$$

where Z is the charge and A the atomic mass number of ion species i. This relation implies $T_i > AT_p$. Since Coulomb collisions would make $T_i \rightarrow T_p$ and $V_i \rightarrow V_p$, wave heating must play a crucial role in lifting the heavy species out of the corona, with low heat transfer occurring in the wave dissipation region. The corresponding unknown heating rates should reflect this requirement, i.e.: $Q_i > AQ_p$. There is ample evidence that waves do preferentially heat heavy ions in interplanetary fast streams as observed by Helios (Marsch, 1991) and Ulysses (von Steiger *et al.*, 1995).

19.3   Solar corona and wind in three dimensions

19.3.1   Types of solar wind

Space missions have revealed that there are three major types of solar wind flows: the steady fast wind which originates on open magnetic field lines in coronal holes; the unsteady slow wind coming from the tips, edges and body of temporarily open streamers, and the transient wind in the form of large coronal mass ejections prevailing during solar maximum. Models for these types of wind have been developed to different levels of sophistication. The majority of the models is concerned with the fast wind which, at least during solar minimum, appears to be the basic or equilibrium mode of the wind. Its properties can be reproduced by using multi-fluid models involving waves. Subsequently, we discuss the constraints imposed on the models mainly by Helios (in-ecliptic) and Ulysses (high-latitude) interplanetary *in-situ* measurements, and by the solar remote-sensing observations of SOHO (Solar and Heliospheric Observatory).

19.3.2   Three-dimensional solar corona

The large-scale solar corona becomes particularly simple near solar minimum activity. Its main features can be seen in Figure 19.1, showing an image in the green Fe XIV 5303 Å emission line combined with a white-light image, both obtained by the LASCO coronagraph on SOHO (Schwenn *et al.*, 1997). The magnetic field can be reasonably well described by a simple model (Banaszkiewicz *et al.*, 1998), involving current-sheet, dipolar and quadrupolar contributions. This model reproduces the observed global properties. It also provides the expansion factors of the magnetic flux tubes or plasma flow tubes, which are required in 1-D solar wind models. Note the

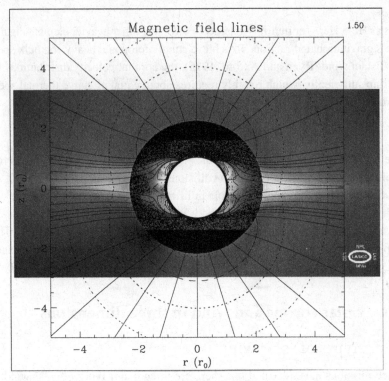

Figure 19.1. The solar corona as seen by the LASCO coronagraph on SOHO (Schwenn *et al.*, 1997) and magnetic field lines of a simple model with dipolar, quadrupolar and current-sheet contributions (Banaszkiewicz *et al.*, 1998). The direction of the magnetic field is indicated by continuous lines and its strength by dotted isocontours.

overall good agreement in Figure 19.1 between the observed electron density distribution, visible as brightness in Thomson-scattered white light, and the morphology of the magnetic field confining the coronal plasma.

The solar corona changes over the 11-year solar cycle, and with it the solar wind flow pattern changes. This is illustrated in Figure 19.2, showing a sequence of coronal images from SOHO together with the associated flow velocity of the solar wind as measured in situ by Ulysses. The fast steams, prevailing at higher latitudes during activity minimum, disappear almost entirely during solar maximum, and the wind becomes slow and highly structured in space and variable in time. The coronal and heliospheric current sheet becomes increasingly warped, highly structured and steeply inclined with changing solar activity. As solar maximum is approached, the magnetic field evolves into a complex multipole field with many active regions (for the present epoch see, for example, the work of Bravo *et al.* (1998). Usually (Hoeksema, 1995), the evolution of the current sheet is visualized by plotting the source field polarity at 2.5 R_s, as constructed from potential field extrapolations of the Sun's surface fields.

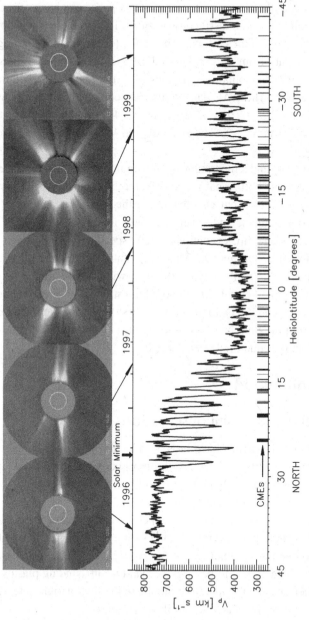

Figure 19.2. The solar corona changing in time as seen by the LASCO coronagraph on SOHO and the associated solar wind flow velocity versus heliographic latitude as measured *in situ* on Ulysses (McComas *et al.*, 2000).

19.3.3 Electron density and temperature

A fundamental parameter in the solar corona is the electron density, radial profiles of which can be inferred from coronagraph images. Global models of the electron density distribution in the corona and interplanetary space can be constructed from remote-sensing combined with *in-situ* observations. This problem is tractable during minimum conditions when the corona is least complex (compare again Figure 19.1 and Figure 19.2). Data from the Skylab and Ulysses missions were used (Sittler and Guhathakurta, 1999) to obtain the profiles shown in Figure 19.3 within a global model fitted to the data. The 1-AU data points were computed by averaging Ulysses plasma densities between $-40°$ ($-5°$) and $-80°$ ($-40°$) in heliographic latitude for the polar (equatorial) estimates and by extrapolation with an r^{-2} dependence to 1 AU before averaging.

The electron density and temperature are of paramount importance, since not only do they characterize the collisional properties and the electron partial pressure (or electric field) in the corona, but also determine the ionization state of Hydrogen, Helium and heavier elements and govern the radiative losses in the EUV and soft X-rays through collisional line excitation. Strictly speaking, it is the electron velocity distribution function that matters. This still cannot be measured in the corona, although its properties are well known in the solar wind, where nonthermal features prevail, such as the suprathermal tail and core-halo structure (see the literature cited in the review by Marsch (1991). Figure 19.4 shows the electron temperature versus radial distance from the Sun. Note how electrons cool off rapidly in coronal holes and more slowly in the equatorial streamer belt.

19.4   Fast solar wind

19.4.1 Coronal and in-situ observations

The recent SOHO observations have brought a wealth of new information on the plasma conditions prevailing in the upper chromosphere, transition region and lower corona. These results define the boundary conditions for the nascent solar wind much more tightly than hitherto (Marsch, 1999).

Coronal hole images suggest that there appear to be two possible source regions of the fast wind: namely polar plumes (with a geometric filling factor of about 4%) and dark lanes (covering the main part, 96%, of a hole). Plumes are conspicuous coronal features which appear to be brighter than the surrounding plasma, because they have a higher density and therefore dominate the line-of-sight intensity. Their basic properties may be summarized as follows. Plumes

- are rooted in bright magnetic (bi-polar) knots in the network,
- show no clear outflow in terms of sizeable Doppler shifts,
- have a lower electron temperature,

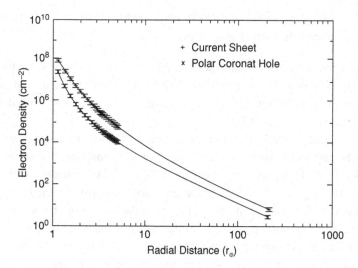

Figure 19.3. The electron density distribution in the solar corona and interplanetary space versus radial distance from the Sun as constructed (Sittler and Guhathakurta, 1999) from Skylab coronagraph data and plasma data measured *in situ* on Ulysses. Profiles with error bars are shown for the polar (asterisks) and equatorial (plus signs) regions.

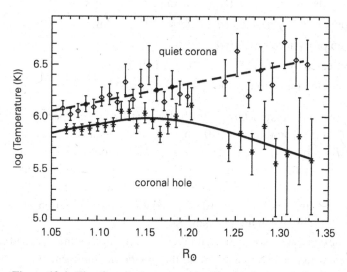

Figure 19.4. The electron temperature profiles for the inner solar corona (David *et al.*, 1998), as obtained from line-ratio diagnostics with the EUV spectrometers on SOHO. The profiles with error bars are shown for the polar coronal hole (diamonds) and equatorial quiet sun (asterisks).

- are possibly FIP-biased (first ionization potential) in abundances,
- reveal outwards moving compressive wave motions apparent in brightness,
- live comparatively short times (for many hours).

The plasma in plumes will eventually be dragged away by the ambient wind, through, for example, shear-flow-driven instabilities. There seem to be signatures of plumes remaining in the near-Sun solar wind (Thieme *et al.*, 1990), where they were detected near the Helios perihelion at 0.3 AU.

Although plumes are prominent coronal features they cannot contribute significantly to the fast wind, which on the average must be supplied at a quite steady rate (Barnes *et al.*, 1995), as the constancy of the mass flux indicates. Even more importantly, the magnetic flux in interplanetary space must also be rooted in the solar wind source. If it were plumes, the field strength at the base of the corona would be about 160 G at the base of the plumes - a value which is much too high.

It is thought that the normal fast wind expands in the dark lanes, corresponding to magnetically open coronal regions (the plumes are also magnetically open in space although they may have a rosette-type field at their base). The fast wind is similar in both the northern and southern hemisphere and appears to contain many micro-streams. In the low corona, the streams

- are associated with dark regions (lanes) in coronal holes,
- emanate from magnetic (mono-polar) funnels between supergranules,
- show significant outflow in terms of Doppler blue shifts,
- show preferentially heated and accelerated heavy ions,
- reveal no pronounced FIP effect,
- are permeated by outward propagating Alfvén waves,
- last a very long time (many months or even years).

Typical fast solar wind parameters (Marsch, 1991) at 1 AU are: Speed: 750 km s^{-1}; density: $n_p = 3\,\text{cm}^{-3}$; proton flux: $2\,10^8\,\text{cm}^{-2}\text{s}^{-1}$; temperatures: $T_p = 2\,10^5$ K (with $T_{p\perp} > T_{p\parallel}$), $T_e = 10^5$ K (in the core), $T_i \geq AT_p$. Heavy ions are observed to "surf" the Alfvén waves with a speed difference $\mathbf{V}_i - \mathbf{V}_p \leq \mathbf{V}_A$. The mass flux is found to be very nearly constant and the same in fast flows in the ecliptic (Helios) during minimum and over the poles (Ulysses) during maximum, an observation which points to globally stable source conditions on the Sun.

By using the conservation equations for the fluxes of mass, momentum and energy, these parameters can be extrapolated from the *in-situ* measurements to the base of the corona. This yields

- a polar magnetic field of about 12 Gauss,
- a particle flux of about $10^{14}\,\text{cm}^{-2}\text{s}^{-1}$,
- an energy flux of $(5 - 10)10^5\,\text{erg cm}^{-2}\text{s}^{-1}$,

as the key input parameters to be considered in any model of the fast solar wind. More details about boundary conditions in the corona and radial gradients of plasma parameters in the solar wind can be found in reviews in the literature (Marsch, 1999). Below we give, from the modelers' point of view, a list of other key parameters (either inferred or obtained directly from observations) at the coronal base:

- number density $\sim (5 - 10)\ 10^7\ \mathrm{cm}^{-3}$,
- flow speed $\sim (10 - 20)\ \mathrm{km\ s}^{-1}$,
- electron temperature $\leq 10^6\ \mathrm{K}$
- Alfvén speed $\geq 2400\ \mathrm{km\ s}^{-1}$,
- wave-turbulence amplitude $\sim 30\ \mathrm{km\ s}^{-1}$,
- sound speed $\sim 150\ \mathrm{km\ s}^{-1}$,
- residence time ~ 1 hour,
- scale height $\sim 50000\ \mathrm{km} \sim 0.07\ R_s$,
- plasma beta ≤ 0.01.

These parameters apply to coronal holes and associated fast flows, which is the basic mode of the quiet solar wind (Axford and McKenzie, 1997). From these parameters we may deduce that the electron partial-pressure gradient (or the electric field) as well as the Alfvén-wave pressure play a minor role, because they act too slowly and cannot lead to rapid acceleration. Furthermore, observationally $T_p > T_e$ and $T_i > T_p$ (for a review, see Marsch, 1999), which implies that the heating process favours the heavy species and would appear to involve high-frequency (e.g., ion cyclotron) waves which provide a discrimination in terms of Z/A, i.e. of the ion gyrofrequency.

19.4.2   Basic model equations

SOHO observations indicate substantial deviations of the coronal plasma from local collisional and thermal equilibrium. Therefore, solar wind models necessarily involve the use of multi-fluid equations, which adequately treat the thermodynamics and internal energy state of the coronal and solar wind plasma. Here we give the basic equations for a 1-D model of the fast solar wind (McKenzie *et al.*, 1995). Because of the low plasma beta in a coronal hole, the plasma is forced to flow along field lines. To describe the differences between the proton and electron temperatures and their anisotropies, anisotropic two-fluid equations are required. The flow-tube geometry for the polar field line may be prescribed by a model profile (e.g., from Banaszkiewicz *et al.*, 1998) of the cross sectional area, $A(r)$, of the flux tube. The polar field line is given by

$$\frac{B(r)}{C} = \frac{2}{r^3} + \frac{D}{r^5} + \frac{E}{a(r+a)^2} .$$

$$(19.3)$$

Here r is the radial distance from the Sun and a, C, D and E are model parameters. Magnetic flux conservation requires

$$B(r)A(r) = B(R_s)A(R_s) .$$ (19.4)

Conservation of mass flux reads

$$\rho(r)V(r)A(r) = \rho(R_s)V(R_s)A(R_s) ,$$ (19.5)

where $\rho = nm_p$ is the proton mass density, and n (we omit the proton index, since $n_e = n_p$ by quasi-neutrality) the proton density. The electrons are assumed to be isotropic and massless, but contribute a partial pressure, $p_e = nk_BT_e$, to the total momentum equation:

$$\rho V \frac{d}{dr}V = -\frac{d}{dr}(p_e + p_w + p_\parallel) - (p_\parallel - p_\perp)\frac{1}{A}\frac{dA}{dr} - \rho\frac{GM_s}{r^2} + \rho A_w .$$

                                                              (19.6)

We omit also the index on the proton pressure and temperature, $p_{\parallel,\perp} = nk_BT_{\parallel,\perp}$. The symbol A_w denotes an acceleration possibly associated with resonant wave-particle interactions. The proton temperature equations are:

$$\frac{dT_\parallel}{dr} + \frac{2T_\parallel}{V}\frac{dV}{dr} = \frac{2v(T_\perp - T_\parallel)}{V} + \frac{Q_\parallel}{nVk_B} ,$$ (19.7)

$$\frac{1}{A}\frac{d(AT_\perp)}{dr} = -\frac{v(T_\perp - T_\parallel)}{V} + \frac{Q_\perp}{nVk_B} ,$$ (19.8)

where v is the collisional rate for temperature anisotropy relaxation. The proton heating rates are denoted by $Q_\perp$ and $Q_\parallel$. For the electrons, heat conduction (with conductivity κ) must be considered, whereas radiative losses can be neglected in coronal holes. We can also neglect collisional temperature equilibriation between electrons and protons sine it is a slow process. For a given heating rate Q_e, the electron temperature is obtained from:

$$\frac{3}{2}\frac{dT_e}{dr} - \frac{T_e}{n}\frac{dn}{dr} = \frac{1}{nVk_B}\left(\frac{1}{A}\frac{d}{dr}\left(A\kappa\frac{dT_e}{dr}\right) + Q_e\right).$$ (19.9)

These equations describe the behaviour of the plasma. The wave pressure is defined by the integral of the power spectrum, $P(f,r)$, of outward propagating waves, and is obtained upon integration over the frequency domain $[f_l, f_u]$ as follows:

$$p_w = \int_{f_l}^{f_u} df\, P(f,r) .$$ (19.10)

The lower, $f_l(r)$, and upper, $f_u(r)$, boundary frequency may vary spatially. Reasonable bounds can be set by the extreme values, $f_l = 2\pi/T_s$, with a solar rotation period, T_s, of 25 days, and the proton gyrofrequency at the coronal base,

$f_u = eB(R_s)/(2\pi m_p c)$, which is in the 10 kHz range. The wave-energy-exchange equation [28] may be written:

$$\frac{1}{A}\frac{d}{dr}\left(\frac{A(V+V_A)^2}{V_A}2p_w\right) = -\left(\frac{V+V_A}{V_A}\right)Q\,, \qquad (19.11)$$

with the Alfvén speed, $V_A = B/\sqrt{4\pi\rho}$, and the total ion heating (negative wave absorption) rate, $Q = (Q_\parallel + 2Q_\perp)/2$. If wave spectral transfer and a turbulent cascade with a flux $F(f, r)$ are to be accounted for, the wave spectral transfer equation [70], namely

$$\frac{1}{A}\frac{\partial}{\partial r}\left(\frac{A(V+V_A)^2}{V_A}2P(f,r)\right) = -\frac{V+V_A}{V_A}\frac{\partial}{\partial f}F(f,r)\,, \qquad (19.12)$$

must be solved. This completes the set of equations required to describe the principal features of the fast solar wind.

19.4.3 Heating functions

Heating of the solar corona is a fundamental problem in solar physics, which cannot be addressed adequately here. However, we briefly discuss some of the heating mechanisms that have been proposed. Since the coronal magnetic field decreases with radial distance, the upper frequency bound, $f_u(r)$ decreases in proportion to the local proton gyrofrequency. For all practical purposes we assume, $f_l = 0$, and $f_u(r) = f_D(r)$, where the latter dissipation frequency is given by $2\pi f_D = eB(r)/(m_p c)$. Integrating equation (19.12) over frequency and comparison with equation (19.11) yields the heating rate

$$Q(r) = F(f_D(r), r) - (V + V_A)P(f_D(r), r)\frac{df_D(r)}{dr}\,. \qquad (19.13)$$

The first term is the energy supplied by the cascade, the second term stems from the inhomogeneity and mimics the effect that an ever increasing fraction of the high-frequency waves is swept up and lost from the spectrum by dissipation. Wave energy cascading and sweeping of the high-frequency domain have been suggested as major mechanisms for ion heating through wave absorption (Tu, 1987).

Other authors, e.g. (Holzer and Axford, 1970) simply assume that some form of mechanical energy is dumped into the corona with a typical exponential damping length, L, so that

$$Q(r) = Q_s \exp\left(-\frac{r - R_s}{L}\right)\,, \qquad (19.14)$$

where L and Q_s have to be appropriately adjusted to fulfill the energy requirements. This combination has the minimum number of free parameters needed to give the observed energy flux and base density. Similar forms of the heating rates have thereafter been employed in many models. To describe extended coronal and interplanetary heating a term Q_{sw}, with $Q_{sw}(r) = Q_{sw}(R_s/(r-R_s))^{3+\delta}$ and $\delta = 0.1$, was included into equation (19.14) (McKenzie et al., 1997). The microphysics of the heating process is probably related to resonant wave-particle interactions, which favour heating and acceleration of the ions by absorption of waves near the cyclotron frequencies. But the processes my be even more complicated than envisaged in this scheme.

General expressions for the heating rates (Marsch and Tu, 2001) and their dependence on the wave spectra, dispersion, polarization and opacity have been derived using kinetic theory. The wave acceleration and heating rates for a given species j can be written as integrals over the wave spectrum and the sum over the wave modes involved:

$$\begin{pmatrix} A_j \\ Q_{j\parallel}/\rho_j \\ Q_{j\perp}/\rho_j \end{pmatrix} = \frac{1}{(2\pi)^3} \int_{-\infty}^{+\infty} d^3k \sum_M \hat{\mathcal{B}}_M(\mathbf{k}) \left(\frac{\Omega_j}{k}\right)^2 \frac{1}{1-|\hat{\mathbf{k}} \cdot e_M(\mathbf{k})|^2}$$

$$\times \sum_{s=-\infty}^{+\infty} \mathcal{R}_j(\mathbf{k}, s) \begin{pmatrix} k_\parallel \\ 2k_\parallel W_j(\mathbf{k}, s) \\ s\Omega_j \end{pmatrix}.$$

$$(19.15)$$

Here the particle's charge is e_j, its mass m_j, density n_j, and the speed along the field is V_j. The mass density of species j is $\rho_j = n_j m_j$. The ion gyrofrequency is $\Omega_j = (e_j B)/(m_j c)$. The s-order parallel resonant speed in equation (19.15) is defined as $W_j(\mathbf{k}, s) = (\omega_M(\mathbf{k}) - k_\parallel V_j - s\Omega_j)/k_\parallel$. It plays a key role in the cyclotron or Landau resonance. The symbol $e_M(\mathbf{k})$ denotes the polarization vector of a given wave mode M with frequency $\omega_M(\mathbf{k})$ and wave vector $\mathbf{k}$. The number s denotes the order of the Bessel function which appears in the resonance function or wave absorption coefficient, $\mathcal{R}_j(\mathbf{k}, s)$. It is given explicitly in Marsch and Tu (2001). The wave spectrum $\hat{\mathcal{B}}_M(\mathbf{k})$, normalized to the background magnetic field density, $B^2/8\pi$, also appears.

The wave-particle terms can only be evaluated once the particle velocity distributions and the power spectral densities are known explicitly. This requires the solution of the full quasi-linear equations (Tu and Marsch, 2001) in order to calculate these quantities self-consistently. In view of these difficulties, no solar wind model to date has included these kinetic features in a consistent way.

19.4.4   Some results from model calculations

The solution procedure for the basic model equations first involves neglecting the wave pressure, p_w, and the acceleration, A_w, in (19.6) entirely. This is justified, because in the corona $V_A \gg V$, and we may conclude from equations (19.7 – 19.11) that the wave pressure gradient is much smaller than the thermal pressure gradient. Also, the resonant acceleration will be equally small when $V_A \gg V$. As a result the plasma and wave equations in this procedure are initially decoupled, except through the heating term $Q(r)$ in (19.11). Once a suitable $Q(r)$ is found, the procedure may be iterated between the plasma and the waves. The source spectrum, involving cascading and dissipation of the waves in the corona, is not known, and any assumption about its properties is, to some extent, arbitrary. The value of Q_s is chosen to satisfy the energetics of the wind, and therefore is constrained by conservation of the mass flux and total energy flux, which, although determined below the coronal base, are conserved quantities and thus known through measurements at 1 AU. In this scheme there is one single free parameter, L, available to fit base densities and temperatures. Typically, this requires $L = 0.35 \ R_s$, which yields rapid acceleration and gives a reasonable fit to the observed density profile. The wave equation (19.11) then gives p_w, and the procedure is repeated until a complete solution of the combined set of equations is obtained. In principle, then (19.12) gives the required wave spectrum compatible with the assumed heating function (19.14).

The results are shown in Figure 19.5, which gives the radial profiles of the plasma density and speed. The acceleration is rapid and the terminal speed is reached within $10 \ R_s$ (McKenzie et al., 1997). This rapid outflow is consistent with the coronal density profile shown in Figure 19.3, which is constructed from the continuity equation (19.5) using semi-empirical velocity profiles (Sittler and Guhathakurta, 1999), and with the UVCS spectroscopic results from SOHO, indicating fast acceleration of the solar wind ions close to the Sun in the polar coronal holes (see, Kohl et al., 1997; and Kohl et al., 1998).

The basic model also gives the radial variation of the proton temperature and anisotropy shown in Figure 19.6. The main characteristic is that the maximum proton

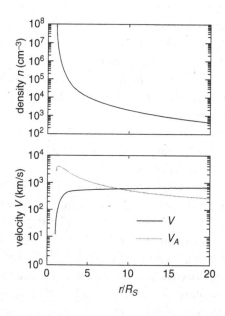

Figure 19.5. Model profiles of density (top) and speed (bottom) of the basic solar wind model (McKenzie et al., 1997) versus radial distance from the Sun.

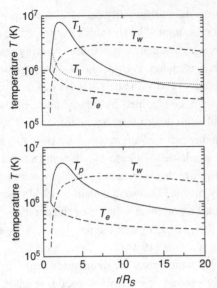

Figure 19.6. Model profiles of coronal temperatures (top, with anisotropy, and bottom the isotropic case) of the basic solar wind model (McKenzie *et al.*, 1997) versus radial distance from the Sun. The wave temperature is also shown for comparison.

temperature is high, of the order of 5 MK. Note that the wave temperature, $T_w = p_w/(k_B \rho)$, is much less than $T_\perp$ in the inner corona (out to about 5 R_s), and this is consistent with the neglect of the wave pressure in the wind dynamics. The proton temperature profile and the corresponding density scale height in the subsonic flow region depend crucially on the heating function, $Q(r)$, for which an exponential e-folding length scale of about 0.35 R_s is required to obtain a base density of 10^8 cm^{-3} or less. This requirement results in the sonic critical point being at less than 2 R_s. Matching the base density, n_s, is the really important constraint and this makes the critical point come closer and the acceleration more rapid as n_s gets smaller. So a correct assumption of the base density is quite crucial.

Several authors (Hansteen and Leer, 1995; Esser *et al.*, 1995) have presented parameter studies of the effects of hot protons on the expanding solar wind in the inner corona. High values of T_p is a characteristic property of all these models, and thus flow velocities of up to 800 km/s are readily obtained within 10 R_s by deposition of heat (or momentum) within a fraction of a solar radius above the Sun's surface. The influence of heavy ions on the fast wind in a three-fluid model has recently been studied in detail (Li *et al.*, 1997), indicating that preferential heating in particular of alpha particles may have an impact on the bulk flow.

Detailed parameter studies of the fast solar wind in models, including proton temperature anisotropy, have been carried out (Li *et al.*, 1997; Hu *et al.*, 1997) and indicate that anisotropy, although it has little influence on the dynamics, imposes requirements on the coronal heating mechanism (see Figure 19.6). The ion and electron temperature profiles in the corona were calculated (Hu *et al.*, 2000) in a four-fluid model including turbulence-driven heating of coronal heavy ions. The results obtained come close to the observations and are shown in Figure 19.7. Note the sizable temperature differences and the very high oxygen temperature in comparison

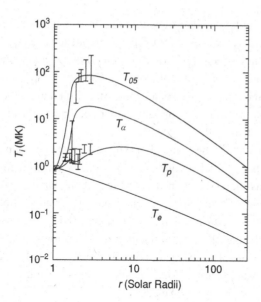

Figure 19.7. The ion and electron temperature profiles in the solar corona as obtained from a four-fluid model including turbulence-driven heating of heavy ions (Hu *et al.*, 2000).

with the cold electrons. The bars indicate SOHO measurements for comparison. As in other models the electrons hardly ever reach a 1 MK temperature (since they are not heated), whereas the protons attain temperatures of 2-5 MK and somewhat higher effective temperatures (including the wave amplitude) of up to 7 MK. Similar results for the heavy ion temperatures were obtained by Czechowski *et al.* (1998), who extended the basic model to include minor ion heating, which was assumed to be in proportion to mass.

In summary: The rapid acceleration obtained in most of the new (wave-driven) models is achieved by a strong thermal pressure gradient, resulting essentially from the steep decrease in ion pressure (temperature) beyond the critical point, after which wave absorption weakens or ceases. The wave dissipation models can account for both the heating of the corona and acceleration of the solar wind. This is because the high-frequency part of the spectrum heats the corona directly via wave dissipation, and thus builds up a strong thermal pressure gradient driving the flow, whereas the lower-frequency part provides an additional push through the Alfvén-wave pressure-gradient force, which acts mainly beyond the Alfvén point at 10 R_s.

19.4.5 The wave spectrum: origin, evolution and dissipation

High-frequency waves (most likely in the Alfvén/ion-cyclotron mode) can provide the necessary coronal heating through rapid dissipation within 1 R_s. This idea (Axford and McKenzie, 1992) was corroborated in a two-fluid turbulence model (Tu and Marsch, 1997), including parametric studies of the wind properties (Marsch and Tu, 1997) in dependence on the average wave amplitude at the coronal base. The wave heating process was shown to work well in the distant interplanetary solar wind (Marsch *et al.*, 1982). The coronal high-frequency Alfvén waves are believed to originate through reconnection (Axford and McKenzie, 1997) from the flaring magnetic network in the lower solar transition region. A key feature of this model is that the damping of Alfvén waves at the cyclotron frequency in a rapidly declining magnetic field (frequency sweeping), provides strong heating close to the Sun. In an empirical model (Cranmer *et al.* 1999) of a polar coronal hole, spectroscopic constraints were placed on the cyclotron-resonance heating. Cranmer (2000) also investigated the ion-cyclotron wave dissipation in the solar corona by consideration of the summed effect of more than 2000 ion species.

In all models assumptions have to be made about the unknown spectrum of the waves injected at the coronal base (which is equivalent to assuming a heating function $Q(r)$). A power-law may be assumed the intensity of which is constrained by extrapolation of *in-situ* measurements. Furthermore the questions of cascading - oblique as well as parallel - remains an open problem. Large-scale MHD structures may preferentially excite perpendicular short-scale fluctuations (Leamon *et al.*, 2000), the dissipation of which may involve strong Landau damping coupled to kinetic processes acting on oblique wavevectors. Whatever the process - it must be more effective for heavy ions than for protons (and electrons).

19.4.6   Critical issues in the models

In the models there are several debated issues which we briefly mention.

1. *Proton heat conduction* reduces the maximum coronal temperature and consequently the initial acceleration. However, high temperatures also yield a long Coulomb mean free path, thus bringing into question the application of the classical heat flux law, particularly in the presence of strong waves which affect the ions on much shorter scales.

2. *Flux tube divergence* is in most models considered by using simple spherical scaling, or an empirical formula (Kopp and Holzer, 1976). This has a direct impact on the wind speed profile and should be based on a more realistic magnetic field model such as that given by equation (19.3), which over the poles tends to increase the wind speed at say 2-3 R_s by about $(20 - 30)\%$, but does not change the asymptotic speed which is the same because of energy conservation.

3. *Base density* is frequently taken to be in the range between $(2 - 4)\ 10^8\ cm^{-3}$ which is rather too high, and therefore yields correspondingly lower speeds near the Sun.

4. *Momentum transfer by resonant waves* at frequencies above 1 Hz may play a role (see equation (19.15)), but can usually be neglected for protons as being small in the corona (of the order of Alfvén Mach number).

5. *Ion cyclotron-resonance heating* is thought to be important, but the problems of wave excitation, propagation, cascading and dissipation involving turbulence and kinetic physics are still not adequately adressed.

19.5   Slow solar wind

19.5.1   Observations of slow flows

The theory of the slow solar wind is much less developed than that of the fast solar wind. For the observed properties of the slow wind see the paper by Schwenn (1990). Although the observational situation is complicated, nevertheless, two kinds of slow wind can be discriminated.

The majority (i.e. between 80 – 90%) type I slow wind connects to variable fields at the coronal base in the magnetic network, where the field lines may continually change their foot points as a result of reconnection in a slowly-changing global coronal field configuration (Axford, 1977). This wind comes from the whole streamer belt (not just its edges) and accelerates slowly in the corona, with the plasma residing there for about a day. The minority (i.e. between 10−20%) type II slow wind comes in the form of small plasmoids or bubbles (Sheeley et al., 1997), and seems to originate from bursty reconnection mostly at field reversals at the top of coronal streamers.

The consequences of this origin are as follows: Type I slow ("junk") wind is highly variable in its properties (not so much in the speed which evens itself out) and never reaches equilibrium with coronal base conditions. Its field lines wander around and form "spaghetti"-like interplanetary structures connected to the Sun. Furthermore in the slow wind

- number density exceeds 10^8 cm^{-3} at the base,
- Coulomb collisions and electron heat conduction are important,
- Alfvén waves have sufficient time to isotropize and cascade,
- proton, heavy ion and electron temperatures are about equal,
- composition resembles that of the closed corona with FIP bias and is variable, sometimes on short timescales.

Type II slow wind has no connection to the base (and hence no information about the hot corona) and seems to be dripping of the tips of streamers like water drops fall from a leaking tap. This origin leads to a highly unsteady wind and has the consequence that

- heat conduction is inhibited, yielding cold and isotropic electrons,
- plasma is deficient in He and heavy ions due to gravitational settling, which depletes the heavy species at great heights in the streamers,
- plasma bubbles are carried by buoyancy and friction with the ambient wind.

The radial flow velocity profiles of such bubbles have been directly measured by the LASCO coronagraph on SOHO (Sheeley et al., 1997). The results are given in the left-hand frame of Figure 19.8, which also shows the measured near-Sun speeds of coronal mass ejections in the right-hand frame. The observations of compositional

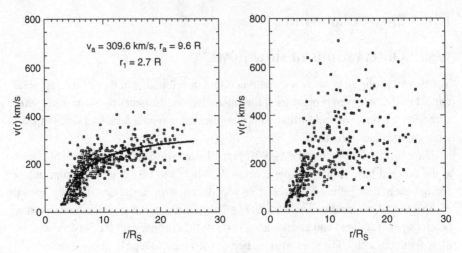

Figure 19.8. The speed profiles in the solar corona of small bubbles in the solar wind (left) and large coronal mass ejections (right), as obtained from difference images taken by the LASCO coronagraph on SOHO (Sheeley *et al.*, 1997).

variations in the solar corona and the processes leading to element fractionation are discussed in the review of (Raymond, 1999).

The third type of solar wind comes in the form of coronal mass ejections which we do not discuss in this paper. It is related to a restructuring of a large fraction of the entire corona, which involves global topology changes by reconnection and often occurs in association with flares and magnetically active regions.

19.5.2 Models of the closed corona and slow wind

The density and temperature structure in the closed field regions of the solar corona has been modelled recently (McKenzie *et al.*, 1999), using a dipole plus current-sheet model for the magnetic field and a heating function of the same type used in fast wind models (see previous sections). The heat equation, describing the redistribution of heat in the presence of sinks through radiative losses, was solved together with the hydrostatic pressure balance. This study shows that it is possible to obtain a temperature distribution consistent with the empirical one in Figure 19.4, and a base density of about $3\,10^8$ cm^{-3}, which is determined mainly by the ratio of heating to cooling.

The global structure of the corona has been modelled by several authors (see, e.g., Linker *et al.*, 1999; Mikić *et al.*, 1999; Usmanov *et al.*, 2000). Suess *et al.* (1999) used stationary two-fluid MHD equations, similar to the basic wind model, with two separate electron and proton temperature equations but with different latitude-dependent heating functions. Also, a momentum source term, with a maximal deposition around 3.5 R_s, was added. Their numerical results reproduced the Ulysses observations (see the model of Guhathakurta *et al.*, 1999) and the latitude dependence

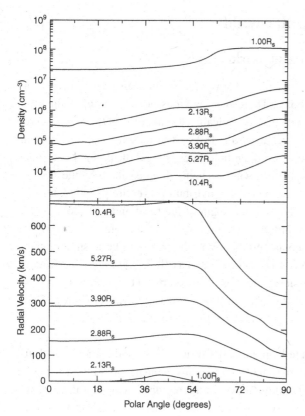

Figure 19.9. The density and speed profiles in the two-dimensional solar corona and wind as obtained from a two-fluid MHD model (Suess *et al.*, 1999). The numerical simulation results (relaxed state after 20 hours) are shown for various radial distances and versus co-latitude.

of the wind parameters consistent with observational constraints. Some results are displayed in Figure 19.9, which shows density and velocity versus polar angle and radial distance. Note the rapid acceleration over the poles (consistent with the basic 1-D model results) and the slow equatorial outflow from the streamer belt.

However, many key observational features of the slow solar wind are not included in these models, and fundamental problems remain unclear or even are not addressed. The slow wind appears to be intrinsically time variable and filamentary in nature, reflecting the unsteady source conditions at the coronal base or the transient changes in the magnetic field structure (in the streamer belt, for example). Topology changes require reconnection, which allows open field lines to form with subsequent plasma release from a previously closed corona. Moreover, the thermodynamics is often treated inadequately, e.g. by using distance-dependent polytropic indices in the equation of state for the ions and without adressing the microscopic physics of the heating process. It may differ considerably from the one in fast wind, as the different microstate (Marsch, 1991) observed in the interplanetary slow wind suggests. We conclude that much more work is required to understand the slow wind, and given the coronal observations, it is inevitable that in the future complex time-dependent theories are required.

19.6   Sources of the solar wind

19.6.1   Chromospheric network

SOHO observations indicate that the solar wind does not emanate directly from plumes in the polar coronal holes. There is no obvious powerful energy source in plumes, and their FIP bias found spectroscopically (Raymond, 1999) is not observed in the *in-situ* solar wind (von Steiger *et al.*, 1997). The reason for this is unclear. Because the plumes appear to be on open field lines beyond the coronal base, they should somehow contribute to the fast wind, as being dragged along by the ambient flow. Perhaps, since their filling is small, their contribution is entirely diluted. Instead, the whole coronal base is involved in the origin of the fast wind, with low plasma density, relatively high upward initial speeds and typical mean fields of 12 G. Below the base, the coronal field is contained mainly in the chromospheric supergranular network, which occupies merely 10% of the base area in holes. The network structure and corresponding EUV emission pattern as observed by EIT (Extreme Ultraviolet Imaging Telescope) and SUMER (Solar Ultraviolet Measurements of Emitted Radiation) on SOHO are shown in Figure 19.10, which gives the full Sun as seen in the emission of iron.

The insert shows the sizable Doppler-shifts of neon ions, indicating blue-shifts, i.e. outflows, at the cell boundaries and lane junctions in the network below the polar coronal hole, and red-shifts, i.e. downflows, in the regions underlying the globally closed corona. In the network the magnetic field pressure dominates the thermal pressure, and the field varies rapidly with height. The network field of (10-100) G, which is rooted in the photosphere in small, kG-field flux tubes ($\sim$100 km in size), expands rapidly with altitude in the transition region and subsequently fills the entire overlying corona.

The high-resolution magnetograms from SOHO show clearly that the magnetic field exists in the network in two side-by-side components, i.e. in uncanceled unipolar flux tubes (funnels) and in closed, multiple loops. The small loops will emerge or contract downwards and collide, thus constituting a permanent source of magnetic energy. The only conceivable energy source for the wind is the dynamic network field itself, because:

- a potential field (open funnel, Gabriel, 1976) has no free energy,
- a complex, though static field (Dowdy *et al.*, 1986), does not envisage magnetic energy release or produce waves,
- only time-varying fields (Axford *et al.*, 1999) undergoing reconnection and allowing flaring can provide energy to the corona.

Theoretical ideas about the origin of the solar wind have been put forward (Axford and McKenzie, 1992; Axford and McKenzie, 1997; Marsch and Tu, 1997) according to which the wind originates in the chromospheric network and draws its energy

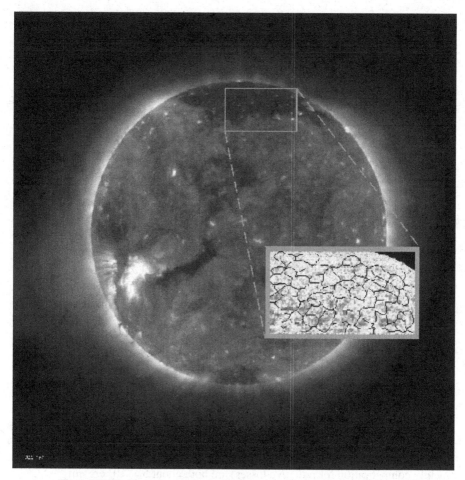

Figure 19.10. The solar corona and polar coronal holes as obtained from EIT and SUMER images on SOHO. The insert shows chromospheric network observations with sizable Doppler blue shifts interpreted as outflow of the nascent solar wind (Hassler *et al.*, 1999) at the base of the polar hole.

from high-frequency waves. The energy generated by magnetic reconnection of the dynamic and complex network fields is presumably released in frequent "pico-flares" with typical energies of about 10^{21} erg. Such small-scale magnetic activity is expected to continually produce waves with a spectrum that may also contain a sizable amount of power at relatively high frequencies. Wave dissipation would occur rapidly within a solar radius and could involve an essentially linear mechanism such as cyclotron damping, as previously discussed already.

19.6.2 Network pico-flares

We may assume that "pico-flares" are similar to other intra-network flares (which we may call "nano-flares" in Parker's definition) in the sense that remnant closed loops containing hot plasma are produced (similar to but smaller than the observed bright points), which can be observed for some time after the flares have released their energy. The dimensions are only 0.01–0.1 arcseconds, and the remnants must still be resolved and seen in the network. Therefore, future high-resolution images, e.g. from the *Solar Orbiter* mission, will be very important. Network "pico-flares" must occur frequently and should have linear dimensions down to the 10–100 km scale.

As concerns the energetics, the required mass and energy fluxes of about 10^{14} cm^{-2}s^{-1} and 10^6 erg cm^{-2}s^{-1} at the coronal base correspond to values of 10^{15} cm^{-2}s^{-1} and 10^7 erg cm^{-2}s^{-1} at the level of a 100 G-field. To achieve this energy flux, a magnetic field of 100 G must be destroyed down to a depth of 250 km about every 15 minutes. A pico-flare with $\sim 10^{21}$ erg might cover an area of only 10 km^2 at an efficiency of 100%, otherwise over an area of 100-1000 km^2. During reconnection, only the "reversed" flux is destroyed, whereas the "escaping" flux is preserved. The flux destroyed by the "magnetic furnace" must be replaced. Magnetoconvection in the photosphere, which is the cause of the granulation and supergranulation, would provide an adequate supply of randomly-looped magnetic flux to the furnace.

19.6.3   Heating of the quiet corona

The corona is hot even when there is no obvious activity. SOHO has shown that the open coronal holes (with their dilute plasma at a few solar radii) are the hottest places on the Sun in terms of proton and heavy ion kinetic temperatures. Observationally, the chromosphere (including the network) looks the same beneath magnetically closed (quiet) corona and the open (hole) corona. The energy flux required to power the solar wind in coronal holes and the energy flux needed to produce the quiet corona (mainly compensation for radiative losses) are both about 5×10^5 erg cm^{-2} s^{-1}.

It is therefore natural to postulate that the source of this energy flux should be more-or-less the same all over the Sun in quiet regions - any additional source in the quiet corona should also appear in the open-field coronal holes. Yet, there is nothing significant. In the Parker model of quiet coronal heating (Parker, 1988) of "nano-flares", low-frequency waves are generated from the twisting and foot-point shuffling of magnetic field lines, whereby the stored energy is released intermittently by coronal reconnections. However, this is equivalent to a flux of very low-frequency waves produced in this way. But the waves are not found in the wind coming from coronal holes with the necessary energy flux. Therefore, this source seems to be inadequate for coronal heating. We may conclude that the quiet corona is also heated by high-frequency waves resulting from pico-flares occurring in the supergranular network.

19.6.4   Some consequences of network flares

The network region, where pico-flares are continuously occurring, is characterised by the following typical parameters: Alfvén speed: $V_A \approx 10\,000$ km s^{-1}; spatial scale: $L = 100$ km, which yields a transit time, $L/V_A \approx 0.01$ s, and therefore Alfvén waves with correspondingly short periods. The associated voltage drop is $V_A B L \approx 10^9$ V. Although the flares are short lived (with the remnants persisting), they can easily accelerate particles up to energies of $1 - 10$ MeV. Waves, reconnection shocks and fast particles are therefore available to heat the corona. Energetic particles precipitating in the solar atmosphere may further produce γ rays, neutrons and rare isotopes through nuclear reactions at the dense solar surface.

19.7   Problems

19.7.1   Problems with the observations

There are difficulties in the interpretation of the coronal observations. For example, what is the role of plumes? They dominate line-of-sight measurements and thus appear to be the corona as seen in the plane of the sky, even during minimum solar activity where they shine prominently in the large polar coronal holes (DeForest et al. 1997). Could it be that there is a significant contribution from plumes, which overlies the low stray-light intensity being genuinely related to the fast wind?

The average outflow speed, inferred first from UVCS through Doppler dimming, indicated a slower acceleration than is obtained in the basic model described above. But the SUMER results (Patsourakos and Vial, 2000) subsequently indicated, in fact, very fast acceleration in the dark lanes. According to the new results obtained from UVCS (Giordano et al., 2000) the situation appears to be clearer. There is strong evidence that the fast solar wind is preferentially accelerated in the interplume lanes and darker background regions surrounding the plumes in the polar coronal holes.

The charge state measurements on SOHO (Ko et al., 1997) seem to suggest a higher electron temperature than the one obtained by spectroscopic diagnostics with SUMER (Wilhelm et al., 1998). Therefore it may be that coronal electrons have a core-halo structure such as found in the interplanetary solar wind. The EUV measurements mainly refer to the bulk electrons in the core ($T_e \approx 1$ MK ≈ 100 eV), whereas the charge states reflect more the halo temperature (~ 5 MK or 500 eV), which would mean that the charge state measurements do not tell us much more than that there is a halo. Recently, Esser and Edger (2000) have done a detailed study indicating how the spectroscopic electron temperature measurements can be reconciled with the in-situ charge state measurements. It turns out that the electron density, temperature and non-Maxwellian distribution as well as the ion differential outflow speeds are all equally important, and that a core-halo-type non-Maxwellian structure must develop within about $2 - 3 R_s$ to obtain model results consistent with observations.

What evidence do we have for the waves that provide the heating? The models imply properties which require observational verification. For example, the waves should:

- disappear beyond $\sim 10\ R_s$ and have relatively short periods < 10 s,
- be fairly linear with $\delta B \ll B$, unless being generated by reconnection,
- preferentially heat heavy ions, discriminating them according to Z/A,
- have frequencies even up to 1 kHz and dissipate quickly within 1 R_s,
- be LH polarized to dissipate by ion-cyclotron absorption,
- be partly also RH polarized to heat the electrons as well.

These properties can only be verified by *in-situ* measurements since remote-sensing or spectroscopic evidence cannot give complete plasma diagnostics. The waves therefore remain a problem from the observational point of view. On the other hand, given our knowledge from solar wind *in-situ* measurements and other accessible space plasmas, such as planetary magnetospheres, we have every reason to believe that high-frequency waves do exist in the corona, because they are, along with the particles, a natural component of any plasma, particularly if it is far from thermal equilibrium.

19.7.2 Problems with the theory

The theories have the weakness that the heating functions are either assumed "*ad hoc*", such as (19.14), or involve the complexity of kinetic physics, such as (19.15), with its associated particle velocity distribution functions and wave spectra, neither of which can be measured directly in the lower corona. Furthermore, a transport theory for a weakly collisional plasma including high-frequency-wave turbulence does not exist. Such a complex situation may require a kinetic treatment, which has recently been attempted for coronal funnels (Vocks and Marsch, 2001). This semi-kinetic model can reproduce some observed features, such as preferential heating of heavy ions by waves, while also including Coulomb collisions.

A model has been developed (Leblanc and Hubert, 1997) which constructs solar wind proton velocity distribution functions by an expansion into higher-order moments. Presently, there is no basic kinetic model of the solar corona and wind. Recently, a model was discussed (Lie-Svendsen *et al.*, 2001) which also included the heat-flux-moment equation explicitly, and in which the equations were integrated all the way out from the chromosphere, through the corona into the inner heliosphere.

Other authors (Maksimovic *et al.*, 1997) have proposed a wind model which relies entirely on suprathermal electrons (kappa velocity distributions) in the corona to accelerate the fast wind. However such a model cannot even explain the major features of coronal holes, namely hot ions and cold electrons, and makes unrealistic assumptions at the coronal base. Substantial progress with kinetic models for the solar wind electrons has been made (Lie-Svendsen *et al.*, 1997), where the fundamental core-halo structure was reproduced by the use of the Fokker-Planck equation for Coulomb collisions, while starting with Maxwellian electrons at the coronal base.

19.8   Conclusions

Within the last few years, considerable progress has been made in characterising the coronal plasma state and in determining the radial profiles of key parameters in the polar coronal holes. The ions were found to be generally very hot and electrons cool in polar coronal holes. These empirical constraints have been taken into account in recent models of the fast solar wind. It seems possible to produce a fast wind on the basis of high-frequency wave dissipation in the lower corona. The fast solar wind ions are mainly driven by their thermal pressure-gradient forces. The implications are that the acceleration is rapid and the protons reach a temperature of several MK. The slow wind is rather unsteady and inherently filamentary. It comes from the streamer belt, being near collisional equilibrium, and also from the predominantly closed corona as a consequence of reconnection.

A theoretical description must involve two-fluid or multi-fluid equations, or even kinetic equations. In the weakly collisional corona, waves and particles appear to be intimately linked through plasma instabilities and wave-particle interactions. These processes involve waves at frequencies (up to the kHz range) much higher than the frequencies of MHD waves which were previously considered in more conventional fluid models.

Yet, many basic questions concerning the corona and wind remain. The coronal base density is crucial for modelling the initial flow speed profile. What is its value? Present estimates range from 2 to 20 times 10^7 cm^{-3}. Do we really know what makes $Q(r)$? It must surely be more than just cyclotron absorption of LH Alfvén/cyclotron waves. What role then do other waves and convected structures or current sheets play, in particular in an oblique turbulent cascade? Are there other micro-processes also contributing to heating of the corona? What is really going on in the network? Pico-flares should be seen there, yet presently they are still below the spatial resolution of optical instruments. How is the particle flux provided, and what causes the ion FIP fractionation seen in the slow wind (it should be related to the FIP effect in the quiet, quasi-static corona), and consequently what determines the solar wind mass flux and coronal base density? How do we adequately describe the unsteady slow wind originating in the transiently open streamer belt?

To answer these questions new observations are required which address the role of plumes, suprathermal and energetic particles, waves in coronal holes, and the flaring and reconnection processes occurring continuously in the network. This requires observations at high temporal and spatial resolution down to the 35 km scale ($<$ 0.05 arcsec at 1 AU) and from new vantage points. These goals could be achieved by spacecraft passing close to the Sun, such as *Solar Probe* (with a low perihelion of 4 R_s) or *Solar Orbiter* (corotating with the Sun at 45 R_s), preferably at solar minimum when there are well-developed coronal holes and high-speed streams. To make further progress in solar wind research, innovative measurements which place new and more complete constraints on the theory are needed.

References

Axford, W.I. (1977). in *Study of Travelling Interplanetary Phenomena*, ed. M.A. Shea, D.F. Smart, and S.T. Wu, (Dordrecht, D. Reidel) 145.

Axford, W.I. and McKenzie, J.F. (1992). in *Solar Wind Seven*, ed. E. Marsch and R. Schwenn, (Pergamon Press, Oxford, England) 1.

Axford, W.I. and McKenzie, J.F. (1997). in *Cosmic Winds and the Heliosphere*, ed. J.R. Jokipii, C.P. Sonett, and M.S. Giampapa, (Arizona University Press, Tucson) 31.

Axford, W.I., McKenzie, J.F., Sukhorokova, G.V., Banaszkiewicz, M.M., Czechowski, A., and Ratkiewicz, R. (1999). *Space Sci. Rev.*, **87**, 25.

Banaszkiewicz, M., Axford, W.I., and McKenzie, J.F. (1998). *Astron. Astrophys.*, **337**, 940.

Barnes, A., Gazis, P.R., and Phillips, J.L. (1995). *Geophys. Res. Lett.*, **22**, 3309.

Belcher, J.W. (1971). *Astrophys. J.*, **168**, 509.

Bravo, S., Stewart, G.A., and Blanco-Cano, X. (1998). *Solar Phys.*, **179**, 223.

Cranmer, S.R., Field, G.B., and Kohl, J.L. (1999). *Astrophys. J.*, **518**, 937.

Cranmer, S. R. (2000). *Astrophys. J.*, **532**, 1197.

Czechowski, A., Ratkiewicz, R., McKenzie, J.F., and Axford, W.I. (1998). *Astron. Astrophys.*, **335**, 303.

David, C., Gabriel, A.H., Bely-Dubau, F., Fludra, A., Lemaire, P., and Wilhelm, K. (1998), *Astron. Astrophys.*, **336**, L90.

DeForest, C.E., Hoeksema, J.T., Gurman, J.B, Thompson, B.J., Plunkett, S.P., Howard, R., Harrison, R.C., Hassler, D.M. (1997). *Solar Phys.*, **175**, 393.

Dowdy, J.F., Rabin, D., and Moore, R.L. (1986). *Solar Phys.*, **105**, 35.

Esser, R., Habbal, S.R., Coles, W.A., and Hollweg, J.V. (1997). *J. Geophys. Res.*, **102**, 7063.

Esser, R. and Edgar, R.J. (2000). *Astrophys. J.*, **532**, L71.

Gabriel, A.H. (1976). *Phil. Trans. Roy. Soc.*, **A281**, 339.

Geiss, J. (1982). *Space Sci. Rev.*, **33**, 201.

Giordano, S., Antonucci, E., Noci, G., Romuli, M., and Kohl, J.L. (2000). *Astrophys. J.*, **531**, L79.

Guhathakurta, M., Sittler, E.C., and McComas, D. (1999). *Space Sci. Rev.*, **87**, 199.

Hansteen, V. and Leer, E. (1995). *J. of Geophys. Res.*, **100**, 21577.

Hartle, R.E. and Sturrock, P.A. (1968). *Astrophys. J.*, **151**, 1155.

Hassler, D.M, Dammasch, I.E., Lemaire, P., et al. (1999). *Science*, **283**, 5040, 810.

Hoeksema, J.T. (1995). *Space Sci. Revi.*, **72**, 137.

Holzer, T.E. and Axford, W.I. (1970). *Ann. Rev. Astron. Astrophys.*, **8**, 31.

Hu, Y.Q., Esser, R., and Habbal, S.R. (1997). *J. Geophys. Res.*, **102**, 14661.

Hu, Y.Q., Esser, R., and Habbal, S.R. (2000). *J. Geophys. Res.*, **105**, 5093.

Jacques, S.A. (1978). *Astrophys. J.*, **226**, 632.

Kohl, J. L., Noci, G., Antonucci, E., et al. (1997). *Solar Phys.*, **175**, 613, 1997.

Kohl, J. L., Noci, G., Antonucci, E., et al. (1998). *Astrophys. J.*, **501**, L127, 1998.

Ko, Y.-K., Fisk, L., Geiss, J., Gloeckler, G., and Guhathakurta, M. (1997). *Solar Phys.*, **171**, 345.

Kopp, R.A. and Holzer, T.E. (1976). *Solar Phys.*, **49**, 43.

Leamon, R.J., Matthaeus, W.H., Smith, C.W., Zank, G.P., and Mullan, D.J. (2000). *Astrophys. J.*, **537**, 1054.

Leblanc, F. and Hubert, D. (1997). *Astrophys. J.*, **483**, 464.

Leer, E. and Axford, W.I. (1972). *Solar Phys.*, **23**, 238.

Leer, E. and Holzer, T.E. (1980). *J. Geophys. Res.*, **85**, 4681.

Li, X., Esser, R., and Habbal, S.R. (1997). *J. Geophys. Res.*, **102**, 17419

Lie-Svendsen, O., Hansteen, V.H., and Leer, E. (1997). *J. Geophys. Res.*, **102**, 4701.

Lie-Svendsen, O., Leer, E., and Hansteen, V.H. (2001). *J. Geophys. Res.*, **106**, 8217.

Linker, J.A., Mikić, Z., Biesecker, D.A., Forsyth, R.J., Gibson, S., Lazarus, A.J., Lecinski, A., Riley, P., Szabo, A., and

Thompson, B.J. (1999). *J. Geophys. Res.*, **104**, 9809.

McComas, D.J., Gosling, J.T., and Skoug, R.M. (2000). *Geophys. Res. Lett.*, **27**, 2437.

McKenzie, J.F., Banaszkiewicz, M., and Axford, W.I. (1995). *Astron. Astrophys.*, **303**, L45.

McKenzie, J.F., Axford, W.I., and Banaszkiewicz, M. (1997). *Geophys. Res. Lett.*, **24**, 2877.

McKenzie, J.F., Sukhorukova, G.V., and Axford, W.I. (1999). *Astron. Astrophys.*, **350**, 1035.

Marsch, E. (1991). in *Physics of the Inner Heliosphere*, ed. R. Schwenn and E. Marsch, (Springer Verlag, Heidelberg, Germany) Vol. II, 45.

Marsch, E. (1999). *Space Science Rev.*, **87**, 24.

Marsch, E. and Tu, C.Y. (1997). *Solar Phys.*, **176**, 87.

Marsch, E. and Tu, C.Y. (1997). *Astron. Astrophys.*, **319**, L17.

Marsch, E. and Tu, C.Y. (2001). *J. Geophys. Res.*, **106**, 227.

Marsch, E., Goertz, C.K., and Richter, K. (1982). *J. Geophys. Res.*, **87**, 5030.

Maksimovic, M., Pierrard, V., and Lemaire, J.F. (1997). *Astron. Astrophys.*, **324**, 725.

Mikić, Z., Linker, J.A., Schnack, D.D., Lionello, R., and Tarditi, A. (1999). *Phys. Plasmas*, **6**, 2217.

Parker, E.N. (1958). *Astrophys. J.*, **128**, 664.

Parker, E.N. (1963). *Interplanetary Dynamical Processes* (Wiley Interscience, New York).

Parker, E.N. (1988). *Astrophys. J.*, **330**, 474.

Patsourakos, S. and Vial, J.-C. (2000). *Astron. Astrophys.*, **359**, L1.

Raymond, J.C. (1999). *Space Science Rev.*, **87**, 55.

Ryan, J.M. and Axford, W.I. (1975). *J. Geophys.*, **41**, 221.

Schwenn, R. (1990). in *Physics of the Inner Heliosphere*, ed. R. Schwenn and E. Marsch, (Springer Verlag, Heidelberg, Germany) Vol. I, 99.

Schwenn, R., Inhester, B., Plunkett, S.P., *et al.* (1997). *Solar Phys.*, **175**, 667.

Sheeley, N.R., Jr., Wang, Y.-M., Hawley, S.H., *et al.* (1997). *Astrophys. J.*, **484**, 472.

Sittler, E. and Guhathakurta, M. (1999). *Astrophys. J.*, **523**, 812.

von Steiger, R., Geiss, J., Gloeckler, G., and Galvin, A.B. (1995). *Space Science Rev.*, **72**, 71.

von Steiger, R., Geiss, J., and Gloeckler, G. (1997). in *Cosmic Winds and the Heliosphere*, ed. J.R. Jokipii, C.P. Sonett, and M.S. Giampapa (Arizona University Press, Tucson), 581.

Suess, S.T., Wang, A.-H., Wu, S.T., Poletto, G., and McComas, D.J. (1999). *J. Geophys. Res.*, **104** 4697.

Thieme, K.M., Marsch, E., and Schwenn, R. (1990). *Ann. Geophys.*, **8**, 713.

Tu, C.-Y. (1987). *Solar Phys.*, **109**, 149.

Tu, C.-Y. and Marsch, E. (1997). *Solar Phys.*, **171**, 363.

Tu, C.-Y. and Marsch E. (2001). *J. Geophys. Res.*, **106**, 8233.

Tu, C.-Y., Pu, Z.X., and Wei, F.S. (1984). *J. Geophys. Res.*, **89**, 9695.

Usmanov, A.V., Goldstein, M.L., Besser, B.P., and Fritzer, J.M. (2000). *J. Geophys. Res.*, **105**, 12675.

Vocks, C. and Marsch, E. (2001). *Geophys. Res. Lett.*, **28**, 1917.

Wilhelm, K., Marsch, E., Dwivedi, B.N., Hassler, D.M., Lemaire, P., Gabriel, A., and Huber, M.C.E. (1998). *Astrophys. J.*, **500**, 1023.

20

Solar observing facilities

B. Fleck

ESA Space Science Department, c/o NASA/GSFC, Mailcode 682.3, Greenbelt, MD 20771, USA

C.U. Keller

National Solar Observatory, 950 N. Cherry Ave., Tucson, AZ 85719, USA

20.1 Introduction

Recent results from solar telescopes in space and on the ground using adaptive optics and other advanced instrumental techniques have produced stunning results that are invigorating solar research and challenging existing models of the Sun. Plans for future space missions, balloon-borne telescopes and new ground-based telescopes and instruments promise to continue this "solar renaissance" in all areas of solar physics, from the solar interior to interplanetary space and from regimes of high-energy to observations requiring high resolution. The next generation of solar telescopes and instruments promises us the ability to investigate solar processes on their fundamental scales, whether sub-arcsecond or global in nature.

Tables 20.1 and 20.2 give overviews over current and planned ground-based facilities and space missions, respectively. In the following sections, these are discussed in more detail. Suborbital rocket and balloon flights are briefly discussed in a separate section.

20.2 Ground-based instruments

The Earth's atmosphere transmits electromagnetic radiation mainly in two wavelength bands: in the visible and infrared part of the spectrum from 300 nm to 28 μm, and at radio wavelengths. Neutrinos are the only particles that make it through the atmosphere directly.

Most of the following discussion on ground-based solar facilities concentrates on the optical spectrum where solar telescopes are unique in the way they are built and

Table 20.1. *Ground-based facilities in operation and under development with apertures larger than 40 cm*

Facility, location	Diameter (m)	Wavelength Coverage
In Operation		
McMath-Pierce, Kitt Peak, USA	1.52, 0.9, 0.9	0.30–28 μm
THEMIS, Tenerife, Spain	0.90	380–1100 nm
Dunn Solar Telescope, Sunspot, USA	0.76	350–2200 nm
Large Solar Vacuum Telescope, Lake Baikal, Russia	0.76	350–2200 nm
Kitt Peak Vacuum Telescope, USA	0.70	350–1100 nm
Vacuum Tower Telescope, Tenerife, Spain	0.70	350–1700 nm
Vacuum Reflector, Big Bear, USA	0.65	350–2200 nm
Domeless Solar Telescope, Hida, Japan	0.60	350–1100 nm
Multi-Channel Telescope, Huairou, China	0.60	390–670 nm
Large Coelostat telescope, Nanjing, China	0.60	380–1100 nm
Einsteinturm, Potsdam, Germany	0.60	350–1700 nm
Tower Telescope, Meudon, France	0.60	350–1100 nm
Large Coronagraph, Tbilisi, Georgia	0.53	500–1100 nm
Large Refractor, Debrecen, Hungary	0.53	656 nm
Large Coronagraph, Wroclaw, Poland	0.53	380–670 nm
Turret Telescope, Pic-du-Midi, France	0.50	380–1100 nm
Vacuum Solar Spectral Telescope, Yunnan, China	0.50	380–1100 nm
Dutch Open Telescope, La Palma, Spain	0.45	350–1100 nm
Gregory Coudé Telescope, Tenerife, Spain	0.45	350–1700 nm
Gregory Coudé Telescope, Locarno, Switzerland	0.45	350–1100 nm
Horizontal Solar Telescope, Tbilisi, Georgia	0.44	350–1100 nm
Vacuum Newton Telescope, Tenerife, Spain	0.40	350–1700 nm
Kanzelhöhe Vacuum Telescope, Austria	0.40	350–1100 nm
Horizontal Telescope, Meudon, France	0.40	350–1100 nm
In Development and Under Study		
Advanced Technology Solar Telescope	4	0.38–28 μm
GREGOR	1.5	0.3–28 μm
New Swedish Vacuum Solar Telescope La Palma, Spain	0.97	high resolution
Solar-C	0.5	coronagraph, 0.5–5 μm
Frequency Agile Solar Radio telescope	100 2–5 m	0.3–30 GHz

Table 20.2. *Past, present, and future space missions*

Mission	Launch	Scientific objectives	Measurements
Past missions (representative)			
OSO 1–8	1962-1978	solar activity, flares	X-ray, UV
Skylab	1973-1974	solar atmosphere, corona	X-ray, EUV, UV, Hα, coronagr.
IMP-8	1973-2001	solar wind plasma	*in situ* particles and fields
HELIOS 1	1974-1984	heliosphere, plasma, particles	*in situ* particles and fields
HELIOS 2	1976-1980	heliosphere, plasma, particles	*in situ* particles and fields
ISEE-3	1978-1983	solar wind plasma	*in situ* particles and fields
SMM	1980-1989	solar activity, flares	X- + γ-ray, UV, total irradiance
Hinotori	1981-1991	solar activity, flares	X-ray imaging + spectroscopy
CORONAS-I	1994-1995	solar interior and activity	X- + γ-ray, UV, total irradiance
In operation			
Ulysses	6 Oct 1990	3-D structure of heliosphere	*in situ* particles and fields
Yohkoh	31 Aug 1991	solar flares	X- and γ-ray imaging + spectrosc.
SAMPEX	3 Jul 1992	energ. particles, cosmic rays	*in situ* particles
Wind	1 Nov 1994	solar wind plasma	*in situ* particles and fields
SOHO	2 Dec 1995	solar interior, coronal heating, solar wind acceleration	helioseismology, irradiance, UV/EUV imaging and spectrosc. coronagraphy, *in situ* particles
ACE	25 Aug 1997	elemental and isotopic comp. of solar wind and energetic part.	*in situ* particles and fields
TRACE	2 Apr 1998	structure and dynamics of TR and corona	high res UV/EUV imaging
GOES-12/SXI	23 Jul 2001	solar flares	soft X-ray images
CORONAS-F	31 Jul 2001	dynamical processes of solar activity; flares	X-ray imaging + spectroscopy UV, coronagraphy, solar irrad.
GENESIS	8 Aug 2001	isotopic and elemental abundances of solar wind	collection of SW sample and return for ground analysis
HESSI	5 Feb 2002	flares / particle acceleration	X- and γ-ray imaging + spectrosc.
In development and under study			
Solar-B	Aug 2005	generation and transport of mag. fields; coronal heating	high res. vis. light magnetometry, EUV spectrosc., X-ray imaging
STEREO	Nov 2005	CME origin and propagation	coronagraphy, UV imaging, *in situ* particles
Space Solar Telescope	2005+	magnetic fields, heating and expansion of corona	1 m optical telescope EUV imaging, coronagraphy X-,γ-ray spectr.
SDO	2007+	nature of solar variability that affect life and society	helioseismology, magnetometry, UV imaging + spectroscopy EUV irradiance, photometry, coronagaphy
Solar Orbiter	2009+	near-Sun heliosphere, dynamics of mag. atmosphere, out-of-ecliptic imaging of polar regions	*in situ* particles and fields, UV/EUV imaging and spectrosc., visible light magnetometry, coronagraphy, radio
Solar Probe	2009+	cornal heating and solar wind acceleration	*in situ* particles and fields, imaging in EUV and vis. light magnetography, coronagraphy
Sentinels	2009+	CMEs and geo-eff. SW	*in situ* and remote sensing

operated. Most general-purpose radio telescopes can be successfully used for solar observations. However, there are specialized solar radio telescopes, which we will list in the following. Due to the lack of space, we will only discuss radio telescopes that are capable of providing two-dimensional images, and we will not discuss neutrino telescopes here.

There is a large number of operational and planned ground-based solar facilities. The selection discussed here is somewhat arbitrary and by no means complete. We group the various facilities into two classes which will be discussed separately: general purpose telescopes that can provide a variety of observing environments, and synoptic facilities that repeat the same type of measurements over years. For general purpose optical telescopes, we set a lower aperture limit of 40 cm. For synoptic instruments to be listed here, the data produced by these instruments need to be accessible over the Internet and observed at least once per day weather permitting. The authors would like to apologize here to anybody whose facility has not been mentioned.

20.2.1 Present

20.2.1.1 General purpose telescopes

The French-Italian THEMIS (Télescope Héliographique pour l'Etude du Magnétisme et des Instabilités Solaires, scc Rayrole 1990) at the Observatorio del Teide, Tenerife, Spain, is the first of a series of next-generation solar telescopes that are compact and axially symmetric, a condition which minimizes instrumentally induced polarization. The main scientific goal of THEMIS is the accurate determination of vector magnetic fields using spectro-polarimetry.

A group of German organizations operate several telescopes on Tenerife. The largest telescope is the Vacuum Tower Telescope (VTT, see Schmidt 1992) with an aperture of 70 cm. To a large degree it is a copy of the Kitt Peak Vacuum Telescope (see below). Postfocus instrumentation includes a vertical Echelle spectrograph, a filter device for the simultaneous observation of solar images in several wavelengths, and an optical laboratory with a Fabry-Pérot interferometer. The evacuated Gregory-Coudé Telescope (Kneer and Wiehr 1989) is an equatorially mounted mirror telescope with a 45-cm aperture that feeds a horizontal Echelle spectrograph. And finally, the Vacuum Newton Telescope is an equatorially mounted evacuated telescope with a 40-cm aperture (e.g. Schröter et al. 1985).

The Dutch Open Telescope (Rutten et al. 1999) at the Observatorio del Roque de los Muchachos, La Palma, Spain, is a novel telescope based on the open telescope concept of Zwaan and Hammerschlag. Wind flushes the primary mirror and the telescope structure continually so that internal seeing is avoided. The mechanical stability was a main driver to allow observing even under strong wind conditions. The open telescope concept has the important advantage that it may be scaled to much larger apertures than is feasible for traditional evacuated or helium-filled telescopes requiring an entrance window.

The Einsteinturm in Potsdam, Germany, in use since the mid-1920s is arguably the architecturally most stunning solar telescope in the world. It contains a coelostat that feeds a 60-cm lens with a focal length of 14 m, and a Littrow spectrograph. The main research topic is vector-spectropolarimetry of sunspots.

The Observatoire de Meudon outside of Paris operates a 60-cm aperture tower telescope with a coelostat that feeds the solar beam to the ground level. A large spectrograph, with a double pass system, can acquire fields of view of 1 by 8 arcmin in up to 9 spectral areas simultaneously with a spatial resolution approaching 1 arcsec.

The Istituto per Ricerche Solari Locarono (IRSOL) was founded and operated by the Universitäts-Sternwarte Göttingen, Germany until 1984, when it was moved to Tenerife along with most of the instrumentation. By 1993 this instrumentation had been rebuilt. The observatory contains an evacuated Gregory-Coudé telescope with a 45-cm primary mirror and Czerny-Turner type spectrograph with a field of view of 200 arcsec.

The Astronomical Institute of Wroclaw University, Poland, operates a 53-cm coronagraph that provides filtergrams and feeds a multi-channel subtractive double-pass spectrograph, and two smaller instruments for studying Hα on the disk and at the limb.

The Abastumani Astrophysical Observatory in Tbilisi, Republic of Georgia, operates a 53-cm multi-purpose coronagraph. Built in 1973, it is the world's largest coronagraph. In addition, it has a 44-cm general-purpose horizontal solar telescope.

The main telescope at the Baikal Observatory is the Large Solar Vacuum Telescope. Constructed in 1980, it looks like an evacuated version of the McMath-Pierce telescope with a 76-cm main lens. A high-dispersion spectrograph is the main post-focus instrument.

The Astronomy Department of Nanjing University operates a ceolostat solar telescope that has an effective aperture of 60 cm. The secondary mirror is scanned for rapid observations of the solar velocity field. It provides monochromatic and white light images of various diameters for CCD and photographic cameras.

The Hida Observatory of the nearby University of Kyoto, Japan, houses the 60-cm Zeiss Domeless Solar Telescope, one of the most versatile solar telescopes in Japan. It has a large vertical spectrograph and tunable filters that are mostly used to obtain high-resolution Hα images and spectra.

The US National Solar Observatory (NSO) operates various telescopes. The McMath-Pierce Solar Telescope Facility (Pierce 1964) on Kitt Peak, Arizona, USA, now almost 40 years old, is still the largest solar optical telescope in the world with an aperture of 1.5 m. Its two auxiliary telescopes are also among the largest solar telescopes in operation. The large light-gathering power, the extended wavelength range from the UV to the far IR, and the well-behaved polarization characteristics of these telescopes make them unique instruments that continue to stimulate a variety of novel investigations of the Sun. The Fourier Transform Spectrometer, housed in the McMath-Pierce complex, is a unique instrument in wide demand by atmospheric physicists and chemists as well as astronomers.

NSO's Dunn Solar Telescope (DST, see Dunn 1969) at Sunspot, New Mexico, USA, now over 30 years old, has played a key role in high-resolution solar physics and influenced the design of subsequent evacuated solar telescopes. With the successful implementation of adaptive optics, the DST is currently providing the highest resolution solar images of any telescope.

The Big Bear Solar Observatory (BBSO) is located in Big Bear Lake, California at an altitude of 2000 m. Originally built by the California Institute of Technology in 1969, it has been operated by the New Jersey Institute of Technology since 1997. The dome contains a single fork mount supporting three solar telescopes. The largest, a 65-cm evacuated reflector, can be used for general-purpose observations and will soon be equipped with adaptive optics.

20.2.1.2 Synoptic telescopes

The Observatoire de Meudon operates various solar synoptic instruments. The magnetograph uses the 40-cm horizontal telescope that is fed by a siderostat. A circular polarisation analyzer allows observations of 2 by 2 arcmin with a resolution of about 1.5 arcsec. For more than 80 years, the 25-cm spectroheliograph has provided full-disk measurements in Hα, and the K3 and K1v parts of the CaII K line with a resolution of about 2 arcsec.

The Kanzelhöhe Observatory, operated by the Institute of Astronomy of the Karl Franzens University of Graz, contains four refractors that are used for sunspot drawings, images of the photosphere and the chromosphere (Hα), respectively. A magneto-optical filter, which was developed at the University of Rome, is attached to the fourth refractor. It operates in the Na-D lines and provides full disk magnetograms, dopplergrams and intensity images. The Kanzelhöhe Vacuum-Telescope (KVT) is a mirror-system with an aperture of 40 cm and an effective focal length of 26 meter with a field of view of a few arc minutes. A horizontal vacuum-spectrograph allows high spectral resolution spectroscopy.

The Catania Astrophysical Observatory on Sicily is operated by the University of Catania. Synoptic solar observations have been conducted there since 1892. The current main products are full disk white light and H-alpha images with emphasis on flare patrol.

Pic-du-Midi is the site of the first coronagraph and is located in southwest France high in the Pyrenees. Synoptic observations have been taken there since the 1940s. The H-Alpha COronagraph (HACO) studies dynamic phenomena of the solar corona's cold area using the Pic-du-Midi coronagraph of 15 cm aperture. Images with a resolution of about 5 arcsec are recorded up to every 30 s.

The Rome Astronomical Observatory has a long tradition of synoptic observations. Currently, they are operating one of the RISE/PSPT instruments (see below) at their observatory in Rome.

Since 1996, the Kiepenheuer Insitute provides full-disk 1024 squared Hα images from a 15-cm refractor located on the building of the VTT at Tenerife. Weather permitting, one image per day is recorded through a 0.05-nm Lyot filter.

The Multichannel Flare Spectrograph installed at the Ondrejov Observatory, Czech Republic, provides daily spectrograms using a horizontal telescope with a 23-cm aperture and a large spectrograph. The telescope creates a slit-jaw image that is imaged in Hα while the spectrograph produces simultaneous spectrograms in CaII 854.2 nm and CaII K. Linear polarization can be measured in Hα. White-light and Hα images of limited fields of view are also provided using a 20.5-cm Clark refractor dating from 1858 and a 21-cm refractor with a 0.06 nm Hα filter.

The Debrecen Observatory is famous for its long record of sunspot positions. White-light images are collected daily. These images are used to determine the position and area of each sunspot. A 53-cm coronagraph is also housed at Debrecen, which is used for Hα studies via a Lyot filter. The location does not allow it to be used as a coronagraph.

Kharkov Astronomical Observatory, Ukraine, provides digital full-disk images in HeI 1083 nm, Hα, and CaII K using a spectroheliograph with 2 arcsec pixel size.

The Hida Observatory (see above) also contains the solar Flare Monitor Telescope, which provides Hα images at video rate in the core and the blue and red wings of Hα.

The National Astronomical Observatory of Japan (NAOJ) has two solar observing sites: Mitaka near Tokyo, with several solar telescopes (started in 1938), and Norikura in the Japanese Alps with three coronagraphs (started in 1950). Mitaka is the site of several instruments including the Solar Flare Telescope, which has two 20-cm and two 15-cm refractors, that provide high-resolution observations of vector magnetic field, velocity, white-light, and Hα data. There are several smaller full-disk telescopes. The Norikura site has one 25-cm and two 10-cm coronagraphs, but is used only in the summer due to access difficulties in heavy snow. The coronagraph can also measure Doppler velocities, a unique feature.

The Hiraiso Solar Terrestrial Research Center is a facility operated by the Communications Research Laboratory of the Ministry of Posts and Telecommunications of Japan. It is located north of Tokyo and serves space weather forecasting. Facilities include a 15-cm refractor for full-disk Hα imaging.

The Huairou Solar Observing Station is located on an island inside an artificial lake outside of Beijing and operated by the Beijing Astronomical Observatory of the Chinese Academy of Sciences. A single spar holds five telescopes: The main telescope is an evacuated 60-cm Gregorian with a 5 by 4 arcmin field of view and a nine-channel birefringent filter working between CaII K and Hα. The evacuated 35-cm refractor has a 4 by 6 arcmin field of view and is used to produce photospheric vector magnetograms and Dopplergrams, and chromospheric longitudinal magnetograms and Dopplergrams in Hβ. A 10-cm full-disk instrument measures vector magnetic fields and velocities. A 14-cm refractor observes in Hα either in a full-disk mode or in a restricted 8 by 8 arcmin mode. And finally, an 8-cm telescope provides full-disk CaII K images.

Yunnan Astronomical Observatory, outside of Kunming, and Purple Mountain Astronomical Observatory, outside of Nanjing, are also operated by the Chinese Academy of Sciences, and use identical instruments built at the Nanjing Optical Factory. There are two parallel vacuum refractors with 26-cm apertures, one observing the photosphere and the other one the chromosphere in Hα. Yunnan Observatory also operates the 50-cm evacuated Vacuum Solar Spectral Telescope, which has a Gregorian design and a spectrograph and a field of view of 100 arcsec. Vector-polarimetry can be performed in FeI 630.2 nm and MgI 517.3 nm. Purple Mountain Observatory has a 9-channel visible light spectroheliograph and an aditional 3-channel near-infrared spectroheliograph operating in HeI 1083 nm, CaII 854.2 nm, and Hα.

The Culgoora Solar Observatory in northeast Australia is operated by the Ionospheric Prediction Service of the Australian government. Their main optical data product is daily full disk Hα images obtained with a 12-cm refractor.

The 25-cm and 65-cm telescopes at Big Bear Solar Observatory observe areas the size of active regions. The 25-cm telescope is equipped with a filter-based magnetograph that can record longitudinal as well as vector magnetograms. The 20-cm telescope monitors the whole Sun in Hα every 30 seconds. It is part of a network with observing stations at the Kanzelhöhe Solar Observatory in Austria and the Yunnan Observatory in China.

The Mees Solar Observatory on Haleakala, Maui (Hawaii) contains various instruments operated by the University of Hawaii. Two coronagraphs observe the corona in Hα and Fe XIV. The Haleakala Stokes Polarimeter produces vector magnetograms. The Mees Stokes Video instrument observes active regions in Hα. The Imaging Vector Magnetograph (IVM) uses a Fabry-Perot filter to scan the FeI 630.2 nm line in all four Stokes parameters. The Mees CCD Imaging Spectrograph (MCCD) repeatedly records the spectra of a region. And finally there is the Mees White Light Telescope providing full-disk solar images.

The Mauna Loa Solar Observatory (MLSO) on the island of Hawaii is operated by the High Altitude Observatory in Boulder, Colorado. Instruments include: the 23-cm refractive Mark-IV K-Coronameter that produces images in white light polarization; the Polarimeter for Inner Coronal Studies (PICS) whose removable occulting disk is used to make observations of the limb or disk in Hα; the Chromospheric Helium I Imaging Photometer (CHIP) that records images of the Sun at 1083 nm, as well as at a number of other nearby wavelengths, using a liquid crystal variable retarder Lyot filter once every three minutes; and finally one of the RISE/PSPT telescopes (see below).

The Marshall Space Flight Center Vector Magnetograph Facility was assembled in 1973 to support the Skylab mission and is housed on a 40-foot tower outside of Huntsville, Alabama. The facility added a co-aligned H-alpha telescope in 1989.

The spatial resolution over the 6 x 6 arcmin field of view is now 0.64 arcsec per pixel. It takes 3 minutes for a full vector magnetogram.

Two solar telescope are located at Mt. Wilson, California, the oldest solar observatory in the USA. The 60-foot tower telescope, operated by USC, is part of a worldwide network monitoring helioseismology. At this very instrument, Robert Leighton discovered the 5-minute oscillations in 1960. The 150-foot tower telescope, operated by UCLA, investigates long-term changes of solar magnetic activity.

At the 60 Foot Solar Tower, dopplergrams are obtained with a sodium-based magneto-optical-filter that is switched between the red and blue wings of the sodium D lines. Along with these images obtained with a 1024 by 1024 pixel CCD camera, a smaller, commercial video CCD records in the light of potassium. The second observing system utilizes a commercial video camera that is controlled by a PC. A similar observing system has been installed at the Crimean Astrophysical Observatory (CrAO) in the Ukraine. Under a joint program with UCLA, full disk white light photographs are made daily to continue the long series of such images started at Mt. Wilson nearly 100 years ago.

The top of the 150-ft tower contains a coelostat and an apochromatic triplet objective lens with a focal length of 150 feet. The effective aperture of the telescope is between 415 and 430 mm in diameter, depending on the time of year. Sunspot drawings, Doppler and magnetogram measurements are performed daily. A 20 or 12.5 arc second square aperture is sliced onto the entrance slit of the Littrow spectrograph, which allows for simultaneous observations of FeI 525.0 nm, the Sodium D doublet, NiI 676.8 nm, and Ca II K.

The NSO at Sunspot has several synoptic telescopes. The 40-cm coronagraph in the Evans Solar Facility acquires daily scans around the solar disk at various limb distances in the green, yellow, and red coronal lines. The Hilltop Dome also contains a real-time Hα feed which is available online. One of the RISE/PSPT stations (see below) is there also.

The synoptic observing program at the NSO Kitt Peak Vacuum Telescope is a joint effort supported by NSO/NSF, GSFC/NASA, and formerly SEC/NOAA. The most significant research result from this facility is the provision of synoptic magnetic and chromospheric data to the solar physics community via the internet. The 70-cm vacuum telescope is equipped with a spectrograph that provides daily full-disk photospheric and chromospheric longitudinal magnetograms and full-disk intensity measurements in HeI 1083 nm.

The San Fernando Solar Observatory (SFO) is operated by the Department of Physics and Astronomy of the California State University at Northridge. Its 28-cm vacuum telescope is used for precise full-disk photometry and spectroscopy.

Wilcox Solar Observatory (WSO) was established in 1975 near the Campus of Stanford University. One telescope produces low-angular resolution synoptic magnetic and velocity field measurements and Sun-as-a-star magnetic field measurements.

The Center for Research on Physics/Universidad de Sonora, Hermosillo Sonora, Mexico operates the Estacion de Observacion Solar and the Observatorio "Carl Sagan", currently the only two solar observatories in the country. They observe active regions in the continuum, Hα, and CaII H and K using two heliostats and a 15-cm refractor. Live Hα pictures are available through the web obtained with a 14-cm Maksutov telescope.

The H-Alpha Solar Telescope for Argentina (HASTA) is installed at an altitude of 2370 m at El Leoncito in the Argentinian Cordillera de los Andes. The 10-cm HASTA provides daily full Sun disk images in Hα using a 0.03-nm Lyot filter and a 1k square CCD camera. The camera acquires images every 5 s. Each image is analyzed in real time in order to detect rapid changes in the overall intensity. If no change is detected, the algorithm stores one image every 1.5 min (patrol mode). On the other hand, if a fast change is detected, the camera automatically switches into the high-speed mode. In this mode, the telescope can take and store full-frame images up to every 3 seconds.

At the same location the Mirror Coronagraph for Argentina (MICA) observes the solar corona from 1.05 to 2.0 solar radii in the green and red coronal lines as well as Hα. It provides high temporal resolution observations of transient phenomena and studies the structure and evolution of solar prominences and coronal streamers. The instrument, with a 6-cm aperture and a resolution of about 8 arcsec, is almost identical to the LASCO C1 instrument on SOHO.

20.2.1.3 Synoptic networks

The Birmingham Solar Oscillations Network (BiSON, see Chaplin *et al.*, 1996) measures the integrated solar velocity using a potassium resonance cell at six stations around the globe (Mt. Wilson, California; Las Campanas, Chile; Tenerife, Spain; Sutherland, South Africa; Carnarvon, Australia; and Narrabri, Australia). Another network called International Research on the Interior of the Sun (IRIS, see Fossat 1991) is similar in concept but uses sodium cells and operates at several places around the globe.

The National Solar Observatory's Global Oscillation Network Group (GONG, e.g. Leibacher 1999) has imaging Michelson interferometers at six stations around the globe (Big Bear, California; Mauna Loa, Hawaii; Learmonth, Australia; Udaipur, India; Tenerife, Canary Islands; Cerro Tololo, Chile). The team has just finished upgrading its instruments from the former 256 by 256 pixel format to 1024 by 1024 CCD cameras running at 60 Hz. In addition, the new system provides magnetograms at this resolution once every minute.

The Taiwan Oscillation Network (TON), built by and operated from the Tsing Hua University, Hsinchu, Taiwan, is a ground-based network to measure solar K-line intensity oscillations. Telescopes are located at Tenerife, Canary Islands, Huairou Solar Observing Station near Beijing, China, Big Bear Solar Observatory, USA, and Tashkent Observatory.

Instruments of the Experiment for Coordinated Helioseismic Observations (ECHO) network are located at Mauna Loa and Tenerife. They are based on a Potassium resonance filter and are operated by the High Altitude Observatory in Boulder.

The US Air Force Solar Observing Optical Network (SOON) is a network of five identical telescopes at selected sites around the world intended to provide real time solar activity information for the prediction of space weather. Data from these telescopes flows in real time to the Space Environment Center of the NOAA in Boulder for space weather forecasts that are also available publically. The main data type is H-alpha images of selected active regions.

The Radiative Inputs of the Sun to Earth (RISE) initiative runs a network of three Precision Solar Photometric Telescope (PSPT) instruments, which produce seeing-limited full disk digital (2048x2048) images in CaII K (393nm $\pm$ 0.3 nm), blue (408-412 nm) and red (605-610 nm) continuum images with 0.1% photometric precision once per hour. They are located at Mauna Loa, Hawaii, Sunspot, New Mexico, and Rome, Italy. The instruments were designed and built at the National Solar Observatory.

The Global H-alpha Network is composed of instruments at Big Bear, California, Kanzelhöhe, Austria, and Yunnan, China. Each instrument is equipped with a 2k by 2k detector that records images once per minute.

20.2.1.4 Synoptic radio telescopes

The Nançay Radio Observatory south of Paris, France, operated by the Observatoire de Paris, houses a radioheliograph, which consists of 19 antennas along a 3200-m east-west baseline and 24 antennas over a 1250-m north-south baseline. It can image the solar corona between 0.6 and 2 m wavelength at up to five images per second.

The Siberian Solar Radio Telescope, located 220 km from Irkutsk consists of two crossed arrays of 128 2.5-m antennas each, giving baselines of 622 m each. Operating at 5.7 GHz, it observes processes in the corona. Observations as short as 14 ms with a spatial resolution of down to 15" can be achieved.

The Nobeyama Radioheliograph in Japan consists of 84 parabolic antennas with 80 cm diameter, sitting on lines 490 m long in the east/west and of 220 m long in the north/south direction. Operational since 1992, it works at 17 and 34 GHz and covers the full disk with a resolution of 5 to 10 arcsec, depending on the frequency. Normally, the temporal cadence is 1 second, but 0.1 seconds can be achieved to study the development of flares.

The Owens Valley Solar Array, California, USA, consists of two large, 27-m antennas and three small 2-m antennas operating between 1 and 18 GHz. It can measure left- and right-hand circularly polarized and linearly polarized radiation within less than 20 ms.

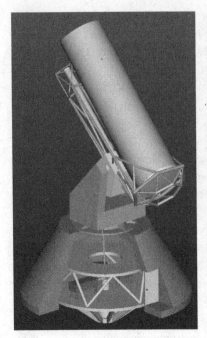

Figure 20.1. Concepts for ATST (left) and GREGOR (right).

20.2.2 Future plans

20.2.2.1 General purpose telescopes

The Advanced Technology Solar Telescope (ATST) (Figure 20.1) is the largest solar facility currently planned. The National Solar Observatory and its partners have just started the design and development phase with first light being planned for late this decade. This 4-m telescope will provide an angular resolution of 0.1 arcsec, or better, a large photon flux for precise magnetic and velocity field measurements, and access to a broad set of diagnostics from 0.3 to 28 μm. A site survey is in progress.

The German 1.5-m GREGOR telescope (Figure 20.1) will replace the Gregory-Coudé telescope at Tenerife. Equipped with adaptive optics and a polarimetry system, it will provide high-precision measurements of the magnetic and velocity fields in the photosphere and chromosphere with high spatial resolution.

The New Swedish Vacuum Solar Telescope, which replaces the old solar telescope at La Palma, should be operational by early 2002. The new telescope has a 97 cm fused silica lens that also serves as the entrance window to the vacuum tube. Various secondary optical systems can be used to optimize for particular observing requirements. The integrated adaptive optics system will allow near-diffraction-limited imaging during a reasonable fraction of the time.

Solar-C is a 0.5-m off-axis reflecting coronagraph that is currently being commissioned on Haleakala adjacent to the Mees Solar Observatory.

Figure 20.2. Concept of SOLIS on top of the Kitt Peak Vacuum Telescope building.

20.2.2.2 Synoptic telescopes

SOLIS, the Synoptic Optical Long-term Investigations of the Sun, is currently being built by the US National Solar Observatory. It is a remote-controlled, autonomous system operating similar to a satellite. It records synoptic core data as well as user-driven observations. It allows event-triggered observing (e.g. flares or rocket flights) and features automatic scheduling of observations with proposal submission and data access over the Internet. Three instruments are attached to a single equatorial mount (see Figure 20.2). SOLIS will be located on top of the present Kitt Peak Vacuum Telescope and become operational in 2002. A network with two additional SOLIS instruments is in the early planning stage.

The Vector Spectro-Magnetograph (VSM) is a 50-cm helium-filled telescope with an active secondary mirror. The entrance slit of the Littrow spectrograph is scanned in declination to provide 2048 by 2048 pixel scans of the full solar disk. Two cameras record over 90 spectra per second, which are then analyzed to obtain the Stokes parameters. The VSM is capable of recording vector magnetograms in FeI 630.2 nm, deep longitudinal magnetograms in FeI 630.2 nm, longitudinal magnetograms in Ca II 854.2 nm, and intensity in HeI 1083.0 nm.

The Full-Disk Patrol (FDP) is a 14-cm refractor with a fast tip-tilt mirror for image motion compensation. It features a 0.25 Å bandpass tunable birefringent filter working from CaII K to Hα and a separate HeI 1083.0 nm filter. Two 2048 by 2048 pixel CCD cameras will provide accurate intensity and Doppler measurements at about one full-disk image per second.

The Integrated Sunlight Spectrometer (ISS) consists of two 8-mm telescopes feeding fibers. One of the two beams goes through an iodine absorption cell for very precise Doppler measurements. The fibers are fed into a commercial 2-m double-pass spectrograph that provides a choice of spectral resolution: 30,000 and 300,000. A 1024 by 1024 pixel CCD camera is mounted onto a movable stage for accurate flat-fielding.

An improved version of the SOON network, ISOON, will transmit Hα images and longitudinal magnetograms in near-real time to the USAF and the Space Environment Center of NOAA for use in space weather forecasting. ISOON is an upgrade of four of the existing SOON facilities with a tunable, dual Fabry-Perot filter system and a 2048 by 2048 CCD camera. The prototype is currently being built at the National Solar Observatory/Sacramento Peak and should become operational in 2002.

20.2.2.3 Radio telescopes

The Frequency-Agile Solar Radiotelescope (FASR) is an imaging radio interferometer operating simultaneously between 0.3 and 30 GHz using an array of about 100 antennas. It will produce high-quality images of the Sun with high spatial resolution (1 arcsec at 20 GHz), high spectral resolution, and high temporal resolution (better than 1 s). In so doing, it will produce a continuous, three-dimensional record of the solar atmosphere from the chromosphere up into the mid-corona. A proposal for a design and development phase has just been submitted in the US.

20.3   Current and planned suborbital missions

The Solar Extreme-ultraviolet Rocket Telescope and Spectrograph (SERTS) has flown once a year since 1995. It obtains spatially resolved spectra and spectroheliograms between 17 and 45 nm with a spectral resolution of about 10,000 and a spatial resolution of 5 arcsec. With a typical suborbital rocket flight, about 7 minutes of observing time are available per flight.

The Flare Genesis Experiment, built by the Johns Hopkins University, has flown twice around the South Pole providing uninterrupted observations of the Sun for almost 2 weeks. The telescope is a 80-cm Cassegrain telescope with a tunable Fabry-Perot filter. While no further flights of this instrument are currently planned, the German-US-Spanish Sunrise project hopes to fly a 100-cm telescope in 2005/2006. Sunrise uses SiC primary and seconday mirrors and a leight-weight

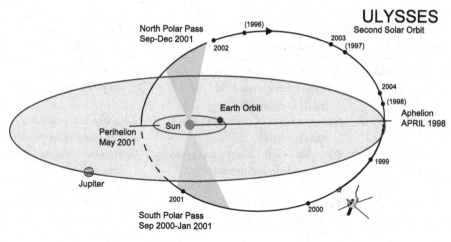

Figure 20.3. The second orbit of Ulysses around the Sun began in April 1998 when Ulysses reached aphelion. The first polar passes occurred in mid 1994 (south pole) and mid 1995 (north pole), following the successful gravity-assist manoeuvre at Jupiter in February 1992.

telescope that feeds two vectormagnetographs, one spectrograph-based and one filter-based system.

20.4 Space missions

20.4.1 In operation

20.4.1.1 Ulysses

The launch on 6 October 1990 of Ulysses[†] (Wenzel *et al.*, 1990), a cooperative mission between ESA and NASA, marked the start of a new era in the study of the heliosphere. For the first time it was possible to explore the uncharted third dimension of the heliosphere by flying over the poles of the Sun (Figure 20.3).

The prime mission objective of Ulysses is to study, as a function of solar latitude, the properties of the solar wind, the structure of the Sun/wind interface, the heliospheric magnetic field, solar radio bursts and plasma waves, solar X-rays, solar and galactic cosmic rays, and the interstellar/inter-planetary neutral gas and dust. To address these objectives, Ulysses carries a set of 9 complimentary *in situ* instruments, which are summarized in Table 20.3.

Ulysses is a spinning spacecraft (5 rpm). It has X- and S-band communication systems and is powered by a Radio-isotope Thermoelectric Generator (RTG). Its total weight is 367 kg (55 kg scientific instruments).

[†] http://helio.estec.esa.nl/ulysses/

Table 20.3. *Ulysses instruments*

Investigation	Measurements	Instrumentation
Magnetic Field	Spatial/temporal variations of helio-spheric mag. field: ± 0.01 to ± 44000 nT	Triaxial vector helium and fluxgate magnetometers
Solar-wind plasma	Energy spectra of solar wind ions and electrons	2 electrostatic analysers with channel multipliers
Solar Wind Ion Composition	Elemental + ionic-charge composition, temperature and mean velocity	Electrostatic analyser with time-of-flight and energy and electrons
Low Energy Ions and Electrons	Energetic ions from 50keV to 5Mev, electrons from 30keV to 300keV	Two sensor heads with five solid-state detector telesc.
Energetic Particle Composition and Interstellar Gas	Composition of energetic ions, interstellar neutral helium	4 solid-state detector telescope
Cosmic Ray / Solar Particles	Cosmic rays and energetic solar particles (0.3–600 MeV/nuc); electrons 4–2000 MeV	5 solid-state detector telesc.; 1 double Cerenkov and semi-conductor telescope for elec.
Unified Radio and Plasma Waves	Plasma waves, radio bursts, electric field: 0–60 kHz + 1–940 kHz magnetic field: 10-500 Hz	72 m radial dipole antenna 8 m axial monopole antenna two-axis search coil
Solar X-rays/ Cosmic Gamma-Ray Bursts	Solar flare X-rays and γ-ray bursts (5–150 keV)	2 Si solid-state detectors 2 CsI scintillation crystals
Cosmic Dust	Particulate matter (10^{-16}–10^{-7} g)	Multicoincidence impact detector with channeltron

20.4.1.2 Yohkoh

Yohkoh[†] ("Sunbeam") is a mission of the Japanese Institute for Space and Astronautical Sciences (ISAS) with participation by NASA and the UK to study solar flares, especially high-energy phenomena observed in the X- and gamma-ray ranges (Ogawara *et al.*, 1991). Named SOLAR-A before launch, it is the second Japanese spacecraft dedicated to flare studies after the Hinotori mission which was launched close to

[†] http://isass1.solar.isas.ac.jp/ and http://www.lmsal.com/SXT/

the maximum of cycle 21 in February 1981. *Yohkoh* was launched on 31 August 1991 from Kagoshima Space Center on board a Mu-IIIs-V rocket and was injected into a slightly elliptical low-earth orbit of about 600 km altitude, 31° inclination, and 97 min period.

The spacecraft has dimensions of about $100 \times 100 \times 200$ cm and weighs about 400 kg. There are four instrument packages on *Yohkoh*: the Soft X-ray Telescope (SXT), the Hard X-ray Telescope (HXT), the Wide Band Spectrometer (WBS), and the Bragg Crystal Spectrometer (BCS).

SXT (Tsuneta *et al.*, 1991) is a grazing incidence telescope with a diameter of 23 cm and 1.53 m focal length which forms X-ray images in the 0.25 to 4.0 keV range (3–60 Å) on a 1024×1024 virtual phase CCD detector. A selection of thin metallic filters located near the focal plane provides the capability to separate the different X-ray energies for plasma temperature diagnostics.

HXT (Kosugi *et al.*, 1991) is a Fourier synthesis telescope providing hard X-ray images simultaneously in four energy bands (15–24–35–57–100 keV) with a temporal resolution up to 0.5 s. It consists of 64 bi-grid modulation subcollimators, each equipped with a NaI(Tl) crystal and a photomultiplier detector. The field-of-view (FOV) of HXT is about 35×35 arcmin, i.e. covering the whole Sun. HXT can therefore detect hard X-rays flares regardless of their position on the Sun without re-pointing the spacecraft. The basic image synthesis FOV of HXT is 2.1 by 2.1 arcminutes with the angular resolution as high as approximately 5 arcseconds.

WBS (Yoshimori *et al.*, 1991) consists of four detectors: the Soft X-ray Spectrometer (SXS), the Hard X-ray Spectrometer (HXS), the Gamma-Ray Spectrometer (GRS), and the Radiation Belt Monitor (RBM).

BCS (Culhane *et al.*, 1991) employs four bent germanium crystals with position-sensitive proportional counters. It views the whole Sun and observes the resonance line complexes of H-like Fe XXVI, Ca XIX, and S XV in four narrow wavelength ranges with a resolving power ($\lambda/\Delta\lambda$) of between 3000 and 6000.

20.4.1.3 Wind

Wind[†] and Polar (Acuna *et al.*, 1995) form the Global Geospace Science (GGS) initiative, which is part of the International Solar Terrestrial Physics (ISTP) programme. The primary scientific objective of GGS is to measure the mass, momentum and energy flow and their time variability throughout the solar wind–magnetosphere–ionosphere system that compromises the geospace environment. Wind's role in this endeavour is to provide complete plasma, energetic particle, and magnetic field input for atmospheric and ionospheric studies and to investigate basic plasma processes occuring in the near-Earth solar wind. Polar, Wind's "sister spacecraft" focuses on plasma processes of the Earth's polar regions inside the magnetosphere.

[†] http://www-istp.gsfc.nasa.gov/istp/wind/

Table 20.4. *Wind instruments*

Investigation	Instrument	Measurements
MFI	Magnetic Field Investigation	d/c magnetic fields
WAVES	Radio and Plasma Wave Experiment	a/c electric/magnetic fields
SWE	Solar Wind Experiment	Mass, energy, direction of low energy ions and e^- (7 eV–22 keV)
3-DP	3-D Plasma	Distribution and energy of ions and e^- (3 eV–30 keV and 20 keV–11 MeV)
EPACT	Energetic Particles: Acceleration, Composition and Transport	Mass, energy, direction of ions (0.2–500 MeV)
SMS	Solar wind/Mass Suprathermal ion compsition studies	Mass, energy, direction of ions (0.5–500 MeV)
TGRS	Transient Gamma Ray Spectrometer	High spectral resolution γ-ray detector in range 15 keV–10 MeV
KONUS	Gamma-Ray Burst Experiment	High time resolution (2 ms) gamma-ray detector

Wind was launched on 1 November 1994 on board a Delta II from Cape Canaveral Air Force Station. It is a cylindrically shaped spinner (20 rpm), approximately 2.4 m in diameter and 1.8 m high. Total mass is 1150 kg.

The scientific payload of Wind includes eight instruments consisting of 24 separate sensors. Table 20.4 provides a summary of these instruments, which are described in more detail in Russell *et al.* (1995).

20.4.1.4 SOHO

The Solar and Heliospheric Observatory[†] (SOHO), is a project of international co-operation between ESA and NASA to study the Sun, from its deep core to the outer corona, and the solar wind (Domingo *et al.*, 1995). It is probably the most comprehensive and complex solar space mission after Skylab, which carried a sophisticated complement of solar telescopes on the Apollo Telescope Mount in 1973 (Tousey, 1977). The three principal scientific objectives of the SOHO mission are:

i) Study of the solar interior, using the techniques of helioseismology

ii) Study of the heating mechanisms of the solar corona, and

iii) Investigation of the solar wind and its acceleration processes.

SOHO carries a complement of twelve sophisticated instruments (Table 20.5), developed and furnished by twelve international PI consortia involving 39 institutes

[†] http://soho.nascom.nasa.gov/ or http://soho.estec.esa.nl/

Table 20.5. *SOHO Instruments*

Investigation	Measurements	Technique
HELIOSEISMOLOGY		
GOLF	Global Sun velocity oscillations ($\ell = 0\text{-}3$)	Na-vapour resonant scattering cell, Doppler shift
VIRGO	Low degree ($\ell = 0\text{-}7$) irradiance oscillations and solar constant	Global Sun and low resolution (12 pixels) imaging, active cavity radiometers
MDI	Velocity oscillations, harmonic degree up to 4500; full disk + high res magnetograms	Michelson interferometer, angular resolution: 1.3 and 4"
SOLAR ATMOSPHERE REMOTE SENSING		
SUMER	Plasma flow characteristics (T, density, velocity) chromosphere through corona	Normal incidence spectrometer, 500-1600 Å, spectral res. 20000-40000, angular res. ≈ 1.3"
CDS	Temperature and density: transition region and corona	Normal and grazing incidence spectrometers, 150-800 Å, spectr. res. 1000–10000, angular res. ≈ 3"
EIT	Evolution of chromospheric and coronal structures	Full disk images (42' $\times$ 42' with 1024 $\times$ 1024 pixels) in He II, Fe IX, Fe XII and Fe XV
UVCS	Ion temperature, velocities and abundances in corona	Profiles and/or intensity of selected EUV lines (Ly α, O VI, etc.) between 1.3 and 6 $R_\odot$
LASCO	Evolution, mass, momentum, energy transport in corona	2 externally occulted coronagraphs (2.5-30 $R_\odot$)
SWAN	Solar wind mass flux anisotropies and its temporal variations	Scanning telescopes with hydrogen absorption cell for Ly-α
SOLAR WIND 'IN SITU'		
CELIAS	Energy distribution and composition (mass, charge, charge state) (0.1-1000 keV/e)	Electrostatic deflection, time-of-flight measurements solid state detectors
COSTEP	Energy distribution of ions (p, He) 0.04-53 MeV/n and electrons 0.04-5 MeV	Solid state and plastic scintillator detectors
ERNE	Energy distribution and isotopic composition of ions (p - Ni) 1.4-540 MeV/n and electrons 5-60 MeV	Solid state and plastic scintillator detectors

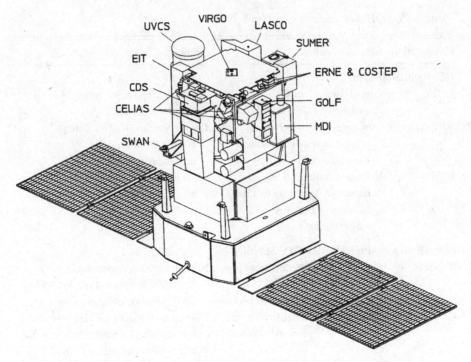

Figure 20.4. SOHO spacecraft schematic view (from Domingo *et al.*, 1995).

from fifteen countries (Belgium, Denmark, Finland, France, Germany, Ireland, Italy, Japan, Netherlands, Norway, Russia, Spain, Switzerland, United Kingdom, and the United States). Detailed descriptions of all the twelve instruments on board SOHO as well as a description of the SOHO ground system, science operations and data products together with a mission overview can be found in Fleck *et al.* (1995).

SOHO is a 3-axis stabilized spacecraft (Figure 20.4) with dimensions of 4.3 × 2.7 × 3.7 m and mass of about 1860 kg (payload 655 kg). It was launched onboard an Atlas II-AS on 2 December 1995 and was injected into a halo orbit around the L1 Lagrangian point on 14 February 1996. From there it has been providing uninterrupted observations of the Sun and the solar wind, 24 hours a day, 7 days a week, except during the SOHO incident in the summer of 1998, when the spacecraft was lost for nearly three months because of ground errors. The successful recovery of the SOHO mission (Vandenbussche, 1999), from confirming its position and determining its role rate with Arecibo radar, to acquiring first short bursts of telemetry and thawing the tank of frozen hydrazine to the successful de-spin and re-acquisition of Sun-pointing attitude is certainly one of the most dramatic rescues in space science. The fact that both the spacecraft as well as all twelve instruments survived this ordeal with minor scars constitutes a great tribute to the skill, dedication and professionalism of the scientists and engineers who designed and built them.

20.4.1.5 ACE

The Advanced Composition Explorer[†] (ACE; Stone *et al.*, 1998) was launched on
25 August 1997 into an orbit around the L1 Lagrangian point at 240 R_E sunward
of the Earth. ACE is a spinning spacecraft (5 rpm). Its main body is a cyclinder 2 m
in diameter and 1.9 m in length. With the solar panels and magnetometer booms
extended, its wingspan measures about 8.3 m. Its total weight is 750 kg, of which
156 kg are scientific instruments.

The prime scientific objective of ACE is to determine and compare the elemental
and isotopic composition of several distinct samples of matter, including the solar corona, the interplanetary medium, the local interstellar medium, and galactic
matter. To address this objective, ACE carries a complement of nine *in situ* instruments, including six high resolution spectrometers that measure the elemental, isotopic, and ionic charge state composition of nuclei from H to Ni ($1 \leq
Z \leq 28$) from solar wind energies (≈ 1 keV/nuc) to galactic cosmic ray energies
(≈ 500 MeV/nuc), and three instruments that provide the heliospheric context for ion
composition studies by monitoring the state of the interplanetary medium. ACE also
provides real-time solar wind measurements to NOAA for use in forecasting space
weather.

20.4.1.6 TRACE

The Transition Region and Coronal Explorer[†] (TRACE, Handy *et al.*, 1999) is a
NASA Small Explorer (SMEX) that images the solar photosphere, transition region
and corona with unprecedented spatial (0.5 arcsec pixel size) and temporal resolution
and continuity. TRACE was lauched on 2 April 1998 from Vandenberg Air Force
Base on a Pegasus-XL launch vehicle, and injected into a sun-synchronous polar
orbit of approximately 625 km altitude and 97.8° inclination.

TRACE carries a single instrument (Figure 20.5). Total mass of the instrument
is 36 kg. The 30 cm aperture TRACE telescope uses four normal-incidence coatings for the EUV and UV on quadrants of the primary and secondary mirrors. The
segmented coatings on solid mirrors form identically sized and perfectly coaligned
images at selected temperatures over the range from 6000 K to 10 MK (Fe IX/X 171Å,
Fe XII/XXIV 195Å, Fe XV 284Å, Lα, C IV 1550Å, UV continuum at 1600 and
1700Å, white light con tinuum at 5000Å). Pointing is internally stabilized to 0.1
arc second against spacecraft jitter. A 1024×1024 CCD detector collects images
over an 8.5×8.5 arc minute field-of-view (FOV). A powerful data handling computer enables very flexible use of the CCD array including adaptive target selection,
data compression, and fast operation for a limited FOV. Science telemetry amounts
to about 700 Mbyte per day.

[†] http://www.srl.caltech.edu/ACE/
[†] http://vestige.lmsal.com/TRACE/

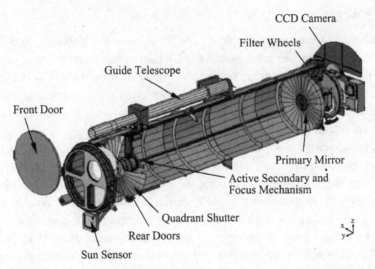

Figure 20.5. Isometric diagram of the TRACE telescope (from Handy *et al.*, 1999).

20.4.1.7 GOES/Solar X-ray imager (SXI)

The next generation of US weather satellites (GOES) have a small X-ray imager mounted on the solar array yoke. SXI[†] provides continuous monitoring of the Sun in X-rays. The data are used by NOAA and the USAF to determine when to issue forecasts and alerts of space weather conditions that may interfere with ground and space systems.

SXI provides full-disk 512 by 512 pixel images in 10 wavelength bands from 6 to 60 Å at one-minute intervals. The expected useful life is 3 years with a goal of 5 years. The first SXI instrument was built by NASA's Marshall Space Flight Center, subsequent models were built by Lockheed-Martin. The first satellite equipped with SXI was launched in the summer of 2001, and subsequent ones should become operational in 2002 and 2003.

20.4.1.8 CORONAS-F

CORONAS-F[‡] was launched on 31 July 2001 on a Cyclone-3 rocket from Russia's Northern Cosmodrome in Plesetsk and was placed into a polar orbit with an altitude of about 500 km and 83° inclination. After CORONAS-I, which was launched in 1994 and provided data until 1995, CORONAS-F is the second mission in the CORONAS project ("Complex Orbital Near-Earth Observations of the Solar Activity"), the main scientific objectives of which are to better understand solar activity, in particular flares, and the solar interior.

[†] http://www.sec.noaa.gov/sxi
[‡] http://www.izmiran.rssi.ru/projects/CORONAS/

CORONAS-F carries a complement of 15 instruments. Among them are 10 X-ray spectrometers and imagers, 2 UV instruments, a radiospectrometer, a coronagraph and several full disk photometers. Total mass of the satellite is 2260 kg, 395 kg of which make up the scientific payload.

20.4.1.9 Genesis

Genesis[†] is a 3-year NASA mission to capture an integrated sample of solar wind material and return it to Earth for ground-based isotopic and elemental analysis. After its launch on 8 August 2001 the spacecraft will spend $\approx$2.5 years at the L1 Lagrangian point collecting solar wind before returning with a capsule-style re-entry in September 2004. Its key scientific objective is to obtain precise measures of solar isotopic abundances and greatly improved measures of solar elemental abundances. In particular, it will measure oxygen isotopes at a 0.1% level (presently known from *in situ* measurements at a 10% level). The hope is that this will allow to distinguish between two important theories for the solar system oxygen heterogenity and thus provide crucial new information about the origin and formation of the solar system.

The Genesis spacecraft has two primary instruments which passively collect solar wind: (a) "collector arrays", a set of panels, each of which can deploy separately to sample different kinds of solar wind ("regimes"); (b) a "concentrator", an electrostatic mirror which concentrates ions of mass 4 through 25 by a factor of 20 by focusing them onto a 6 cm diameter target.

20.4.1.10 HESSI

The High Energy Solar Spectroscopic Imager[‡] (HESSI; Lin, 2000) is a NASA SMEX mission to explore the basic physics of particle acceleration and explosive energy release in solar flares. It was launched on 5 February 2002 and was placed in a low-earth orbit of 600 km altitude and 38° inclination. It is a spin-stabilized spacecraft (15 rpm), weighing 296 kg, 130 kg of which make up its single instrument, a Fourier synthesis hard X-ray and Gamma-ray imaging spectrometer. It provides high resolution images of flares over a broad range from soft X-rays (3 keV) to gamma-rays (20 MeV), and at the same time also provides high spectral resolution information on a point-by-point basis. To accomplish this, it uses two new complementary technologies: fine grids to modulate solar radiation for the imaging, and germanium detectors to measure the energy of each photon very precisely.

The imaging is achieved with nine pairs of widely spaced, fine grids made of tungsten and molybdenum that modulate the solar X-ray flux as the spacecraft rotates at 15 rpm. HESSI always sees the full Sun ($\approx$ 1°). Spatial resolution varies from 2 arcsec at 100 keV to 36 arcsec at 1 MeV. Temporal resolution is as good as

[†] http://genesismission.jpl.nasa.gov/
[‡] http://hesperia.gsfc.nasa.gov/hessi/

tens of milliseconds for a basic image. A detailed image takes 2 s (half a rotation of the spacecraft).

High-resolution spectroscopy is achieved with nine segmented, hyperpure (n-type) cooled (75 K) germanium cyrstals, 7.1 cm in diameter and 8.5 cm long. Spectral resolution is about 1 keV for photons up to 100 keV, 3 keV up to energies of 1 MeV, increasing to 5 keV at 20 MeV.

20.4.2 In development and under study

20.4.2.1 Solar-B

Solar-B[†] is an ISAS mission with US and UK participation as follow-on to the highly successful *Yohkoh* mission (Antiochos *et al.*, 1997). Target launch date is August 2005. Its key scientific objectives are to study the generation and transport of magnetic fields and their role in heating and structuring the chromosphere and corona, and in eruptive events and flares. The 3-axis stabilized satellite (total mass ≈ 900 kg) will be launched into a polar sun-synchronous orbit with inclination 97.9°, altitude approximately 600 km and a nominal lifetime in orbit of three years.

Its scientific payload comprises three instruments. At the heart of Solar-B is a diffraction limited 50-cm aperture Solar Optical Telescope (SOT), with its Focal Plane Package (FPP) designed for high resolution photospheric and chromospheric imaging and spectro-polarimetry. In addition there are two coronal instruments, the X-Ray Telescope (XRT) and the Extreme-ultraviolet Imaging Spectrometer (EIS).

The SOT will preserve image quality and ≈ 150 km diffraction limited resolution from 3880 – 6600 Å. It will feed the FPP which consists of a narrow-band tunable birefringent filter imager, a broad-band interference filter imager, and a spectro-polarimeter (SP), essentially a space version of the HAO Advanced Stokes Polarimeter. The image will be stabilized to better than 0.02 arcsec by a correlation tracker and active tilt mirror. The SP will make vector magnetic measurements from Stokes spectra of the Fe I lines 6301 and 6302 Å, with 0.16 arcsec pixels and FOV up to 164×328 arcsec. The broad-band system takes diffraction-limited images with 0.05 arcsec pixels in the Ca II H line, CN and G bandheads, and continuum bands. The narrow-band system makes filtergrams, magnetograms, Dopplergrams, and Stokes images in several photospheric lines, Mg b, and Hα. It has 0.08 arcsec pixels and FOV same as that of the SP. The SP and filter imagers will usually observe simultaneously on the same target region.

XRT is an enhanced version of the SXT instrument on *Yohkoh*, providing atmospheric images at wavelengths from 2 to 60 Å. The image scale is 1 arcsec/pixel which is a factor 2.5 better than that of SXT, and it will respond to a broader range of plasma temperatures. XRT is a grazing-incidence (GI) modified Wolter I X-ray telescope, of 35 cm inner diameter and 2.7 m focal length. The 2048x2048 back-illuminated CCD

[†] http://science.msfc.nasa.gov/ssl/pad/solar/solar-b.stm

has 13.5 μm pixels, corresponding to 1.0 arcsec and giving full Sun field of view. A small optical telescope using the same CCD will provide visible light images for coalignment with the SOT.

EIS consists of a multilayer-coated off-axis telescope mirror (150 mm aperture) and a multilayer-coated toroidal grating spectrometer. It uses two back-thinned CCDs with 13.5 μm pixels as detectors. The mirror and spectrometer combined will have a spatial resolution capability of 2 arcsec while the plate scale is 1"/pixel. The spectral resolution is good enough to allow the measurement of velocity to ± 3 km/s. Half of each optic is coated to optimize reflectance at 170 – 210 Å, and the other half to optimize reflectance at 250 – 290 Å. Each wavelength range is imaged onto a separate back-thinned CCD detector of high quantum efficieny ($\approx 80\%$) for the chosen wavelenghts.

20.4.2.2 STEREO

The principal scientific objective of NASA's Solar-TErrestrial RElations Observatory[†] (STEREO; Rust *et al.*, 1998) is to understand the origin and consequences of coronal mass ejections (CMEs). By using two identically instrumented spacecraft, one drifting ahead of Earth and one behind, it will provide new perspectives on the structure of the solar corona and CMEs. It will obtain simultaneous images of the Sun from the two spacecraft and build a 3-D picture of CMEs and the complex structures around them. It will also study the propagation of disturbances through the heliosphere and their effects at Earth orbit. STEREO is a multilateral international collaboration involving participants from France, Germany, the United States, and the United Kingdom. The two spacecraft carry four identical instrument suites:

- SECCHI, the Sun Earth Connection Coronal and Heliospheric Investigation, which includes four instruments: two white-light coronagraphs (covering the range 1.25–4 $R_\odot$, and 2–15 $R_\odot$), an EUV imager (full-disk, 1" pixels) and a heliospheric imager (an externally occulted coronagraph that can image the heliosphere from 12 $R_\odot$ to beyond Earth's orbit).
- STEREO/WAVES (SWAVES), an interplanetary radio burst tracker that will trace the generation and evolution of traveling radio disturbances from the Sun to the orbit of Earth.
- IMPACT, the In situ Measurements of Particles and CME Transients investigation, which will measure the 3-D distribution and plasma characteristics of solar energetic particles and the local vector magnetic field.
- PLASTIC, the PLAsma and SupraThermal Ion and Composition experiment, which will provide plasma characteristics of protons, alpha particles and heavy ions.

[†] http://sd-www.jhuapl.edu/STEREO/

The STEREO Mission spacecraft are expected to be launched in November of 2005. The two spacecraft will slowly drift apart in ecliptic longitude with a total separation of 45° after the first year and 90° after the second year. The mission is expected to last two years at minimum, hopefully five years.

20.4.2.3 Space Solar Telescope – SST

The Space Solar Telescope (SST) was proposed in 1992 by Guoxiang Ai in China, and has subsequently been considered for a bilateral collaborative project between China and Germany (Schmidt and Ai, 1998). Its key scientific objectives are to measure magnetic fields in the photosphere and chromosphere at very high spatial resolution, to determine their 3-D topology, and to understand the heating, cooling and expansion of the corona in the context of the underlying magnetic field structure. The principal instrument of SST will be a 1-m diffraction limited optical telescope for high resolution magnetometry. It will be complemented by a full disk Hα and white light telescope, a bundle of high resolution EUV imagers, a coronagraph, a wide band X-ray and Gamma-ray spectrometer, and a solar and interplanetary radio spectrometer. Total spacecraft mass is estimated to be 2000 kg. Target orbit is a sun-synchronous, nearly polar orbit of 730 km altitude and 98.3° inclination.

20.4.2.4 SDO

The Solar Dynamics Observatory[†] (SDO, Hathaway *et al.*, 2001) is the first cornerstone mission in NASA's Living With a Star (LWS) programme, which is an initiative "to develop the scientific understanding necessary to effectively address those aspects of the coupled Sun-Earth system that directly affect life and society". SDO's objective in this programme is to understand the nature and source of solar variability that affects life and society. Key questions to be addressed by SDO are: What mechanisms drive the quasi-periodic 11-year cycle of solar activity? How is active region magnetic flux synthesized, concentrated, and dispersed across the solar surface? Where do the observed variations in the Sun's total and spectral irradiance arise, and how do they relate to the magnetic activity cycle? Is it possible to make accurate and reliable forecasts of space weather and climate?

SDO will carry three instrument suites:

- Helioseismic and Vector Magnetic Imager (HVMI), a high resolution helioseismograph/vectormagnetograph that will provide stabilized 1"-resolution full-disk Doppler velocity and line-of-sight magnetic flux images every 45 seconds, and vector-magnetic field maps every 90 seconds. HVMI has significant heritage from the SOHO MDI instrument with enhancements to provide higher resolution (e.g. 4k×4k detectors), higher

---

[†] http://lws.gsfc.nasa.gov/lws_missions_sdo.htm

cadence, and the addition of a second channel to provide full Stokes polarization measurements.

- Solar-Heliospheric Activity Research and Prediction Program (SHARPP), which comprises two closely integrated instrument suites: the Atmospheric Imaging Assembly (AIA) and a white light coronagraph (KCOR). KCOR is identical to the STEREO/SECCHI/COR2 coronagraph (2.5–15 $R_\odot$), itself an advanced version of the LASCO/C2 instrument. The AIA package contains seven telescopes to simultaneously image the full solar disk over a temperature range from 0.02 to 3 MK with high spatial resolution (1.3", 4k×4k detectors) and with very fast cadence (10 sec). The 7 bandpasses are: Ly-α 1215 Å, He II 304 Å, O V 629 Å, Ne VII 465 Å, Fe XII 195 Å (includes Fe XXIV), Fe XV 284 Å, and Fe XVI 335 Å. The 7 telescopes are grouped by instrumental approach: (1) MAGRITTE filtergraphs (all normal incidence telescopes): 5 multilayer "EUV channels", with bandpasses ranging from 195 to 500 Å, and one Ly-α channel; (2) SPECTRE: a zero-dispersion spectroheliograph to image the transition region in O V 629 Å ("soft EUV channel").

- EUV Variability Experiment (EVE), which will measure the solar EUV irradiance with unprecedented spectral resolution, temporal cadence, accuracy, and precision. EVE consists of three subsytems: The Multiple EUV Grating Spectrograph (MEGS) measures the 40–1200 Å spectral irradiance with 1 Å spectral resolution and with 10-second cadence. The Optics Free Spectrometer (OFS), being ionization cells, provides daily inflight calibrations for the MEGS channels. The EUV Spectrophotometer (ESP) completes the spectral coverage at 1–50 Å and 1190–1250 Å. Collectively, EVE will measure EUV irradiance from 1 to 1250 Å with 7% accuracy and 4% long-term precision.

Target launch date for SDO is August 2007. It will fly in an inclined geosynchronous orbit, which satisfies the requirements for a high scientific data rate well in excess of 100 Mbps and nearly continuous observations with a single dedicated ground station.

20.4.2.5 Solar Orbiter

Solar Orbiter[†] (Marsch *et al.*, 2000, 2001) was selected by ESA's Science Programme Committee (SPC) in October 2000 as a Flexi-mission, to be implemented after the BepiColombo cornerstone mission to Mercury before 2013. The key mission feature of the Solar Orbiter is to study the Sun from close-up (45 solar radii, or 0.21 AU) in an orbit tuned to solar rotation in order to examine the solar surface and the space above from a co-rotating vantage point at high spatial resolution, and to provide images

[†] http://solarsystem.estec.esa.nl/index.htm

of the Sun's polar regions from heliographic latitudes as high as 38°. To reach its novel orbit (Figure 20.6), Solar Orbiter will make use of low-thrust solar electric propulsion (SEP) interleaved by Earth and Venus gravity assists.

The scientific goals of the Solar Orbiter are

- to determine *in situ* the properties and dynamics of plasma, fields and particles in the near-Sun heliosphere,
- to investigate the fine-scale structure and dynamics of the Sun's magnetised atmosphere, using close-up, high-resolution remote sensing,
- to identify the links between activity on the Sun's surface and the resulting evolution of the corona and inner heliosphere, using solar co-rotation passes,
- to observe and fully characterise the Sun's polar regions and equatorial corona from high latitudes.

The strawman payload encompasses two instrument packages: Solar remote-sensing instruments: EUV full-Sun and high resolution imager, high-resolution EUV spectrometer, high-resolution and full-sun visible light telescope and magnetograph, EUV and visible-light coronagraphs, radiometer. Heliospheric instruments: solar wind analyzer, radio and plasma wave analyser, magnetometer, energetic particle detectors, interplanetary dust detector, neutral particle detector, solar neutron detector.

20.4.2.6 Solar Probe

The primary objective of the Solar Probe[†] mission (Gloeckler *et al.*, 1999) is to understand the processes that heat the solar corona and produce the solar wind. The present mission profile foresees the Probe to arrive at the Sun along a polar trajectory perpendicular to the Sun-Earth line with a perihelion of 4 solar radii from the Sun's center. It will first travel to Jupiter for a gravity assist, leave the ecliptic plane, fly over the Sun's poles to within 8 solar radii, and reach perihelion over the equator at 4 solar radii. Two perihelion passages are planned, separated by about 5 years.

To achieve its scientific objective, it carries a complement of *in situ* and remote sensing instruments. The strawman payload includes for the *in situ* science package: vector magnetometer, solar wind ion composition and electron spectrometer, energetic particle composition spectrometer, plasma wave sensor, and a fast solar wind ion detector. The remote sensing package includes a magnetograph/helioseismograph, a high resolution XUV imager, and an all-sky, 3-D coronagraph imager. Mass and power allocations for the whole payload are 19 kg and 16 W, respectively.

An Announcement of Opportunity to propose instruments was issued by NASA in September 2000. Unfortunately, due to budgetary pressures the selection process for these packages is currently on hold and the prospects for the mission were unclear at the time of writing.

[†] http://www.jpl.nasa.gov/ice_fire/sprobe.htm

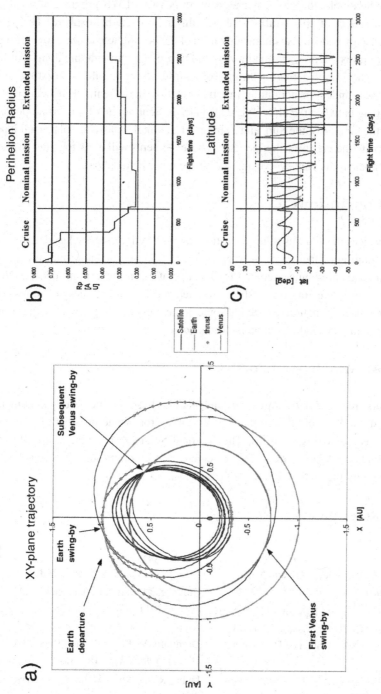

Figure 20.6. Orbit design of Solar Orbiter: Ecliptic projection of the spacecraft trajectory (a); perihelion radius (b) and spacecraft heliographic latitude (c) versus mission flight time in days.

20.4.2.7 Solar Sentinels

The Solar Sentinels, the next solar mission in NASA's LWS program after SDO, consists of a fleet of spacecraft distributed throughout the heliosphere. Their key objective is to provide global characterization of the heliosphere leading to improved accuracy of transient propagation models and to provide tomographic images of the Sun. While not yet well defined at the time of writing, a Sentinels fleet consisting of six spacecraft was under consideration: four Inner Heliospheric Sentinels in heliocentric orbits ranging between 0.5 and 0.95 AU, a FarSide Sentinel in a 1 AU orbit opposite Earth, on the far side of the Sun, and a single L1 Sentinel to provide solar wind input information to the geospace components. These elements will work together to track solar disturbances as they evolve and transit the inner heliosphere.

20.5   Conclusions

Major advances in solar observing facilities have ocurred during the last decade, and a next generation of facilities on the ground, on balloons and suborbital rockets as well as in space are being prepared to start operating in the coming decade. New instruments and detectors on existing facilities and completely new facilities will provide crucial information for solving some of the long-standing scientific questions addressed in the preceeding chapters of this book.

Acknowledgements

The authors are grateful to J.W. Harvey and W.C. Livingston for carefully reading the manuscript. CUK would like to thank J.W. Harvey for providing a list of synoptic, ground-based solar observatories. The National Solar Observatory is operated by the Association of Universities for Research in Astronomy, under a cooperative agreement with the National Science Foundation.

References

Acuña, M.H., Ogilvie, K.W., Baker, D.N., Curtis, S.A., Fairfield, D.H., and Mish, W.H. (1995). *Space Science Rev.*, **71**, 5.

Antiochos, S., Acton, L., Canfield, R. *et al.* (1997). *The SOLAR-B Mission.* Final Report of the NASA Science Definition Team.
http://science.msfc.nasa.gov/ssl/pad/solar/sdt-rpt.stm

Chaplin, W.J., Elsworth, Y., Howe, R., Isaak, G.R., McLeod, C.P., Miller, B.A.,

van der Raay, H.B., Wheeler, S.J., and New, R. (1996). *Solar Phys.*, **168**, 1.

Culhane, J.L., Hiei, E., Doschek, G.A. *et al.* (1991). *Solar Phys.*, **136**, 89.

Domingo, V., Fleck, B., and Poland, A.I. (1995). *Solar Phys.*, **162**, 1.

Dunn, R.B. (1969). *Sky and Telescope*, **38**, 368.

Fleck, B., Domingo, V., and Poland, A.I. (eds.) (1995). *Solar Phys.*, **162**.

Fossat, E. (1991). *Solar Phys.*, **133**, 1.

Gloeckler, G. *et al.* (1999). *Solar Probe: First Mission to the Nearest Star.* Report of the NASA Science Definition Team for the Solar Probe Mission.
http://www.jpl.nasa.gov/ice_fire/
SP_SDT_Report.htm

Hathaway, D. *et al.* (2001). *Solar Dynamics Observatory.* Report of the NASA Science Definition Team.
http://lws.gsfc.nasa.gov/
lws_mission_sdo.htm

Handy, B.N., Action, L.W., Kankelborg, C.C. *et al.* (1999). *Solar Phys.*, **187**, 229.

Kneer, F. and Wiehr, E. (1989). The Gregory-Coudé-Telescope at the Observatorio Del Teide Tenerife, *NATO Advanced Science Institutes (ASI) Series C,* **263**, 13.

Kosugi, T., Makishima, K., Murakami, T., Sakao, T., Dotani, T., Inda, M., Kai, K., Masuda, S., Nakajima, H., Ogawara, Y., Sawa, M., and Shibasaki, K, (1991). *Solar Phys.*, **136**, 17.

Leibacher, J. (1999). *Adv. Space Res.*, **24**, 173.

Lin, R.P. and the HESSI Team (2000). *ASP Conf. Ser.*, **206**, 1.

Marsch, E. *et al.* (2000). *Solar Orbiter - A High-Resolution Mission to the Sun and Inner Heliosphere.* Assessment Study Report, ESA-SCI(2000)6.

Marsch, E., Harrison, R., Pace, O. *et al.* (2001). In *Solar Encounter: The First Solar Orbiter Workshop*, ESA SP-493, pp. xi–xxvi.

Ogawara, Y., Takano, T., Kato, T., Kosugi, T., Tsuneta, S. Watanabe, T., Kondo, I., and Uchida, Y. (1991). *Solar Phys.*, **136**, 1.

Pierce, A.K. (1964). *Appl. Opt.*, **3**, 1337.

Rayrole, J. (1990). *L'Astronomie*, **104**, 390.

Russel, C.T. (ed.) (1991). *Space Sci. Rev.*, **71**, Nos. 1–4.

Rust, D. *et al.* (1998). STEREO. Science Definition Team Report,
http://sd-www.jhuapl.edu/STEREO/
Sci/teamReport.html

Rutten, R.J., Hammerschlag, R.H., Bettonvil, F.C.M. (1999). *Solar and Stellar Activity: Similarities and Differences, ASP Conference Series 158*, ed. C. J. Butler and J. G. Doyle, 57–60.

Schmidt, W. (1992). *Sterne und Weltraum*, **31**, 167.

Schmidt, W. and Ai, G. (1998). In *A Crossroads for European Solar and Heliospheric Physics*, ESA SP-417, 189.

Stone, E.C., Frandsen, A.M., Mewaldt, R.A., Christian, E.R., Margolies, D., Ormes, J.F., and Snow, F. (1998). *Space Sci. Rev.*, **86**, 1–22.

Schröter, E.H., Soltau, D., and Wiehr, E. (1985). *Vistas Astron.* **28**, 519–525.

Tousey, R. (1977). *Appl. Opt.*, **16**, 825.

Tsuneta, S., Acton, L., Bruner, M., Lemen, J., Brown, W., Caravalho, R., Catura, R., Freeland, S., Jurcevich, B., Morrison, M., Ogawara, Y., and Owens, J. (1991). *Solar Phys.*, **136**, 37.

Vandenbussche, F.C. (1997). *ESA Bull.*, **97**, 39.

Wenzel, K.-P., Marsden, R.G., and Battrick, B. (eds.) (1990). *ESA Bull.*, **1050**.

Yoshimori, M., Okudaira, K., Hirasima, Y., Igarashi, T., Akasaka, M., Takai, Y., Morimoto, K., Watanabe, T., Ohki, K., Nishimura, J., Yamagami, T., and Ogawara, Y. (1991). *Solar Phys.*, **136**, 69.

Index